Power Systems

Z. Lubosny

Wind Turbine Operation in Electric Power Systems

Springer

*Berlin*
*Heidelberg*
*New York*
*Hong Kong*
*London*
*Milan*
*Paris*
*Tokyo*

Z. Lubosny

# Wind Turbine Operation in Electric Power Systems

## Advanced Modeling

With 191 Figures and 34 Tables

Springer

Dr. Zbigniew Lubosny
Gdansk University of Technology
Dept. Electrical Power Systems
Narutowicza Street 11/12
80-952 Gdansk
POLAND

Cataloging-in-Publication Data applied for

Bibliographic information published by Die Deutsche Bibliothek
Die Deutsche Bibliothek lists this publication in the Deutsche Nationalbibliografie;
detailed bibliographic data is available in the Internet at <http://dnb.ddb.de>.

ISBN 3-540-40340-X Springer-Verlag Berlin Heidelberg New York



Springer-Verlag Berlin Heidelberg New York
a member of BertelsmannSpringer Science + Business Media GmbH

http://www.springer.de


Printed in Germany



Typesetting: Digital data supplied by authors
Cover-Design: de'blik, Berlin
Printed on acid-free paper 62/3020 xv 5 4 3 2 1 0

# Preface

The total capacity of wind turbine generator systems (WTGSs) operating in the world reached 27 600 MW in Autumn 2002. In the European Union (EU) countries, wind generators capacity of 20 500 MW were operational. Most of them were located in Germany (10 650 MW), Spain (4100 MW) and Denmark (2500 MW). The Danish electric power system has the largest ratio of the capacity of WTGSs to the capacity of all sources of energy and it reaches about 30%. The mean value of the rated power of WTGSs sold in 2000 was 930 kW. Today, wind generators with a rated power equal to 2 MW have become standard.

The exponential growth of the number of WTGSs installed in power systems in the EU results from the policy which forces the sale of electric energy from renewable energy sources. The EU plan is to achieve a total capacity of WTGS of 60 000 MW by 2010 and 150 000 MW by 2020. In practical terms, such an increase is possible by utilizing the wind as a source of energy. Other systems, e.g. solar or thermal, cannot yield such levels of energy. Utilization of water as a source of energy (from new power plants) is also limited, for various reasons. This means that energy from the wind is a reasonable prospect for the years to come.

Unlike classical sources of energy, WTGSs are variable, which is why during normal operation (below rated power) the real power introduced to the network varies. At the same time, in some types of WTGS, the production of real power is correlated with the consumption of reactive power. This variation of power produced/consumed causes variations of voltages with all the usual consequences for electric power systems and customers (e.g. flicker). Using power electronic systems (converters) introduces voltage and current harmonics into the power system. As uncontrollable (at partial load) sources of energy, the WTGSs can introduce problems of voltage stability and transient stability into power systems. The first problem results from the location of wind turbines and wind farms, often at seaside areas, which are often relatively weak networks. The second problem is also related to the one mentioned before, but it can become an especially serious problem in the power system when the number of WTGSs operating in the system increases significantly.

The rapid increase of the number of WTGSs operating in power systems on the one hand and the problems that are introduced into power systems by WTGS operation on the other hand show the importance and necessity of their operation in power system analysis.

Today, modeling is the basic tool for analysis (optimization, project, design, etc.). In any particular case, because of the various types of analysis that have to be made, various types of models have to be used. The models utilized for steady-state analysis are extremely simple, while the dynamic models of WTGS (for stability analysis) are not easy to develop. What is the reason? A model can be considered as a structure (e.g. set of equations) together with data. Only appropriate data with the appropriate set of equations give the proper model (of the object under consideration).

In the case of power systems with classical sources of energy analysis, the modeling is relatively simple because the models of objects and controllers are well known and even standardized (IEEE standards). The data – especially the object data – are available. But in the case of WTGS modeling, researchers meet problems related to the lack of data and lack of control-system structures due to strong competition among the WTGS manufacturers, who, for obvious reasons, do not want to share their new ideas and constructions. This leads to the situation in which many researchers model the WTGSs in relatively (even extremely) simple form, almost neglecting the control systems, which, of course, significantly influences the reliability of the analytical results.

Therefore, the author's intention is to present a relatively comprehensive approach to the modeling of WTGSs and their operation in electric power system analysis. The author does realize that the rapid development of WTGSs means that the ideas and considerations presented here may become outdated quite soon (this is especially true of control systems).

The structure of the monograph is as follows. Chapters 1 and 2 present the history and state-of-the-art of modern wind turbines. In Chap. 3 standards related to the assessment of wind turbine generator systems on the electric power system and on the power quality are considered. Chapter 4 deals with some aspects related to wind turbine operation in the power system in the steady state. WTGS modeling (as a dynamic object) in the analysis of its operation in the grid is also presented in Chap. 4. The results of time domain-domain analysis based on various types of modern WTGSs (with synchronous and asynchronous generators) are presented and discussed in Chap. 5. Finally, a summary is given in Chap. 6.

This research was supported through a European Community Marie Curie Fellowship, Contract No. HPMF-CT-2000-01046, Deutscher Akademischer Austauschdienst (DAAD) and the Rector of Gdansk University of Technology.

The author would like to acknowledge Professor Zbigniew Styczynski from the Otto-von-Guericke-Universitat, Magdeburg for his support during the project realization.

Magdeburg–Gdansk, Spring 2003

Zbigniew Lubosny
zlubosny@ely.pg.gda.pl

# Frequently Used Symbols and Abbreviations

## Notation

a bar at the bottom of a symbol denotes a phasor or a complex number (e.g. $\underline{V}$, $\underline{\boldsymbol{Y}}$);
lower-case symbols normally denote instantaneous values (e.g. $v$, $i$);
upper-case symbols normally denote per unit, rms or peak values (e.g. $V$, $I$);
bold face denotes a matrix or a vector (e.g. $\boldsymbol{Y}$, $\boldsymbol{V}$, $\boldsymbol{I}$).

## Symbols

| | |
|---|---|
| $B$ | susceptance |
| $C$ | capacitance |
| $D$ | damping coefficient |
| $E$ | electromotive force (emf) or energy |
| $f$ | frequency |
| $G$ | conductance |
| $H$ | constant of inertia |
| $i$, $I$ | current |
| $J$ | rotor moment of inertia |
| $K$ | coefficient |
| $l$ | length |
| $L$ | inductance |
| $M$ | magnetizing inductance |
| $n$ | revolution per minute |
| $p$ | number of poles |
| $P$ | real power |
| $Q$ | reactive power |
| $R$ | resistance |
| $s$ | shaft slip or Laplace operator |
| $S$ | apparent power |
| $t$ | time |
| $T$ | time constant |
| $v$, $V$ | voltage |
| $W$ | work |
| $X$ | reactance |
| $Y$ | admittance |
| $Z$ | impedance |
| $\alpha$ | firing angle |
| $\gamma$ | instantaneous position of $d$-axis relative to phase A |
| $\delta$ | power (or rotor) angle |
| $\Delta$ | deviation |
| $\eta$ | static drop of the turbine-governor characteristic |
| $\varphi$ | power-factor angle |

| | |
|---|---|
| $v$ | wind velocity |
| $\tau$ | torque |
| $\upsilon$ | transformation ratio |
| $\omega$ | angular velocity in electrical radians |
| $\Psi$ | flux |
| $\Theta$ | angular displacement |

## Subscripts

| | |
|---|---|
| a | armature |
| A, B, C | natural axes |
| cr | critical |
| d, q | axes of two-axis (orthogonal) reference frame |
| D, Q | *d*- and *q*-axis damper winding |
| e | electromagnetic |
| E | excitation system |
| f | field |
| G | synchronous generator |
| GB | gearbox |
| D, Q | damper winding in *d* and *q* axes |
| l | leakage |
| L | load |
| m | magnetizing or mechanical |
| M | asynchronous motor (generator) |
| n | nominal (rated) |
| N | nominal system voltage (line voltage) |
| R | rotor |
| ref | reference |
| s | synchronous |
| S | stator |
| Q | power system |
| T | transformer |
| W | wind wheel (wind turbine) |
| $\alpha, \beta, \gamma$ | axes of multi-axis arbitrary reference frame |

## Superscripts

| | |
|---|---|
| ‘ | transient, or rotor variables and/or parameters referred to stator winding |
| “ | subtransient |

## Abbreviations

| | |
|---|---|
| AC | alternative current |
| ASM | asynchronous motor |
| CB | circuit breaker |
| DC | direct current |
| HV | high voltage |
| LV | low voltage |
| MV | medium voltage |
| PCC | point of common coupling |
| rms | root-mean-square |
| SG | synchronous generator |
| WTGS | wind turbine generator system |

# Contents

# 1 Development of Wind Turbines

The utilization of wind energy has a very long tradition. Some historians suggest that wind turbines (windmills) were known over 3000 years ago. They also maintain that remains of stone windmills were discovered in Egypt near Alexandria. But the first reliable information, from historical sources, points to the utilization of windmills in 640 AD, on the Persian–Afghan border. These Persian windmills were characterized by a vertical axis of rotation. The windmills were usually used for milling grain.

Another historical source – dated a few ages later – shows that the Chinese also utilized wind wheels with a vertical axis of rotation. These wind wheels were used for draining rice fields. There is no information whether the Persians or the Chinese invented the wind turbines (wind wheels).

Unlike China and Persia, Europe utilized windmills with a horizontal axis of rotation. Historians say that such a type of windmill was invented in Europe. The first reliable information about the windmill use in Europe is from the year 1180 from Duchy in Normandy. The source describes a windmill, the so-called post mill, which is a typical European windmill.

The most common and most popular windmills built and utilized in Europe were of the following types:

- Post windmill – the main feature is that the whole wooden millhouse is located on the central post (usually reinforced by four diagonal cross-beams) and can revolve. The wind wheel has four blades covered by fabric (in central Europe) or by wood (in Northern and Eastern Europe). The millhouse can be turned into the wind direction by the help of a "tail" fixed to the back wall. This type of windmill was usually used for grain milling.
- Hollow post mill – used since the 15th century for pumping water, which influenced its construction. In general, the construction of this type of windmill is similar to that of the post mill (e.g. four blades) but in the case of the hollow post mill, the millhouse is located on the pyramid-shaped tower. The yawing millhouse is smaller than the post windmill because it contains the "gear box" only (wind-wheel bearing with cogwheel and wallower). The rest of the mechanism (machine), i.e. the water-pump drive, is located in the windmill tower.
- Tower windmill – this was widespread in the Mediterranean regions. The millhouse had the shape of a round stone tower. Originally, the wind-wheel could not be yawed. In later constructions, the possibilities of wind wheel yawing

were added. Greek tower windmills differed from other windmills because they often had framed sail blades (more than ten).

- Dutch windmill – this has a small yawing cap (roof) housing, the wind wheel and the "gear box". The millhouse is fixed and usually is built from bricks (19th century) and wood. This construction enabled locating more powerful (and heavier) machinery inside.
- Paltrock mill – from 16th or 17th century, with wooden construction similar to that of post windmills. The millhouse is located a few metres above ground or water. These mills were utilized usually as wood-saw mills. Then the sawing machinery installed was on the ground or water level. Location on the water resulted from the utilization of water transport for wood transportation.

Windmills were very popular and commonly used for driving various types of machines up to the middle of the 19th century. It is estimated that about 200 000 windmills were operational in Europe at that time. But as usual in history the use of this invention eventually came to an end. The beginning of the end of the windmill era came with the beginning of the steam era. The internal combustion engine contributed to this process as well. But it was the electrification of rural areas that marked the real end of the utilization of windmills.

In the United States windmills started to be used in the early 19th century when, in Europe the utilization of windmills reached its peak. European windmills could not be used in America, especially by settlers of the middle and western part of the country. The settlers needed rather light windmills (easy to move to other places) with low speed and relatively high initial torque because they were usually used for water pumping purposes.

The first windmill was constructed in about 1850 by Richard Halladay, and the next one – also widely used – by Leonhard R. Wheeler:

- Halladay's windmill was a windmill with the wind wheel assembled from numerous blades. The blades, divided into six sections, were not connected directly to the shaft but suspended loosely on a ring. The movement of the ring enabled the blade pitch angle to be changed. The movement of the ring was triggered by centrifugal forces. The control system kept the pitch angle relatively small at low winds and increased it greatly at high winds.
- Wheeler's windmill – the main concept of this windmill was similar to that of Halladay's windmill, but the wind wheel was not divided into sections. Wheeler's windmill was equipped with an additional wind vane. The enabled proper orientation of the whole windmill to the wind and turning it out of the wind when the wind velocity become too high. This type of windmill – named Eclipse – became the standard design of American wind turbines.

In America, the windmill era also began to end with the electrification of the rural areas of the country.

Looking back into history one can say that the windmills were supplanted by electric energy. It was only a temporary process, but now windmills are coming back. Ironically enough, they are coming back with "electric energy" supporting the electric power systems.

But that comeback – on a massive scale – has been neither fast nor easy. The first attempts to use windmills for producing electricity date back to the 19th century. In 1891, Poul La Cour in Askov (Denmark) built an experimental wind turbine driving a *dynamo* (DC system). In 1897, he built his next, larger test station. The production of commercial wind turbines of rated power 10–35 kW resulted from his work. By 1918 about 120 wind turbines were in operation. Wind generators were usually used to feed the small islanded electric grids with the help of diesel engines driving electric generators. The first Danish wind turbines used four-bladed rotors with shutter blades. Poul La Cour was also the first to tackle the problem of energy storage. He used wind turbines for electrolysis and stored hydrogen for later energy conversion.

In 1941–1942 the F. L. Smith Company produced wind turbines – called the *Aeromotor* – with two-bladed wind wheels, of diameter 17.5 m. Their rated power was about 50 kW at a wind velocity of 11 m/s. The blades were made of laminated wood. Unfortunately, some problems with the blades and turbine dynamics appeared in that type of wind turbine and the company switched into production of wind turbines with three-bladed wind wheels (rated power 70 kW). A concrete tower was used for the first time for that type of turbine.

The wind turbines of both constructions were equipped with DC machines.

In the following years the process of wind turbines projecting, constructing and testing was carried on. Some of the constructions of that time are presented in Table 1.1.

**Table 1.1.** Wind turbine generator systems in the middle of 20th century

| Name | Wind wheel diameter [m] | Rated power [kW] |
|---|---|---|
| WIMF D-30, Balaklava, Crimea, Russia, 1931 | 30.0 | 100 |
| John Brown Company, Orkney Island, 1950 | 15.0 | 100 |
| Andreau-Enfield, St. Albans, England, 1956 | 24.4 | 100 |
| Best Romani, France, 1958 | 30.1 | 800 |
| Nypric-Vadot, Saint-Remy-de-Provence, France, 1962 | 21.1 | 800 |
| Gedser, Denmark, 1957 | 24.0 | 200 |
| W-34, Stotten, Germany, 1959-1968 | 34.0 | 100 |

There was a tendency to construct high-power wind turbines (with rated power above 1 MW) because of the "scale effect", where the effects achieved (e.g. energy sold) to the outlay costs looked more profitable. Many projects of that type were considered. Some of these "big" projects were not implemented, e.g. MAN-Kleinhenz with a 10 MW turbine, from 1942; Russian project (Arctic Sea) with a 5 MW wind turbine from the 1940s, because of the war.

The first wind turbine with rated power equal to 1.25 MW (at a wind velocity of 13.2 m/s) was installed in USA in 1941. The turbine was destroyed four years later, when the blade broke off. A German project called GROWIAN 1 (GROsse Windenergie Anlage) was also unsuccessful. Wind turbine with wind wheel of a diameter of 100.4 m and rated power 3 MW (at a wind velocity of 11.8 m/s) was built in 1982 on the North Sea seaside near Brunsbutel. The wind turbine was stopped soon afterwards because of the cracks on the blades. Similar projects were carried out in a few other countries as well. A short summary of such large projects is presented in Table 1.2.

Today, many technical and technological problems related to the construction of high-power wind turbines have been overcome, but wind turbines are still the subject of intensive investigations. The typical rated power of present wind turbines is 1–2 MW. But wind turbine generator systems (WTGS) with rated power above 2 MW (e.g. 3.5–4.5 MW) are also produced. Wind power plants with higher rated power are built as the wind farms consisting of teen to tens WTGSs of rated power, usually up to 2 MW.

**Table 1.2.** Examples of high-power wind turbine generator systems

| Name | Wind wheel diameter [m] | Rotor height [m] | Rated wind velocity [m/s] | Rated speed [rpm] | Rated power [MW] |
|---|---|---|---|---|---|
| Putnam (USA, 1941) | 58.0 | 37.0 | 13.12 | 29.0 | 1.25 |
| Nypric-Vadot (France, 1964) | 35.0 | – | – | – | 1.0 |
| MOD 2 (USA, 1977) | 91.5 | 61.0 | 12.5 | 17.5 | 2.5 |
| GROWIAN (Germany, 1982) | 100.4 | 96.6 | 11.8 | – | 3.0 |

# 2 Wind Turbine Generator Systems

Modern wind turbine generator systems (WTGSs) are constructed mainly as systems with a horizontal axis of rotation, a wind wheel consisting of three blades, and high speed (1500/750 rpm) asynchronous generators and gear boxes (with ratio grater than 60); see Fig. 2.1. Asynchronous machines are used because of their advantages, such as simplicity of construction, possibilities of operation at various operational conditions, and low investment and operating costs. Asynchronous generators with squirrel-cage rotors and with wounded (slip-ring) rotors are used. In the second case, power electronic converters or variable resistors are connected to the rotor side of the machine, which makes it possible for the machine to operate at various speeds.[1]

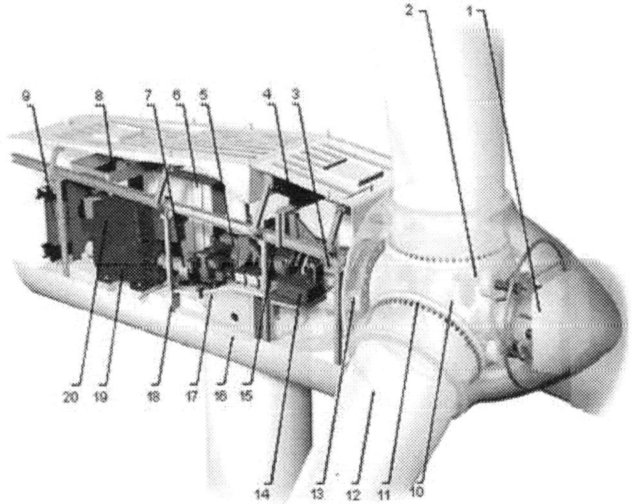


**Fig. 2.1.** WTGS with asynchronous generator (V 80-2.0 MW; 1 – hub controller, 2 – pitch cylinder, 3 – main shaft, 4 – oil cooler, 5 – gearbox, 6 – generator controller with converter, 7 – parking brake, 8 – service crane, 9 – step-up transformer, 10 – blade hub, 11 – blade bearing, 12 – blade, 13 – rotor lock system, 14 – hydraulic unit, 15 – hydraulic shrink disk, 16 – yaw ring, 17 – machine foundation, 18 – yaw gears, 19 – generator, 20 – generator cooler)

[1] Unlike the WTGS with a squirrel-cage rotor asynchronous machine, which operates at constant speed (so-called constant speed system).

The machine configuration called the doubly-fed asynchronous machine (DASM) is becoming very popular. In this case, the electronic converter is connected between the rotor and stator side of the machine.

The generators of WTGSs' are usually machines with a switched (usually between 6 and 4) number of poles. Often two separate generators are located in a single machine. When the machine operates at a rated speed of 750 rpm, the so-called small generator is utilized. The generator operates at small wind speed velocity. When the wind velocity is high, then the machine operates at a rated speed 1500 rpm and the big generator is switched on.

The utilization of such a high-speed machine forces the utilization of a gearbox, which is located between the generator and the wind wheel, because of the difference between the machine rotor speed and the wind-wheel speed (up to 40 rpm). This relatively low wind-wheel rotation results mainly from the need for WTGS optimal operation. Optimal operation means maximum power extraction from the wind. Optimal operation – for three-bladed wind wheels – is achieved for a tip-speed ratio equal to about 6–8. In spite of the relatively slow rotation of the wind wheel, the blade-tip linear speed is often high, usually higher than 60 m/s (216 km/hour) during steady-state operation and slightly higher for transients.

The WTGS usually operates at a wind velocity between 3–5 m/s up to 25 m/s. The rated power is achieved at a wind velocity equal to about 12–16 m/s.

Asynchronous generators – even the ones with rated power up to 2 MW – utilized today in WTGSs are low-voltage machines with a rated voltage equal to 690 V. To connect the generator to the electric power system, step-up transformers with high-voltage side equal to 10–40 kV are usually used. Step-up transformers are located in the WTGS nacelle (units with rated power equal to and above 2 MW) and in or near the WTGS tower (smaller units).

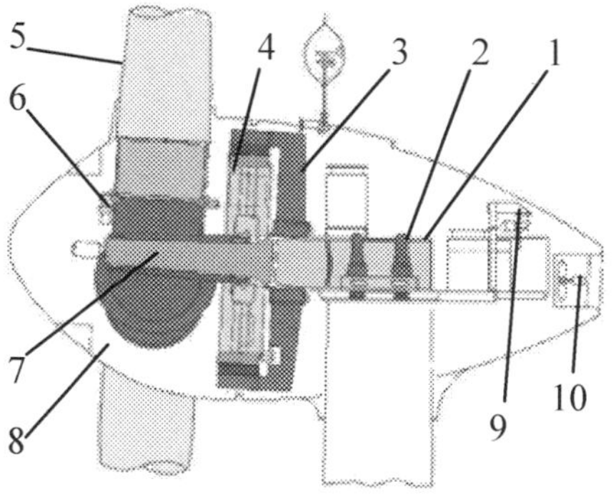


**Fig. 2.2.** WTGS with synchronous generator (ENERCON E 66 wind turbine; 1 – machine foundation, 2 – yaw gears, 3 – generator stator, 4 – generator rotor, 5 – blade, 6 – blade positioning gear, 7 – main shaft, 8 – hub, 9 – service crane, 10 – cooler)

Anther type of commercial WTGS is that equipped with a synchronous generator; see Fig. 2.2. In such a construction[2], a synchronous generator is connected to

[2] In fact, there is only one manufacturer of such a commercial WTGS.

the electric power system through a power electronic converter. The generator here is connected to the wind wheel directly, i.e. without a gearbox. Therefore the WTGS, which operates with the wind wheel speeds up to 40 rpm, has to be equipped with a synchronous generator with an appropriate high number of poles. The synchronous generators utilized in such constructions are machines with a high number of poles – about 80, in the case of the ENERCON wind turbines).

WTGSs with synchronous or asynchronous generators are usually equipped with a blade pitch angle control system, which enables the power generated by the wind turbine to be controlled. Some wind turbine generator systems (usually the constant speed system with smaller rated power) utilize the stall effect to "control" (in fact to limit) the power extracted from the wind.

The number of types of modern wind turbine generator systems is huge. Some WTGS constructions with higher rated power are presented in Table 2.1 as an example.

Apart from the above-mentioned types of WTGS, wind turbines with a vertical axis of rotation are also utilized. The number of such units operating in the power system is today small. Because of some disadvantages of such a construction, manufacturers concentrate their efforts on turbines of the horizontal axis of rotation type.

The new trend in WTGS construction, aiming at improving wind turbine efficiency (to increase the amount of power extracted from the wind), is the utilization of diffusers. The diffusers are tested for horizontal-axis turbines. But so far such constructions have not been sold commercially.

**Table 2.1.** Modern constructions of wind turbine generator systems

| Name | Gen. type | Wind wheel diameter [m] | Hub height [m] | Rated wind velocity [m/s] | Operational speed range [rpm] | Rated power [MW] |
|---|---|---|---|---|---|---|
| NM 52/900 | A | 52.2 | 61.5–73.8 | 15 | 14.9/24.4[3] | 0.9 |
| EW 900 | A | 52–57 | 45–70 | 15 | 15.0–28.0[4] | 0.9 |
| N 60/1300 | A | 60 | 46–85 | 15 | 12.8/19.2 | 1.3 |
| EW 1500 | A | 70.5 | 64.7–100 | 12 | 11.0–22.0 | 1.5 |
| NM 64c/1500 | A | 64 | | 16 | 17.3 | 1.5 |
| MD 77 | A | 77 | 61.5–100 | 13 | 9.2–17.3 | 1.5 |
| V66-1.75MW | A | 66 | 60–70 | 16 | 10.5–24.5 | 1.75 |
| E 66 | S | 70 | 65–114 | 12 | 10.0–22.0 | 1.8 |
| NM 72/2000 | A | 72 | | 14 | 12.0/18.0 | 2.0 |
| V80-2.0MW | A | 80 | 60–100 | 15 | 9.0–19.0 | 2.0 |
| MM 70 | A | 70 | 65–100 | 14.5 | 10.0–23.3 | 2.0 |
| N 80/2500 | A | 80 | 60–100 | 15 | 10.9/19.1 | 2.5 |
| NM 92/2750 | A | 92 | 70–100 | 14 | 15.6 | 2.75 |

A – asynchronous generator, S – synchronous generator

[3] Two-speed machine (constant speed system).

[4] Variable-speed system.

Because of their operation in electric power system, WTGSs are equipped with devices and systems to allow them to operate in the power system. The generic scheme of the electrical part of the WTGS, presented in Fig. 2.3, consists of four sections:

- generator section,
- low-voltage (LV) section,
- step-up transformer section,
- medium-voltage (MV) section.

The generator-section configuration depends on the generator type. For squirrel-cage rotor asynchronous generators, this section includes an asynchronous generator only. For doubly-fed asynchronous generators, a converter is switched between the rotor and stator through the transformer or reactor. For wounded rotor asynchronous generators with variable resistors connected to the rotor windings, appropriate power electronic switches are utilized.

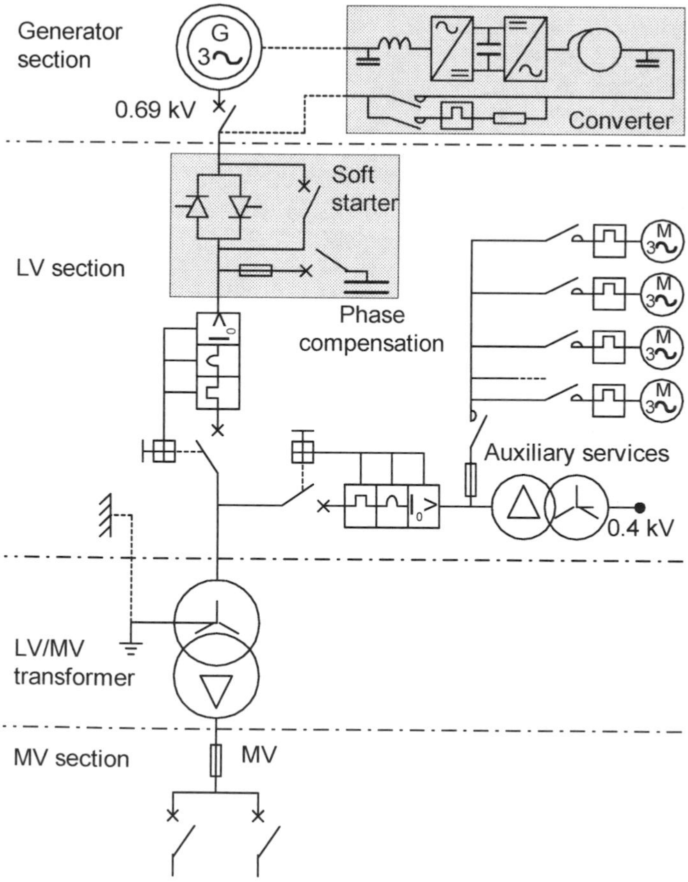


**Fig. 2.3.** General configuration of WTGS electrical equipment

The low-voltage section comprises protection systems, auxiliary services, soft-start system, and phase-compensating condensers:

- The soft-start system is used to smooth the connection and disconnection of the generator to the power system. It helps to avoid high changes of voltage and currents in the grid to which the WTGS is connected, and simultaneously protects the mechanical part of the WTGS (e.g. shaft and gearbox) against high torques and forces. The thyristors, which are used in the soft starter, waste about 1–2% of the energy running through them. Therefore, the wind turbines are equipped with the so-called bypass switch (a mechanical one) which is activated after the turbine has been soft started.
- The phase-compensating system – in the form of a bank of condensers – is utilized in WTGSs with squirrel-cage rotor asynchronous generators and with wounded-rotor generators with variable resistors connected to the rotor winding. Usually, the no-load reactive power consumed by the asynchronous machine is compensated. In the case of doubly-fed asynchronous generators, the reactive power compensation is realized by the converter – through the machine rotor current control. For such a WTGS, the soft start is realized by the converter and resistor (with bypass switch) limiting the converter current at the generator start up.
- The auxiliary services of the wind turbine comprise drives of actuators, pumps, and fans, but also supply of the control and communication systems, heaters, lights etc.
- This section is also equipped with protection systems, which comprise the grid-protection systems (over-current, under- and over-voltage, etc.) and the technological protection systems (e.g. speed, temperature, etc.).

The step-up transformer section comprises the step-up transformer with appropriate protection systems, while the medium-voltage section comprises buses, switches, fuses, and voltage and current transformers for protection and measuring purposes. Electrical energy measuring systems are usually located at the medium-voltage section, but sometimes they are located at the low-voltage side of the step-up transformer.

The wind turbine generator system operation is permanently determined by the velocity and variations of the wind. The following four basic operating states can be distinguished:

- Standstill of the turbine – as a result of the wind velocity value falling below the cut-in velocity $v < v_{\text{cut-in}}$.
- Partial load – operation with maximum energy extraction from the wind, when the wind velocity $v$ is within the range $v_{\text{cut-in}} \leq v \leq v_{\text{n}}$, where $v_{\text{n}}$ is the rated wind velocity. The WTGS generates the rated power at the rated wind velocity.
- Full load – operation with constant and rated load when the wind velocity is higher than the rated wind velocity $v_{\text{n}} < v \leq v_{\text{cut-out}}$ and simultaneously lower than the maximum one. The cut-out wind speed is usually $v_{\text{cut-out}} = 25$ m/s.
- Standstill of the turbine – because of too high wind velocity $v > v_{\text{cut-out}}$.

These operating states of the wind turbine are usually presented in the form of the power versus wind velocity characteristics of the wind turbine. An example of such a characteristic is shown in Fig. 2.4. The dashed line here shows the power in the air stream and the continuous line shows the power extracted from the wind and transferred to the electric power system.

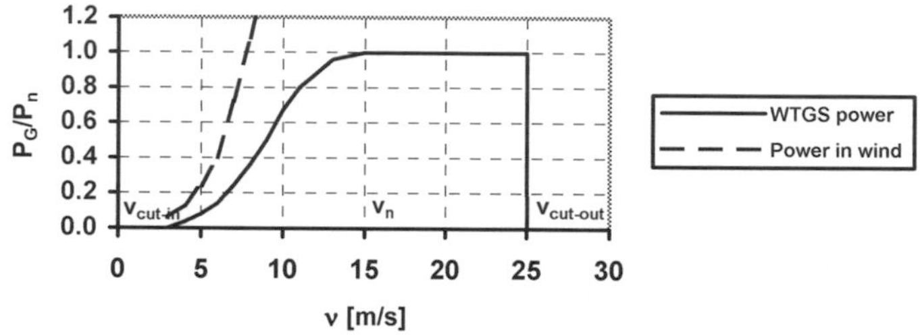


**Fig. 2.4.** Power versus wind velocity characteristics of a WTGS

The WTGS passes from one operating state to another and the WTGS performance during operation at partial and full load influences the electric power system operation.[5] This influence can have positive and/or negative aspects (character). In general, the utilization of wind turbines can be considered as positive for a number of reasons (e.g. ecological – renewable energy). But there is some negative influence related to WTGS operation in power systems as well. This negative influence on the electric power system is the subject of the considerations here. The influence of the wind turbine on the power system depends on wind variations and on the wind turbine construction. But the original reason for the most inconvenient effects is the wind velocity variation. Additionally, in the case of wind turbines equipped with power electronic converters, the converters are the source of some "problems" for the power system.

The negative influence of the WTGS on the electric power system, according to the problem source, can be as follows:

1. Wind variations as the source of torque, power, voltage and current variations:
   - stochastic process with spectrum frequencies from 1 to 10 Hz and amplitude up to 10% of the mean value of the wind velocity,
   - wind gusts with amplitude up to 20% of the mean wind velocity and with a time duration up to 30 s,
2. Wind turbine construction as the source of torque, power, voltage and current variations:
   - Shadow effects as a result of the wind wheel blades passing in front of WTGS tower. This effect is manifested as torque fluctuations. The frequency of these variations is proportional to the rotor speed and to the number of blades, e.g. 1–1.5 Hz for a three-bladed wind wheel. The amplitude of the

[5] Also the WTGS standstill can be considered as a state influencing the power system, e.g. by lack of electric energy production.

torque variations depends, highly non-linearly, on the WTGS load and can reach even 30% of the torque mean value.

- Wind velocity and turbulence unevenly spatially distributed over the rotor-swept area. The wind velocity is usually higher at higher altitude. The effect of that can be seen in the torque. The frequency of the torque fluctuations is proportional to the rotor speed, e.g. about 0.3–0.5 Hz. The amplitude of this fluctuation is up to a few percent of the torque mean value.
- Complex oscillations of turbine tower, rotor shaft, gearbox and blades (there are a few types of possible oscillations), which can be seen as torque fluctuations. The frequency of these fluctuations is from ten-hundredth up to several Hz. The amplitudes of these processes – which are extremely nonlinear – can reach up to ten percent of the mean value. Resonance in the drive train can occur.
- Current and voltage harmonics generated during converter operation.
- Soft-start system operation.

The above-mentioned torque variations and converter operation lead to many effects, which appear in the power system. These effects can be presented as groups of the following problems:

1. power quality:
   - voltage variations and sags,
   - additional reactive power consumption that can lead to decrease of the nodal voltages,
   - flicker,
   - switching on/off causing dips and sags,
   - harmonics,
   - power-flow variations,
2. voltage stability,
3. transient stability,
4. protection and control:
   - protection systems co-ordination,
   - voltage and frequency control.

Because of these factors (mentioned above) – disturbances and effects – the WTGS should be considered as a highly variable source of energy. The level of variation of the power generated by the wind turbine, but also the voltage and current variations, is usually high, and it depends on many factors. As an example, WTGS operation (power generated) at full load and partial load for various types of wind turbines is presented in Figs. 2.5–2.10. Figures 2.5 and 2.6 show the power variation of a WTGS equipped with a squirrel-cage rotor asynchronous generator. Figures 2.7 and 2.8 present the power variation of a WTGS equipped with the doubly-fed asynchronous generator. And Figs. 2.9 and 2.10 present the power variation of a WTGS equipped with a synchronous generator and converter located between the generator and the power system.

In general, the figures show[6] that WTGS output power varies highly – with all the consequences for the electric power system. If compared to partial load, at full load the real power variations are lower for all types of WTGS. It is a result of the control system operation, which at partial load (when the WTGS extracts the maximum power from the wind) cannot "effectively damp" the power variations, but at full load can do so.

The figures also show that the WTGSs equipped with a doubly-fed asynchronous machine or with a synchronous generator (Figs. 2.7 and 2.9) at full load damp the power variations more effectively than the WTGSs equipped with squirrel-cage rotor asynchronous generators (Fig. 2.5).

At partial load, the superiority of WTGSs equipped with a doubly-fed asynchronous machine over other types of wind turbines is not as high as at full load.[7] The power variation damping is only slightly higher here (in comparison with other types of wind turbines).

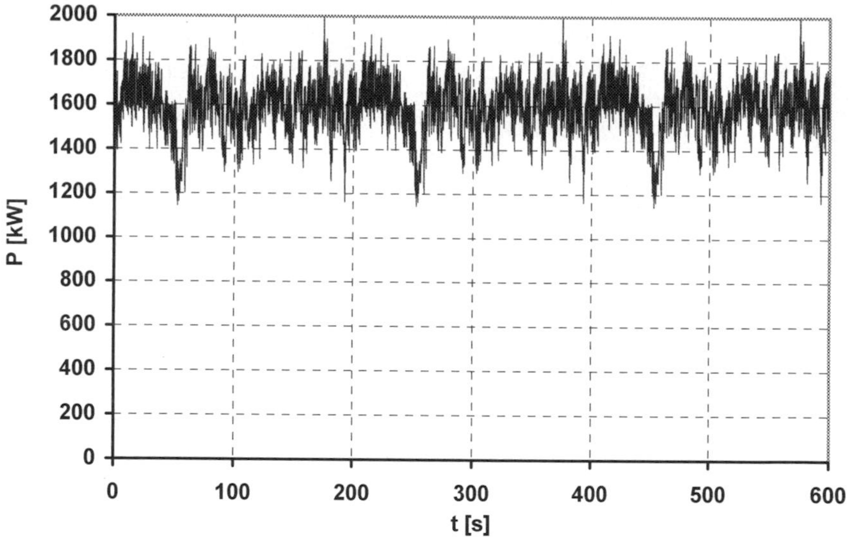


**Fig. 2.5.** Real power generation at full load by Nordtank NTK1500 wind turbine [27]

[6] Please note that the figures show the WTGS power production measured for various wind conditions for each turbine. Therefore, the curves (e.g. amplitudes of power variations) cannot be directly compared. The curves should be treated here as typical for the given type of wind turbine only.

[7] Overall costs, effectiveness and reliability of wind turbines are not considered or compared here.

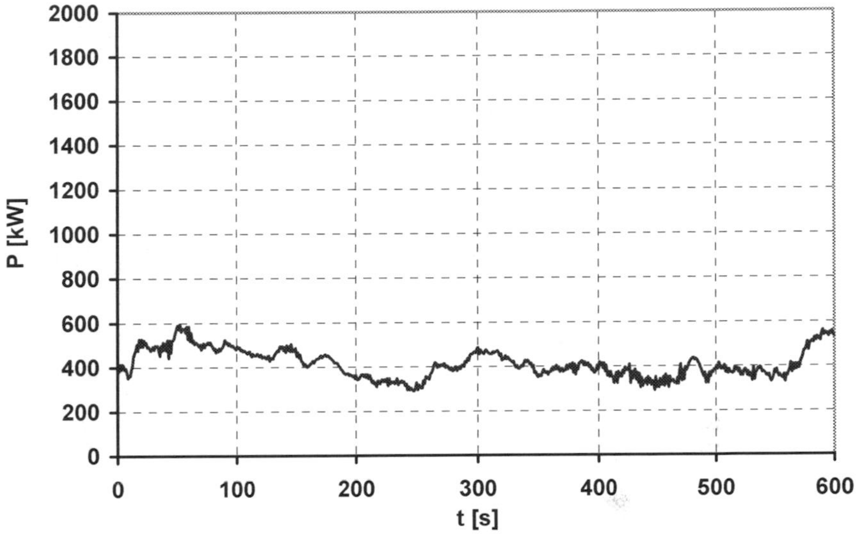


**Fig. 2.6.** Real power generation at partial load by Nordtank NTK1500 wind turbine [27]

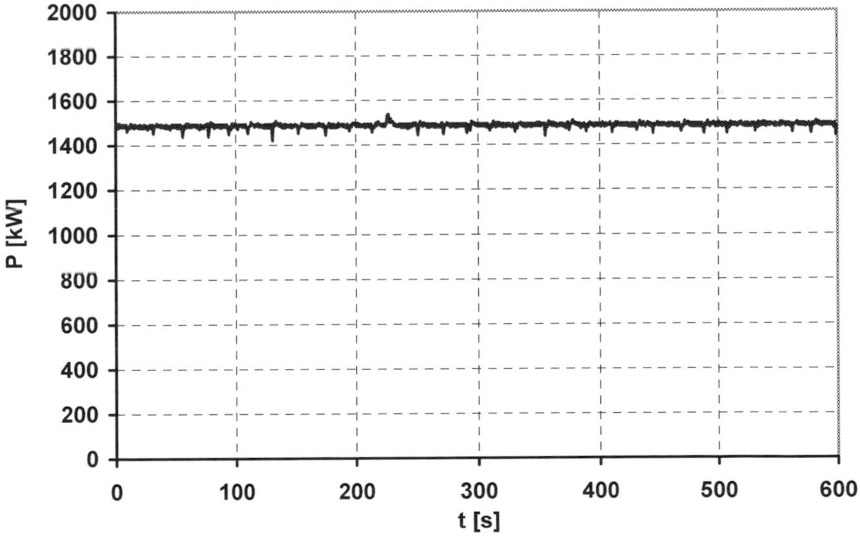


**Fig. 2.7.** Real power generation at full load by Vestas V63-1500 wind turbine [27]

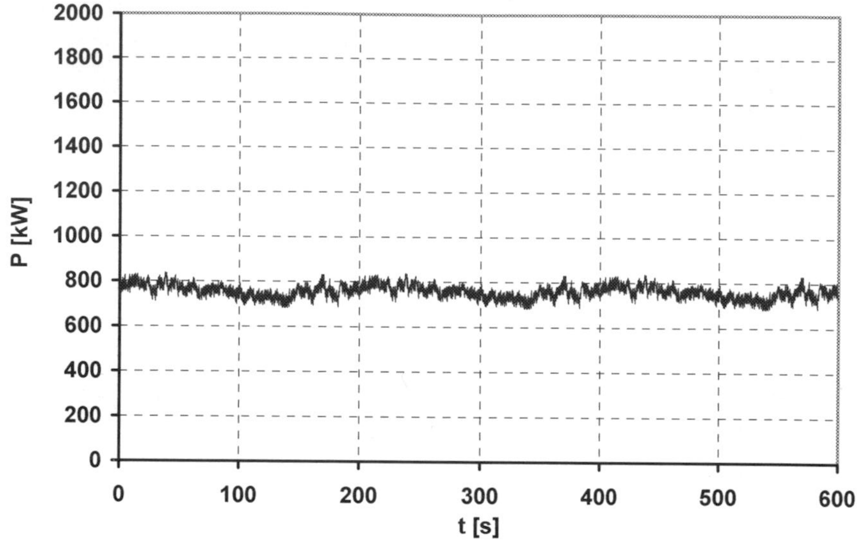


**Fig. 2.8.** Real power generation at partial load by Vestas V63-1500 wind turbine [27]

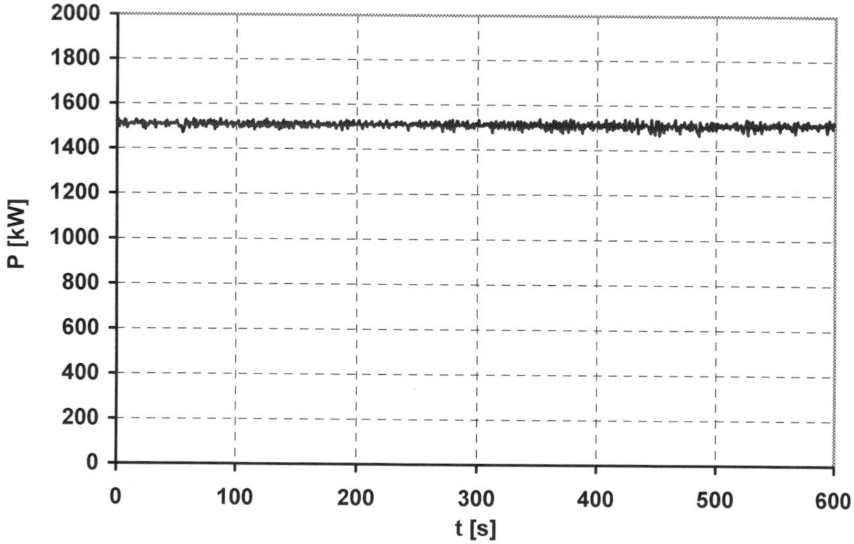


**Fig. 2.9.** Real power generation at full load by Enercon E66-1500 wind turbine [27]

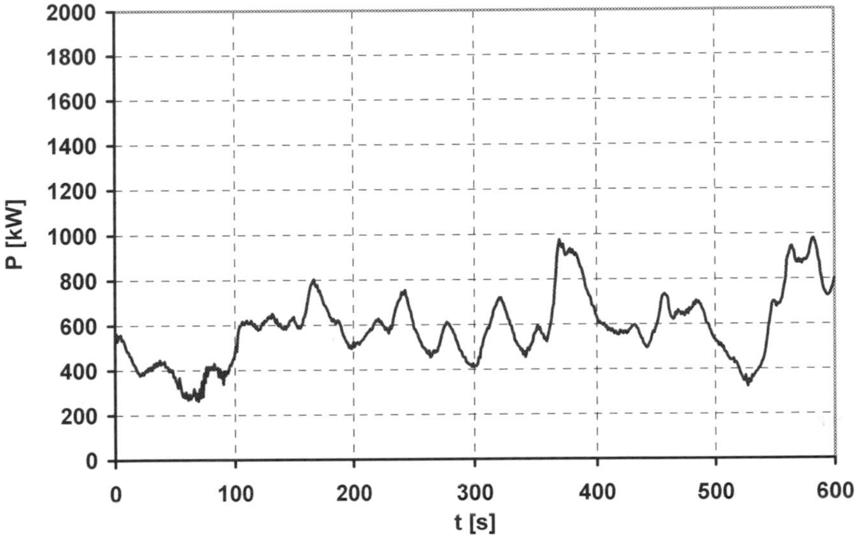


**Fig. 2.10.** Real power generation at partial load by Enercon E66-1500 wind turbine [27] (the high power variation results from high wind velocity variation)

The need for increased power production from the wind and economic reasons, when the rated power of today's wind turbines is still relatively small (2 MW units are now typical), makes it necessary to group wind turbines into so-called wind farms. A typical structure of the electrical connections of a wind farm is presented in Fig. 2.11. Wind turbine generator systems in a farm are usually divided into a few groups, with up to 20 units in the group. In each group, the WTGSs are connected to the transmission line (usually cable).

The wind farm is connected to the power system through a MV/HV transformer. Because of the rated power of the wind farm – tens to a hundred MW – connection to the HV power system is necessary. The auxiliary services and the supervisory control system (if any) of the farm are fed from the MV side of MV/HV transformer.

The power produced by the wind farm is the sum of the power produced by each WTGS. Therefore the wind farm performance (i.e. power variations, etc) is the combination of the performance of each individual wind turbine, which means summing up the output power, voltage and current variations. Fortunately, due to the fact that the wind variation at each unit site is uncorrelated, the wind farm power variation is not a simple sum of the unit power variations but is lower.[8] This can be treated as a positive feature.

[8] The wind farm output power variations can be computed as $\sqrt{n}\Delta P$, where $\Delta P$ is the *i*th turbine output power variation and *n* is the number of wind turbines.

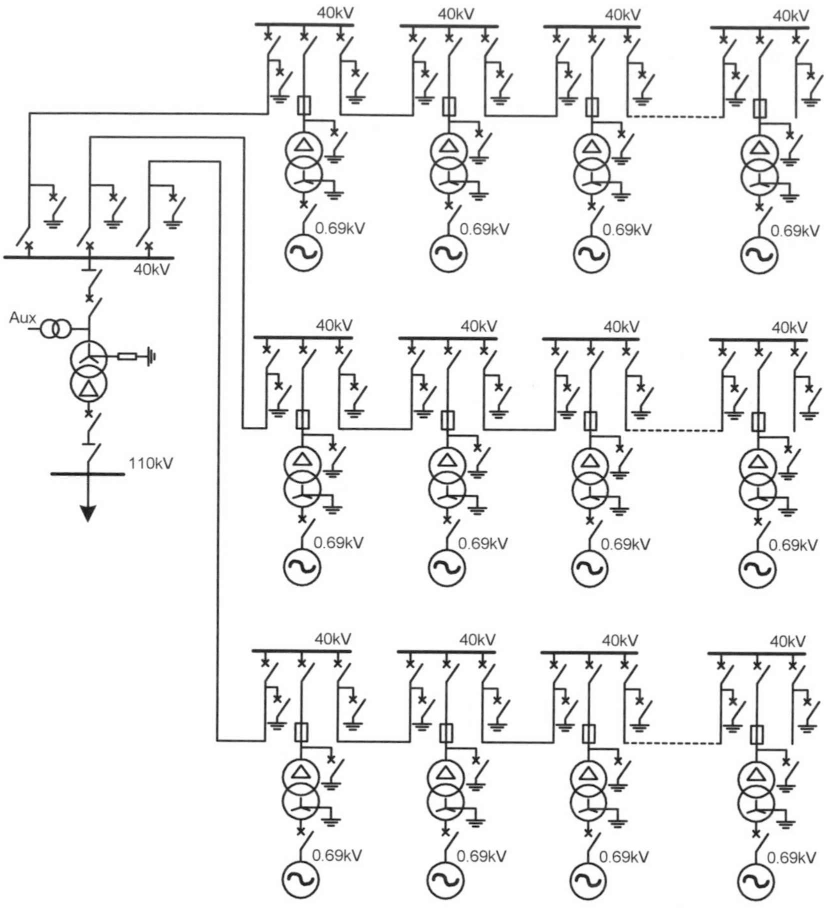


**Fig. 2.11.** Typical structure of a wind farm[9]

An example of 24 MW wind farm operation (power production) during 20 minutes at "full" load and partial load is presented in Figs. 2.12 and 2.13. The curves shown in the figures (typical for wind farms) indicate that the wind farm performance is similar to the single WTGS performance. It means that at partial load, the output power is characterized by higher variations. At full load, the power variations are significantly smaller.

Wind farms can operate with or without supervisory control. Small farms, consisting of a few up to several wind turbines, operate without supervisory control. Large farms are usually equipped with a control system. Supervisory control systems enable formulating the reference power for the whole wind farm (by setting the power reference for each wind turbine). Then, it is possible to control the wind

[9] An MV voltage equal to 40 kV is given as an example here. Usually an MV voltage equal to 10–30 kV is used.

farm power production. Of course, it is possible only to limit the power produced by the wind farm below the maximum power that can be extracted from the wind in given wind conditions (below the power capability). But it is also the only way today to influence the dynamic performance of the wind farm. An example of such control and its effectiveness is shown in Fig. 2.14. The curves in the figure show that by decreasing the power production below the power capability it is possible to keep the power production at the required level. It is also possible to change (relatively fast) the wind farm power production.

Wind farms are built on land, but in recent years there has been (and will probably be in the future) a strong trend towards locating them offshore [16]. There are many reasons for locating wind farms offshore, including but not limited to the following:

- Lack of suitable wind turbine sites on land. It is particularly the case in densely populated countries or regions and in areas of protected landscape.
- Higher wind speeds, up to 20% at some distance from the shore, which gives up to 50% more energy extracted from the wind than on a land site.
- More stable winds. At sea, periods of complete calm are extremely rare and short. Then the effective use of wind turbine capacity is higher than on land.
- Huge offshore wind resources. Wind energy resources in the seas of the European Union with water depths up to 50 m are several times larger than total European electricity consumption.
- Low surface roughness. Because of the smooth water surface, the wind speeds do not increase as much with height above sea level as they do on land. This implies that lower (and cheaper) towers for wind turbines can be utilized.
- Lower turbulence – as a result of relatively small temperature variations during the day. The lifetime of turbines located offshore is predicted to be 5–10 times longer than that of those located on land.

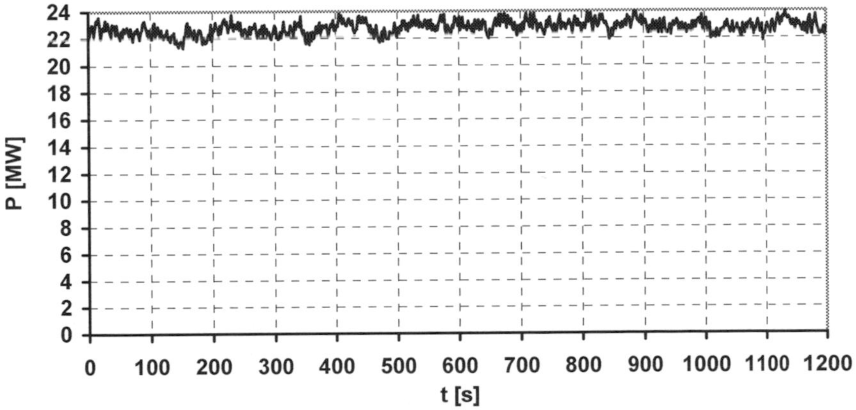


**Fig. 2.12.** Real power generation at "full" load by a 24 MW wind farm [27]

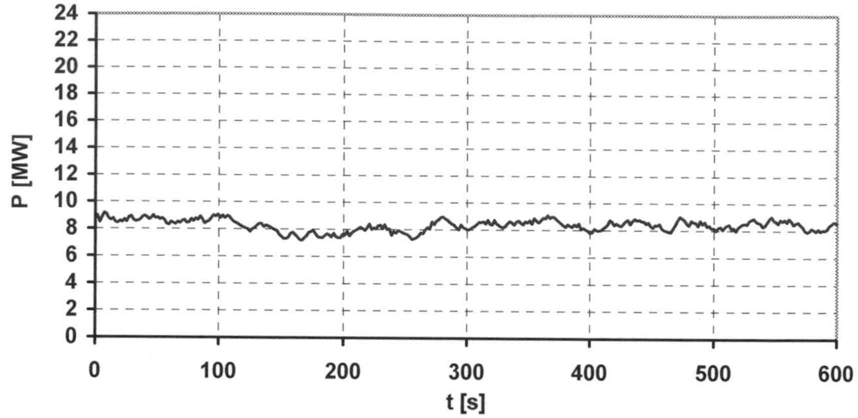


**Fig. 2.13.** Real power generation at partial load by a 24 MW wind farm [27]

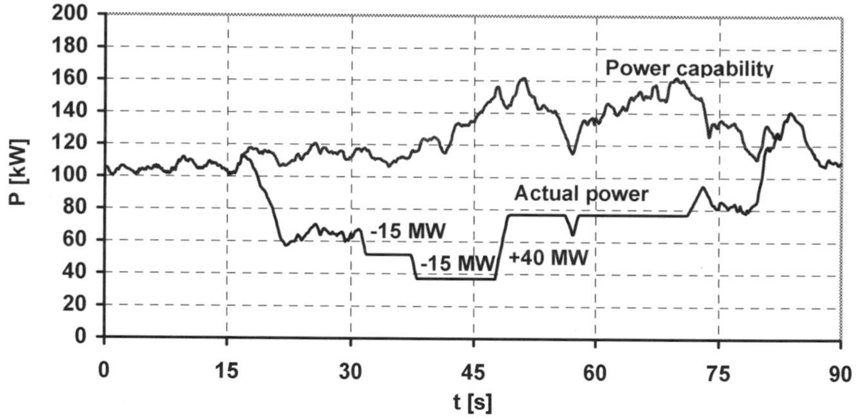


**Fig. 2.14.** Real power generation control by a supervisory farm controller [27]

The wind farm site (onshore or offshore) does not influence the farm performance. Wind farms connected to the electric power system by a DC transmission line are an exception here, but in a limited form. In this case, the power, voltage and current variations which appear at the farm AC side can be almost fully eliminated at the DC connection and in the grid (power system). This is of course true when the power transmitted to the power system is lower than the wind farm power capability. Then the only source of harmonics introduced to the grid is the DC/AC converter.

# 3 Power Quality Standards and Requirements

The aim of an electric power system is to produce and deliver to the consumers electric energy of defined parameters (power quality), where the main quantities describing the electric energy are the voltage and frequency. The power system (in fact the distribution companies) should ensure the continuous supply of the consumers' terminals, where the voltage should be a sinusoidal wave with nominal amplitude (e.g. 311 V in a LV system) and a frequency of 50 Hz.

Unfortunately, it is impossible because of the electric power system structure and features (physical interactions and phenomena). During normal operation of the system, the frequency varies as a result of the variation of the real power generated and consumed at a given time. The voltages (amplitude) vary as a result of the variation of the reactive power generated and consumed. At the same time, because of voltage drops in the transmission lines and transformers it is impossible to keep voltages at the nominal level in all the nodes of the power system. It is also impossible to keep an ideal sinusoidal shape of the voltage or current wave as a result of nonlinearities (e.g. saturation effect) in many devices used for electric energy generation, transmission and consumption. This means that the electric power system (excluding disturbances) operates in a quasi-steady state and that continuous efforts should be (and in fact are) made to keep the quantities defining the electric energy near (in defined limits) the nominal one.

The above-mentioned features of the power system have been well known since power systems were established. But new problems related to the power quality[1] have appeared in recent years. The problems are related to the load equipment and devices used in electric energy generation, transmission and distribution that have become more sensitive to power-quality variations than those used in the past. Many new devices contain microprocessor-based controls and electronic power elements that are sensitive to many types of disturbances. Some of them cause disturbances. As highly variable sources of energy, wind turbines belong to this group of devices. And, of course, the WTGSs are the sources of these problems, too.

These problems, the increased awareness of power quality issues by the consumer, and at the same time, the readiness (and wish) of the electric utilities to deliver a good product cause a need for standardization and evaluation of performance criteria and measurement concepts. This has resulted in many IEEE [40–44], IEC [33–35] and European (CENELEC) [23] standards. The European standard

[1] Power quality problem is defined [20] as any problem manifested in voltage, current or frequency deviations that result in failure or misoperation of customer equipment.

EN 50160-1999 (*Voltage characteristics of electricity supplied by public distribution systems*), which has become a national standard in many European countries, can be treated as basic for further considerations related to power quality. Another standard (draft) IEC 61000-21 (*Wind turbine systems*) [35] can be considered as a base for WTGS influence on the power system assessment.

The standards define the power quality term as *a set of parameters defining the properties of the power supply as delivered to the user in normal operating conditions in terms of continuity of supply and characteristics of voltage (symmetry, frequency, magnitude, waveform)*.

The set of parameters defining the quality of the power supplied to the consumers based on standard EN 50160-1999 refers to voltage variations, harmonics and power frequency variations. The standard gives limits for some of the components of the above mentioned variations. For each of these variations, the value is given. It should not be exceeded for 95% of the time. The length of this window is 10 minutes for most variations, which means that very short timescales are not considered in the standard.[2] The following limits for the low-voltage supply[3] are given:

- *Voltage magnitude*: 95% of the 10-minute averages during one week should be within ±10% of the nominal of voltage 230 V.
- *Voltage magnitude steps*: should not exceed ±5% of the nominal voltage, but changes up to ±10% can occur a number of times per day.
- *Voltage fluctuations*: 95% of the 2-hour long-term flicker (voltage fluctuation leading to light flicker) severity values obtained during one week should not exceed 1. Long-time flicker severity value $P_{\mathrm{lt}}$ is defined as:

$$P_{\mathrm{lt}} = \sqrt[3]{\sum_{i=1}^{12} \frac{P_{\mathrm{st}i}^3}{12}} \tag{3.1}$$

  where $P_{\mathrm{st}i}$ is a short-time (10-minutes average) flicker severity value.
- *Harmonic distortion*: there are given values of harmonic voltage components up to order 25 that should not be exceeded during 95% of the 10-minute averages obtained in one week. The permissible values of harmonics are presented in graphic form in Fig. 3.1.

  Total harmonic distortion, THD, is a measure commonly used to indicate harmonics. It can be calculated for either voltage or current. THD is defined as:

$$THD = \sqrt{\sum_{h=2}^{40} V_h^2} \tag{3.2}$$

  where $V_h$ is the rms value of the harmonic component $h$ of the quantity $V$ (here voltage). The THD calculated for voltage shall not exceed 8%.

---

[2] Neglecting the short-time phenomena in many power quality standards (also in EN 50160 standard) is pointed out as a limitation of the standards.

[3] Most of the consumers are connected to the power system at low voltage.

- *Voltage unbalance*: the ratio of negative and positive sequence voltage should be obtained as 10-minute averages, 95% of those shall not exceed 2% during one week.
- *Signaling voltages*: 95% of the 3-second averages during one day should not exceed 9% for frequencies up to 500 Hz, 5% for frequencies between 1 and 10 kHz, and a threshold decaying to 1% for higher frequencies.
- *Frequency*: 95% of the 10-second averages should not be outside the range 49.5–50.5 Hz.

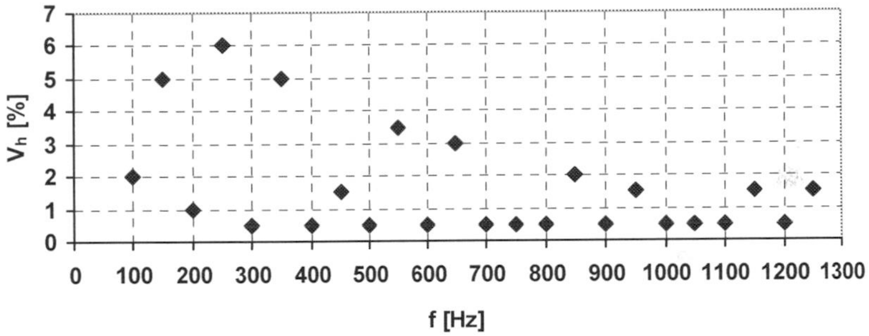


**Fig. 3.1.** Permissible level of voltage harmonics in electricity supplied

EN 50160 standard also mentions event-type phenomena – not giving their limits – such as:

- *Voltage sags*: the frequency of occurrence is between a few tens to one thousand events per year. The duration is mostly less than 1 s, and the voltage rarely drops below 40%. For a system with wind turbines, the sags can occur relatively frequently as a result of switching the WTGS on and off.
- *Short interruptions*: occur between a few tens and several hundred times per year. The duration is in about 70% of the cases less than 1 s.
- *Voltage swells*: (short overvoltages) occur under certain circumstances. Overvoltages due to short-circuit faults elsewhere in the system will generally not exceed 1.5 kV rms in a 230 V system.

The "area of interest" of the standard in relation to voltage magnitude is presented in graphic form in Fig. 3.2. The question marks show regions for which the power quality limits are not precisely defined. An example of voltage sags and swells is presented in Fig. 3.3.

For the medium voltage supply, the power quality parameters limits are defined similarly to the ones presented above, i.e. for low-voltage supply. The difference is practically formal and consists, for a few parameters, in more narrowly defined limits.

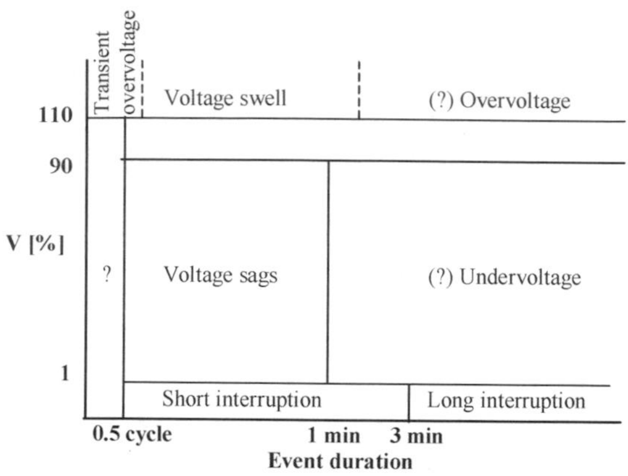


**Fig. 3.2.** Definition of voltage-magnitude events used in EN 50160 [12]

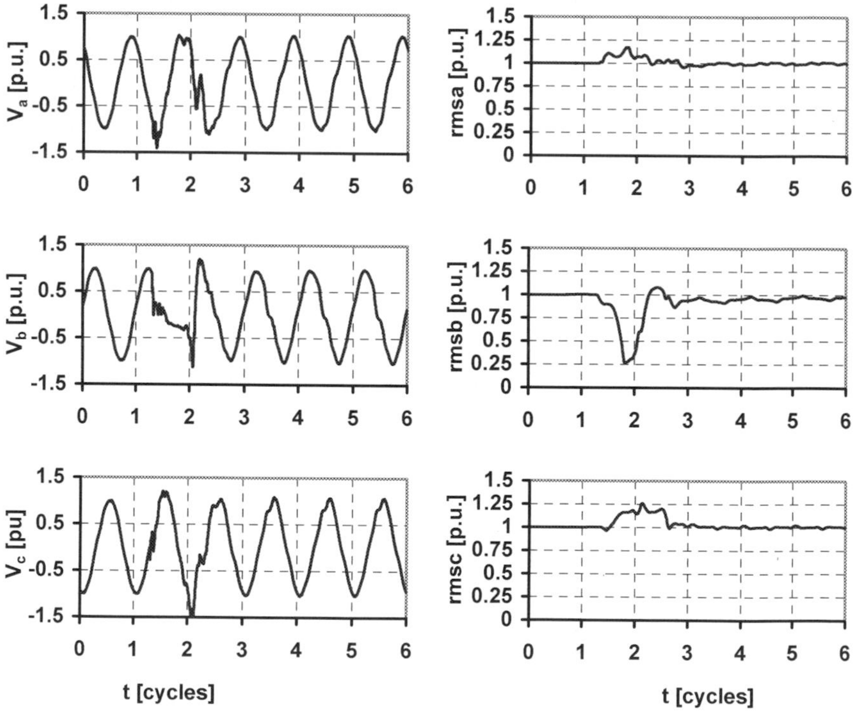


**Fig. 3.3.** Example of voltage sags and swells in a 3-phase system [12]

The standard EN 50160 is criticised in [12] – especially because of using relatively long time averages (10 min) with reference to 95% of the time of energy supply. In the following text and usually in study reports on the assessment of WTGS connection to a given point in the electric power system, the power quality parameter limits defined by standard EN 50160 are taken directly, i.e. without counting averages and without considering the 95% time of supply. This condition is more severe but is also easier to apply in analyses based on mathematical models of power systems. Additionally, by fulfilling the severe conditions in (and by) a modeled system – usually with some simplifications – we can guarantee (by assumption) fulfilling the EN 50160 standard conditions in the real power system.

According to the standard, the long and short-time flicker severity values ($P_{lt}$, $P_{st}$) should be measured by using the flicker meter whose measuring algorithm is described in the standards IEC 61000-4-15:1997, EN 61000-4-15:1998. Utilization of flicker severity values according to standard IEC (EN) 61000-4-15 (by measurement) for voltage fluctuation evaluation is relatively easy, but only when the measured object (e.g. the wind turbine connected to the grid) exists. On the projecting stage (e.g. during assessment of the WTGS influence on the electric power system) these factors are not so useful because it is not easy or even not possible to do exact computing when using typical models of the WTGS and power system. Utilization of the frequency characteristics seems to be an easier way of tackling the flicker problem evaluation. For example, the IEEE standard 141-1993 considers the flicker problem by defining curves that show the observable and objectionable flicker level, as shown in Fig. 3.4. Such an approach (use of frequency characteristics) can be much more useful during the project design stage, because it is relatively easy to obtain the frequency characteristics of modeled objects. A similar characteristic is presented also in IEC standard 61000-3-7[4] in the form of the so-called "$P_{st}$ = 1 curve", which presents the rectangular voltage changes for which the long-term flicker severity value $P_{st}$ equals 1.

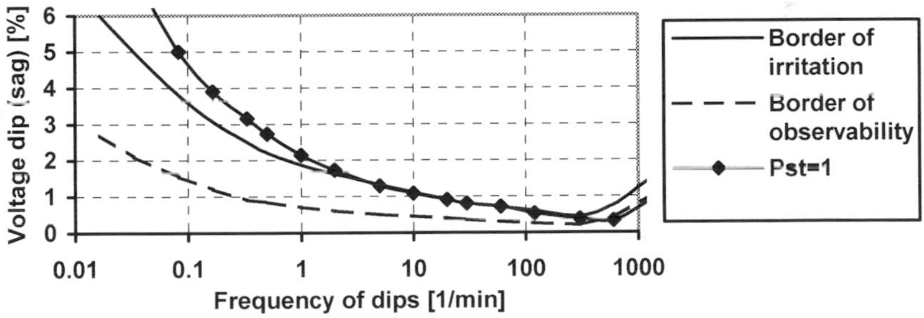


**Fig. 3.4.** Range of observable and objectionable flicker: IEEE standard 141-1993 and curve $P_{st}$ = 1 for LV (230 V) grid from IEC standard 61000-3-7

Today, the measurement and assessment of the power quality characteristics of grid-connected wind turbines is defined by IEC standard 61400-21 *(Wind turbine systems)* prepared by IEC Technical Committee 88. The standard has not been finally released yet but it is widely used by wind turbine manufacturers, the electric

[4] This characteristic is shown as an example for an LV system. In the standard, the flicker severity levels assessment procedure does not in general use the characteristic in an explicit form.

utilities and other commercial organizations involved in "power generation from the wind". The power quality is assessed here from the flicker values given in the approval certificate (wind test) for the wind turbine (of a given type) in combination with values of the known short-circuit apparent power level and the network impedance angle $\psi_k = \arctan(X_k/R_k)$ at the point of common coupling (PCC). The standard, in the case of voltage fluctuations, refers to the standard IEC 61000-3-7, while in the case of harmonics and interharmonic currents it refers to standard IEC 61000-3-6.

The standard IEC 61000-3-7 *(Assessment of emission limits for fluctuating loads in MV and HV power systems)* gives voltage fluctuation limits and simultaneously enables the computation of the flicker emission for a given load that is planned to be connected to the given power system. The power quality limits presented in the standard are usually utilized for evaluating influence of the WTGS on the power system. The standard defines the following quantities (and their limits or way of assessment) related to the flicker levels: compatibility levels, planning levels and emission levels. These are as follows:

- *Compatibility levels* are reference values of $P_{lt}$ and $P_{st}$ factors utilized for coordinating the emission and immunity of the equipment which is part of, or supplied by a supply network in order to ensure electromagnetic compatibility (EMC) *in the whole system*. The levels are based on 95% probability levels of the entire systems using distributions which represent both time and space variations of disturbances. The compatibility levels in LV and MV power system are defined as equal to $P_{st} = 1$ and $P_{lt} = 0.8$.
- *Planning levels* of $P_{lt}$ and $P_{st}$ are used for planning purposes in evaluating the impact of the supply system of *all consumers' loads*. Planning levels are equal to or lower than compatibility levels. Indicative values of planning levels are given in the standard and the levels are equal to: $P_{st} = 0.9$, $P_{lt} = 0.7$ in MV power systems, and $P_{st} = 0.8$, $P_{lt} = 0.6$ in HV (and EHV) power systems.
- *Emission level* of a *fluctuating load* is the flicker level, which would be produced in the power system if no other fluctuating loads were present. The standard describes the method of emission level evaluation.

Details of flicker of the calculation of severity levels according to the standard are presented in Sect. 4.4.

The standard IEC 61000-3-6 *(Assessment of emission limits for distorting loads in MV and HV power systems)* has a formal structure and suggested computing method for connecting large distorting loads (producing harmonics and/or interharmonics) analogues to presented in the IEC 61000-3-7 standard. The part of the standard that considers harmonics and interharmonic currents, is relatively short and practically limited to giving the current limits. Details of the harmonic and interharmonic current level computation, in the case of assessment of WTGS operation in the grid, are presented in Sect. 4.5.

The last group of "standards" (which are rules rather than standards) related to the connection and operation of WTGSs in electric power systems are guidelines prepared by electric utilities [17, 22]. The guidelines consist of various requirements, conditions and limits resulting from various standards, e.g. from those mentioned above and from utility practice. The Polish Power Grid Company's *"Instruction of the traffic and use of the transmission network"*, ENERGA's utility *"Instruction of the traffic and use of the distribution network"*, or German "*Eigenerzeugungs-anlagen am Mittelspannungsnetz"* are examples of such "standards". Usually this type of guideline does not impose more rigorous conditions than those resulting from relevant standards. Examples of such conditions are equations (4.66)–(4.70) in Sect. 4.6.

# 4 WTGS Operation in Power Systems

## 4.1 Introduction

The connection of the wind turbine generator system to parallel operation with the electric power system influences the system state (operating point) and influences the load flow (real and reactive power). At the same time, nodal voltages and power losses themselves change, too. These quantities, defining the electric power system state, can be assumed to be basic but in fact the spectrum of factors influencing the electric power system as a result of the WTGS connection to the grid is much wider. The factors, which are considered in the following sections in a more detailed form, can be generally characterized as follows:

- Location of the wind turbine in the electric power system. The way of connecting it to the electric power system (which in general is an AC system) highly influences the impact of the WTGS (or wind farm) on the power quality. As a rule, the impact on power quality at the consumer's terminal for the WTGS located close to the load (i.e. connected to an MV system) is higher than one connected "electrically" far away (i.e. connected to an HV or EHV system).
- Voltage variations of amplitude and frequency. The variations mainly result from the wind velocity, but other factors influencing the generator torque play an important part, too. Voltage variations are directly related to real and reactive power variations. The WTGSs equipped with an asynchronous generator with a squirrel-cage rotor or with a resistor connected to the rotor (in wounded-rotor machines) are consumers of reactive power, which can cause additional negative problems for the grid.

  Voltage changes appear also as a result of switching the WTGS on and off. The biggest voltage changes are caused by the process of switching the fully loaded WTGS off.
- Flicker – which occurs on the consumer load (lamp bulbs). This results directly from voltage variations.
- Harmonics – which result mainly from the operation of power electronic systems (e.g. converters). The effects of harmonics (voltage or currents) in the power system can lead to degradation of power quality at the consumer terminals, increase of the power losses, overload of condensers by higher harmonics, communication systems malfunction, etc.

- Short-circuit currents and protection systems. The wind turbine connection to the power system increases the short-circuit current level and changes the character of the currents. The short-circuit currents of asynchronous machines are highly damped in time (decay fast) but their initial value can be high. Then, as a minimum, the impulse current cannot be neglected in short-circuit considerations.

  WTGSs are equipped with their own protection systems that usually, relatively soon after the disturbance, cut off the turbine from the power system. It protects the wind turbine but under some circumstances can be bad for the grid. Wind turbine protection systems settings should be co-ordinated with the settings of the power system protection.
- Stability. The WTGS influence on the voltage stability, steady-state stability and dynamic stability can be considered. Wind turbines equipped with asynchronous generators (excluding doubly-fed machines), because of their $Q=f(V)$ characteristics, decrease the voltage stability margin in nodes to which they are connected. In the case of steady-state stability, WTGSs with asynchronous generators in general improve the stability margin.[1] In the case of dynamic stability one can say that connection of the wind turbines (wind farms) to the power system can cause the decrease of the stability margin.
- Self-excitation of a WTGS with an asynchronous generator. This can take place after disconnection of a WTGS with local loads from the power system (islanding process). The risk of self-excitation arises especially when WTGSs equipped with compensating capacitors are used. Compensating systems with condensers and grid capacitance can also increase the risk of self-excitation. In such cases, the voltage and/or frequency usually increase in the islanded part of the grid. Usually, WTGS protection and control systems enable overvoltages and overfrequencies to be avoided.
- Real power losses in power networks with WTGSs. These usually change after connection of the wind turbine (wind farm) to the power system. The effect achieved, i.e. increase or decrease of the power losses, depends on the network configuration, the load flow before connection and the power introduced by the WTGS to the grid.

## 4.2 WTGS Location in the Electric Power System

Wind turbine generator systems (WTGSs) are often located in the regions that have favourable wind conditions and where their location is not burdensome. These regions are low urbanised, which means that the distribution power network in these regions is usually weakly developed. This state is typical for all countries developing a wind power industry.

The point of common coupling (PCC) of the WTGS and the power network,

[1] Of course, improperly developed control systems of WTGSs can decrease the natural – for asynchronous generators – high damping of any electromechanical oscillations.

including the parameters of the power network, the parameters of the WTGS and the structure of the grid (the way that the WTGS or wind farm is connected to the power system) is of essential significance in further operation of the WTGS in the power system and its influence on that system.

WTGSs can be connected to MV networks and to HV networks. The second way is utilized when at least a few WTGSs (or wind farms) are connected to one PCC. A WTGS can be connected to an MV power network in the following ways:

- The WTGS is connected to the existing MV transmission line, which feeds existing customers as well. A schematic diagram of such a network is presented in Fig. 4.1. The distance between the WTGS and the PCC is usually relatively small – up to a few kilometres. This type of connection is also relatively cheap (in comparison with other types of connection).

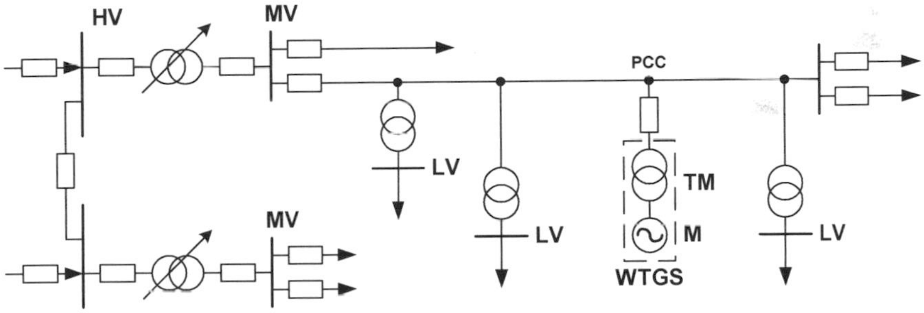


**Fig. 4.1.** WTGS coupled to an MV transmission line

- The WTGS is connected to an MV bus in feeding an HV/MV substation through a separate transmission line. A schematic diagram of such a network is presented in Fig. 4.2. This type of connection is sometimes neccssary when the MV network is weak (which leads to too big a negative influence of the WTGS on the power quality) or the distance between the WTGS and the feeding substation is less than that between the WTGS and the existing MV transmission line. This type of connection is more expensive than the one presented above, but it has some advantages related to the low influence of the WTGS on the customer's load (power quality).

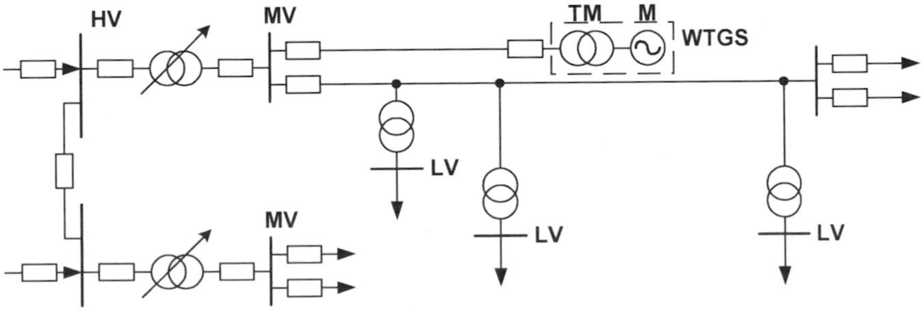


**Fig. 4.2.** WTGS coupled to a separate MV transmission line

- The WTGS is connected to an HV bus in feeding an HV/MV substation through a separate transmission line and its own HV/MV transformer. A schematic diagram of such a network is presented in Fig. 4.3. This type of connection is sometimes necessary when the relatively large (in rated power) WTGS has to be connected to the power network in the region where the MV network is weak (long lines, small diameters of lines, etc.) and the rated power of the feeding transformer is also low. This is the most expensive type of connection than those presented above because of the cost of the HV/MV transformer with its necessary equipment (switches, protection systems, etc.).

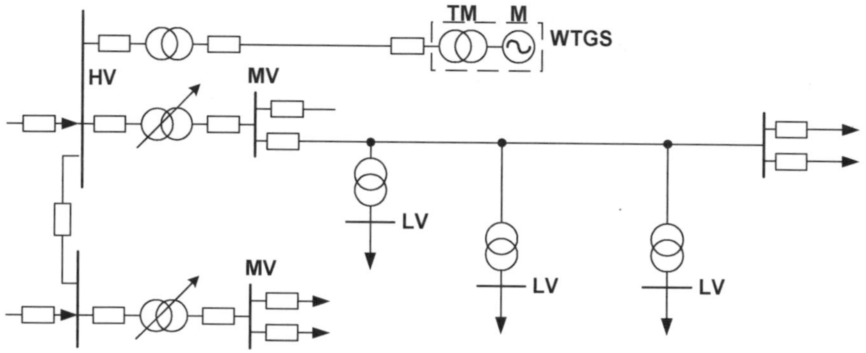


**Fig. 4.3.** WTGS coupled to an HV bus through a separate transmission line and an HV/MV transformer

When the number of WTGSs in a small area is high (in some systems three and more) and their rated power is high or when it is necessary to connect so-called wind farm, the ways of connecting them can be as follows:

- A wind farm is connected to the HV transmission line through its own feeding HV/MV transformer. A schematic diagram of such a network is presented in Fig. 4.4. The transformer is located directly near the HV transmission line. The connection between the WTGSs and the feeding transformer is usually made as an MV cable network. The operation of the wind farm and its influence on the power system in such a case depends on the structure of the adjoining power network. In particular, the structure and existence (or not) of such power elements as synchronous generators (power plants), transformers connecting to the transmission system, etc., that are located near the HV bus to which the farm is connected has a crucial meaning.
- A wind farm is connected directly to the bus of an HV substation. The connection is made through its own feeding HV/MV transformer and MV transmission line(s). A schematic diagram of such a network is presented in Fig. 4.5. This type of connection can be made when the wind farm is located relatively close to the existing HV substation. From the point of view of the operation of the wind farm in the power system, there is a small difference between the two types of connection.

- A wind farm is connected with an HV power system through a DC link. This type of connection is considered for off-shore wind farms and sometimes for islanded systems (also located on geographic islands) that should operate separately from the bulk system controlling both the voltage and frequency. A schematic diagram of such a network is presented in Fig. 4.6.

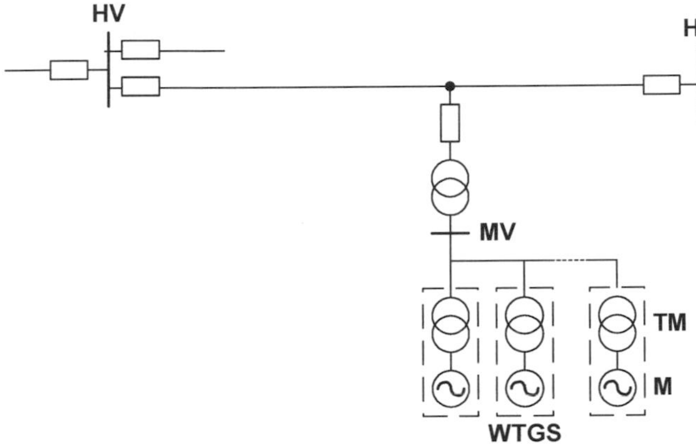


**Fig. 4.4.** Wind farm coupled to an HV transmission line

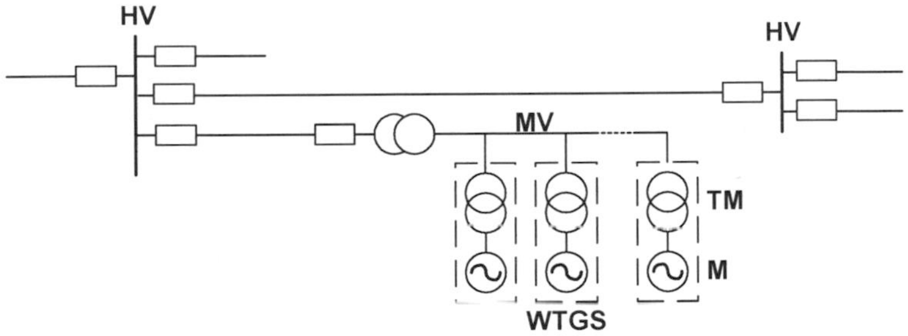


**Fig. 4.5.** Wind farm coupled to an HV bus through a separate transmission line

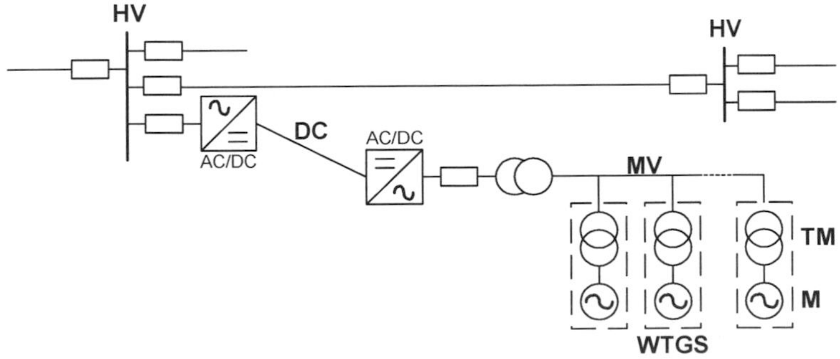


**Fig. 4.6.** Wind farm coupled to an HV power system through a DC transmission line

## 4.3 Voltage Variation

The problem of voltage variation as a result of WTGS operation in an electric power system is a basic one. Let us divide the considerations related to the problem into two cases: voltage variation in a radial network, and voltage variation in a meshed network.

For the first case let us assume a radial structure for the network with the WTGS coupled at the end of the last (from the feeding bus) branch. The network is presented in Fig. 4.7. The wind turbine, generating real power $P_{WTGS}$ and consuming reactive power $Q_{WTGS}$, is connected to node $N$.

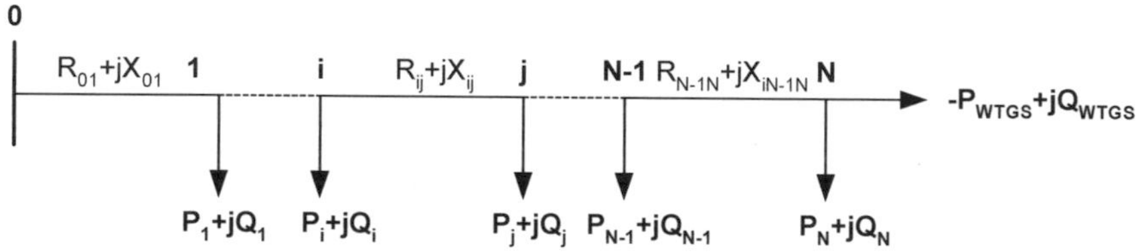


**Fig. 4.7.** Structure of a typical radial power network

If the apparent power flowing in the $ij$th element to the $j$th node is defined as that

$$\underline{S}_{ij}^{(j)} = \sqrt{3}\underline{V}_j \underline{I}_{ij}^{*} = P_{ij}^{(j)} + jQ_{ij}^{(j)}, \tag{4.1}$$

the $ij$th branch current is defined as

$$\underline{I}_{ij} = \frac{\underline{S}_{ij}^{(j)*}}{\sqrt{3}\underline{V}_j^{*}}. \tag{4.2}$$

The voltage losses in the $ij$th branch (defined by serial impedance $\underline{Z}_{ij} = R_{ij} + X_{ij}$), assuming $\underline{V}_j = V_j e^{j0}$, is equal to

$$\delta\underline{V}_{ij} = \delta V'_{ij} + j\delta V''_{ij} = \frac{P_{ij}^{(j)} R_{ij} + Q_{ij}^{(j)} X_{ij}}{V_j} + j\frac{P_{ij}^{(j)} X_{ij} - Q_{ij}^{(j)} R_{ij}}{V_j}. \tag{4.3}$$

For transmission lines in an electric power system the shunt voltage loss $\delta V''$ is usually small. The voltage drop, defined as $\Delta V_{ij} = V_i - V_j$, is often very close to the serial voltage loss $\delta V'$ (real part of voltage loss). Then one can write the voltage drop in the $ij$th branch as equal to

$$\Delta V_{ij} \approx \delta V'_{ij} = \frac{P_{ij}^{(j)} R_{ij} + Q_{ij}^{(j)} X_{ij}}{V_j}. \tag{4.4}$$

The real and reactive power flowing through the $ij$th branch to the $j$th node is equal to

$$P_{ij}^{(j)} = P_j + P_{j,j+1}^{(j)}$$
$$Q_{ij}^{(j)} = Q_j + Q_{j,j+1}^{(j)}, \qquad (4.5)$$

where the power consumed (generated) by the shunt element of the branch is added to the power $Q_j$ consumed at the $j$th node. Then the power flowing into the $N$th node is equal to

$$P_{N-1,N}^{(N)} = P_N$$
$$Q_{N-1,N}^{(N)} = Q_N \qquad (4.6)$$

when WTGS is not connected to the network. It is equal to

$$P_{N-1,N}^{(N)} = P_N - P_{\mathrm{WTGS}}$$
$$Q_{N-1,N}^{(N)} = Q_N + Q_{\mathrm{WTGS}} \qquad (4.7)$$

when WTGS operates in the power system. In general, the power flowing in the $ij$th branch can be defined as

$$P_{ij}^{(j)} = \sum_{p=j}^{N} P_p + \sum_{p=j}^{N-1} \Delta P_{p,p+1} - P_{\mathrm{WTGS}}$$
$$Q_{ij}^{(j)} = \sum_{p=j}^{N} Q_p + \sum_{p=j}^{N-1} \Delta Q_{p,p+1} + Q_{\mathrm{WTGS}}, \qquad (4.8)$$

where $\Delta P_{p,p+1}$, $\Delta Q_{p,p+1}$ are power losses in the $p,(p+1)$th branch.

The voltage drop in branches from node 0 to node $j$ in the case without WTGS is equal to

$$\Delta V_{0j} \approx \sum_{l=0}^{j-1} \frac{P_{l,l+1}^{(l+1)} R_{l,l+1} + Q_{l,l+1}^{(l+1)} X_{l,l+1}}{V_{l+1}} =$$
$$\sum_{l=0}^{j-1} \frac{(\sum_{p=j}^{N} P_p + \sum_{p=j}^{N-1} \Delta P_{p,p+1}) R_{l,l+1} + (\sum_{p=j}^{N} Q_p + \sum_{p=j}^{N-1} \Delta Q_{p,p+1}) X_{l,l+1}}{V_{l+1}}. \qquad (4.9)$$

The voltage drop in the same branches in the case of WTGS operation in the power system is equal to (4.10)

$$\Delta V_{0j}^{\mathrm{WTGS}} \approx \sum_{l=0}^{j-1} \frac{(\sum_{p=j}^{N} P_p + \sum_{p=j}^{N-1} \Delta P_{p,p+1} - P_{\mathrm{WTGS}}) R_{l,l+1} + (\sum_{p=j}^{N} Q_p + \sum_{p=j}^{N-1} \Delta Q_{p,p+1} + Q_{\mathrm{WTGS}}) X_{l,l+1}}{V_{l+1}}.$$

The voltage drop in the *j*th node as a result of the WTGS connection to the power system is equal to

$$\Delta V_j = \Delta V_{0j} - \Delta V_{0j}^{\mathrm{WTGS}} . \tag{4.11}$$

Assuming that the nodal voltages $V_l$ are close to the rated voltage $V_n$ and, simultaneously, that the power losses in both cases are equal to each other[2], (4.11) takes the following form:

$$\Delta V_j = \frac{P_{\mathrm{WTGS}} \sum_{l=0}^{j-1} R_{l,l+1} - Q_{\mathrm{WTGS}} \sum_{l=0}^{j-1} X_{l,l+1}}{V_n} . \tag{4.12}$$

Equation (4.12) can be used for evaluation of the voltage drop in a given node as a result of the power generation change[3], and for computing the maximum rated power of the WTGS that can be connected to the power system in the given node *N*, satisfying voltage drop limit.

In general, the voltage drop in each node, as a result of the WTGS power-generated change should be less than the voltage drop limit $\Delta V_{\max}$. This requirement can be described as

$$\underset{j \in \{1 \ldots N\}}{\forall} \Delta V_j \le \Delta V_{\max} . \tag{4.13}$$

By rearranging (4.12) and (4.13) it is possible to relate the voltage drop limit and the WTGS power injection in node *N*:

$$\frac{P_{\mathrm{WTGS}} \sum_{l=0}^{j-1} R_{l,l+1} - Q_{\mathrm{WTGS}} \sum_{l=0}^{j-1} X_{l,l+1}}{V_n} \le \Delta V_{\max} . \tag{4.14}$$

Equation (4.14) can also be used for evaluation of the maximum (e.g. rated) power injection. In the considered case, for the given power factor $\tan\varphi_{\mathrm{WTGS}}$ of the wind turbine (connected to node *N*), the WTGS real power change (and simultaneously the rated power[4]) should be less than

$$P_{\mathrm{WTGS}} \le \frac{\Delta V_{\max} V_n}{\sum_{l=0}^{N-1} R_{l,l+1} - \tan\varphi_{\mathrm{WTGS}} \sum_{l=0}^{N-1} X_{l,l+1}} . \tag{4.15}$$

Equation (4.15) can be used for determining the rated power of the wind turbine that can be connected to the given network in node *N*.

---

[2] This is a simplification only. Power losses are considered in Sect. 4.8.

[3] The power generated by WTGS, i.e. $\underline{S}_{\mathrm{WTGS}} = -P_{\mathrm{WTGS}} + \mathrm{j}Q_{\mathrm{WTGS}}$ can be also considered as a change of power generated by the wind turbine during its operation.

[4] The maximum power change will take place after disconnecting the fully loaded $P_{\mathrm{WTGS}} = P_{\mathrm{nWTGS}}$ wind turbine from the power system.

Equation (4.12) enables us to formulate a "control" rule allowing the WTGS to keep the voltage drops equal to zero [83]. This could be possible for generating units whose consumed/generated reactive power can be controlled.[5]

Then, assuming the voltage drop $\Delta V_N$ at the PCC (node $N$) equal to zero, the reactive power consumed by the WTGS should be equal to:

$$Q_{\text{WTGS}} = P_{\text{WTGS}} \frac{\sum_{l=0}^{N-1} R_{l,l+1}}{\sum_{l=0}^{N-1} X_{l,l+1}} = P_{\text{WTGS}} \frac{1}{\tan \psi_{\text{k}}}, \tag{4.16}$$

where $\psi_{\text{k}}$ is the short-circuit impedance angle.

Equation (4.16), after multiplication by the rated voltage and the short-circuit impedance at $N$th node and after rearranging, can take the following form:

$$\frac{\Delta V_N}{V_{\text{n}}} = \frac{cP_{\text{WTGS}}}{S''_{\text{k}N}} \frac{1 - \tan \varphi_{\text{WTGS}} \tan \psi_{\text{k}}}{\sqrt{1 + (\tan \psi_{\text{k}})^2}}, \tag{4.17}$$

where $c$ is the voltage factor, and where $S''_{\text{k}N}$ is the apparent short-circuit power at the $N$th node.

According to (4.17) the rated power of the wind turbine connected to the considered network at node $N$ should fulfil the following condition:

$$\frac{P_{\text{nWTGS}}}{S''_{\text{k}N}} \leq \frac{\Delta V_{\max}}{cV_{\text{n}}} \frac{\sqrt{1 + (\tan \psi_{\text{k}})^2}}{\left|1 - \tan \varphi_{\text{WTGS}} \tan \psi_{\text{k}}\right|}. \tag{4.18}$$

In many publications, the minimum value of the quotient of the apparent short-circuit power $S''_{\text{k}N}$ and the WTGS apparent rated power $S_{\text{nWTGS}}$ is presented for computing the possibilities of connecting the wind turbine to the given network. The quotient from (4.18) can be computed as:

$$\frac{S''_{\text{k}N}}{S_{\text{nWTGS}}} \geq \frac{cV_{\text{n}}}{\Delta V_{\max}} \frac{\left|1 - \tan \varphi_{\text{nWTGS}} \tan \psi_{\text{k}}\right|}{\sqrt{1 + (\tan \psi_{\text{k}})^2} \sqrt{1 + (\tan \varphi_{\text{nWTGS}})^2}}. \tag{4.19}$$

In Fig. 4.8, based on (4.19), the values of the $S''_{\text{k}N} / S_{\text{nWTGS}}$ quotient as a function of the tangent of the short-circuit impedance angle $\tan \psi_{\text{k}}$ for chosen values of the power factor $\tan \varphi_{\text{WTGS}}$ and the voltage drop limit $\Delta V_{\max}$ are presented.

[5] Today, such a type of control is not utilized in wind turbine generator systems. For WTGSs with doubly-fed asynchronous machines and synchronous generators, the reactive power is usually kept equal to zero. For other WTGSs with asynchronous generators, the reactive power consumption results from the machine characteristics (including the compensation capacitor bank).

High values of the short-circuit impedance angle are typical for sites located near transformers and in an HV/EHV network. For these sites, the apparent short-circuit power can be relatively small in comparison with the wind turbine rated power.
Small values of the short-circuit impedance angle are typical for sites located far away (in terms of the impedance) in the MV network (especially behind transmission lines characterized by a small cross-sectional area of the conductor and cables). For these sites, the apparent short-circuit power should be relatively high in comparison with the wind turbine rated power.

Figure 4.8 also shows that the WTGS reactive power consumption (in limited amount) decreases the voltage variations and mitigates the requirement for minimal short-circuit power at the point of the wind turbine connection to the network.

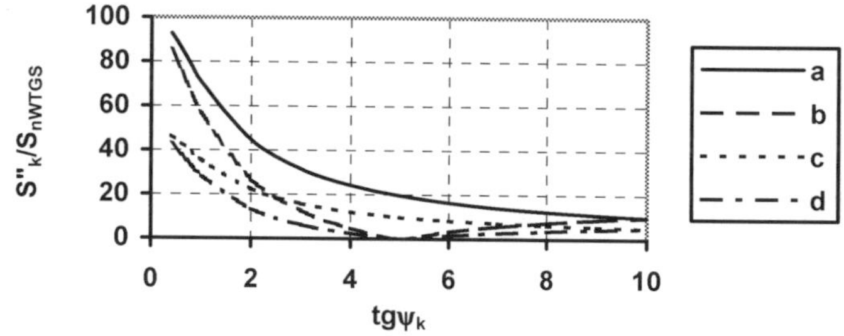


**Fig. 4.8.** Quotient of the apparent short-circuit power and WTGS rated power fulfilling a given voltage drop limit (a: $\tan\varphi = 0$, $\Delta V_{max} = 1\%$; b: $\tan\varphi = 0.2$, $\Delta V_{max} = 1\%$; c: $\tan\varphi = 0$, $\Delta V_{max} = 2\%$; d: $\tan\varphi = 0.2$, $\Delta V_{max} = 2\%$)

The analysis of the voltage variation in a meshed network in analytical form is more complicated than in a radial network. Therefore for this type of analysis, load-flow computer programs are usually utilized. An analytical analysis in this case is possible for a doubly-fed transmission system, such as the one presented in Fig. 4.9.

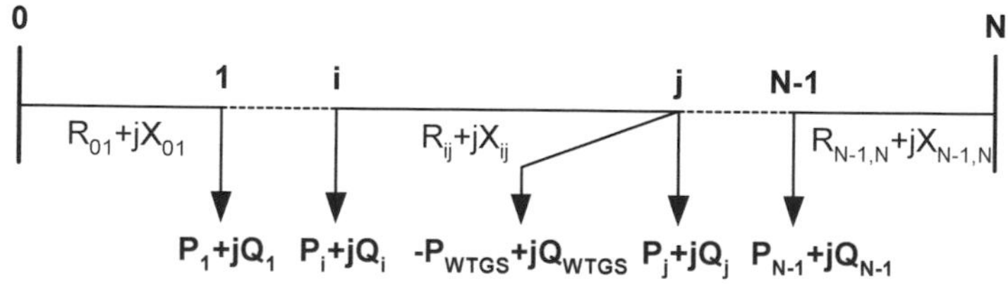


**Fig. 4.9.** Structure of a typical doubly-fed power network

For the following considerations, let us assume that the WTGS is coupled to the network at the $j$th node. The current flowing in the $ij$th branch can be computed as a sum of two currents: the circulating current and the forced current. The first current depends on the difference of voltages at nodes 0 and $N$, and is the same in each branch:

$$\underline{I}_{\text{cir}} = \frac{\underline{V}_0 - \underline{V}_N}{\sqrt{3}\underline{Z}_{0N}} = \frac{\underline{V}_0 - \underline{V}_N}{\sqrt{3}\sum_{p=0}^{N-1}\underline{Z}_{p,p+1}}. \tag{4.20}$$

The second current depends on the loads located in the transmission system. The forced current in the $ij$th branch (when the loads are defined by currents $\underline{I}_k$) in a system without an WTGS can be computed as

$$\underline{I}_{ij} = \frac{-\sum_{k=1}^{i}\underline{I}_p\underline{Z}_{0,p} + \sum_{p=j}^{N-1}\underline{I}_p\underline{Z}_{p,N}}{\sum_{p=0}^{N-1}\underline{Z}_{p,p+1}}. \tag{4.21}$$

The forced current in the system after WTGS coupling is equal to

$$\underline{I}_{ij} = \frac{-\sum_{p=1}^{i}\underline{I}_p\underline{Z}_{0,p} + \sum_{p=j}^{N-1}\underline{I}_p\underline{Z}_{p,N} + \underline{I}_{\text{WTGS}}\underline{Z}_{jN}}{\sum_{p=0}^{N-1}\underline{Z}_{p,p+1}}. \tag{4.22}$$

The nodal voltage in the $i$th node is equal to

$$\underline{V}_i = \underline{V}_0 - \sqrt{3}\underline{I}_{\text{cir}}\sum_{p=0}^{i-1}\underline{Z}_{p,p+1} - \sqrt{3}\sum_{p=0}^{i-1}\underline{I}_{p,p+1}\underline{Z}_{p,p+1}. \tag{4.23}$$

Then the voltage change in the $i$th node, where $i = 1, 2,..., j$-1, as a result of the WTGS connection to the network at the $j$th node is equal to

$$\underline{V}_i - \underline{V}_i^{\text{WTGS}} = \delta\underline{V}_i = \frac{\sqrt{3}\underline{I}_{\text{WTGS}}\sum_{p=0}^{i-1}\underline{Z}_{p,p+1}\sum_{p=j}^{N-1}\underline{Z}_{p,p+1}}{\sum_{p=0}^{N-1}\underline{Z}_{p,p+1}}. \tag{4.24}$$

According to (4.24), the voltage change in the $j$th node after substituting WTGS current $\underline{I}_{\text{WTGS}}$ by the apparent power $\underline{S}_{\text{WTGS}}$ and the nodal voltage $\underline{V}_j$ is equal to

$$\delta\underline{V}_j = \frac{\underline{S}^*_{\text{WTGS}}\sum_{p=0}^{j-1}\underline{Z}_{p,p+1}\sum_{p=j}^{N-1}\underline{Z}_{p,p+1}}{\underline{V}^*_j\sum_{p=0}^{N-1}\underline{Z}_{p,p+1}} = \frac{\underline{S}^*_{\text{WTGS}}}{\underline{V}^*_j}\underline{Z}_{\text{k}}, \tag{4.25}$$

where $\underline{Z}_{\text{k}}$ can be treated as the short-circuit impedance at the $j$th node, when nodes 0 and $N$ are infinity busses.

The real part of (4.25) is equal to the formula presented in (4.12). Therefore, the considerations related to (4.13)–(4.19) and Fig. 4.8 for a radial network are also valid for a doubly-fed network (transmission system).

The consideration is based on the power network simplification, which consists in the replacement of the network by the short-circuit impedance. Unfortunately, such an approach gives limited computational accuracy. Therefore, computation of the nodal voltage variation is usually made by using a complex (and exact) mathematical model of the grid. This ensures that the results of the computation are valid for the real system.

In general, while evaluating the voltage variation, two types (steps) of analysis are possible:

- Computation of the voltage change in each node as a result of the power injection to the given node (node of planned WTGS connection). This step of the computation can be made without any knowledge about the WTGS characteristics. Only the knowledge about the network structure and the parameters is needed.
  In fact, only the network capacity can be computed here, where the network capacity means the maximum power (real and reactive[6]) which, when injected into the given node, makes a given change of voltage.
- Computation of the voltage change in each node as a result of the power injection into the node of the planned WTGS connection, taking into account the WTGS characteristics. In this step, one of the following characteristics should be known: the reactive power consumption versus the real power and the terminal voltage $Q = f(P, V)$; the power factor as a function of the terminal voltage and the real power $\tan\varphi = f(P, V)$ or $\cos\varphi = f(P, V)$.

An example of an analysis based on a detailed model of the grid is presented below. Let us assume that the planned WTGS will be connected to the network[7] presented in Fig. 6.1 at node 5. Let us consider two types of network: the first consisting of overhead transmission lines and the second consisting of cable lines only. In both cases, the lengths of the lines are the same.

The nodal voltage change as a function of the power injected to node 1 (first type of analysis) is presented in Figs. 4.10 and 4.11. Figure 4.10 shows the voltage change at node 8 for the network with overhead lines and Fig. 4.11 shows the same voltage change achieved in the network with cables. The voltage change equal to zero at point $(P, Q) = (0, 0)$ shows the initial state of the network – a state without the WTGS.

The change of voltage at node 8 only is presented here because the highest change of voltage usually (but not always) takes place at the point of the WTGS connection to the network (which will be shown for this network later). An addi-

[6] The real power is injected into the network node but the reactive inductive power is consumed in the case of the wind turbine.

[7] The network structure, data of transmission lines, transformers and loads are presented in Sect. 6.1.

tional reason is that in the real system most of the loads are located in the low-voltage network.

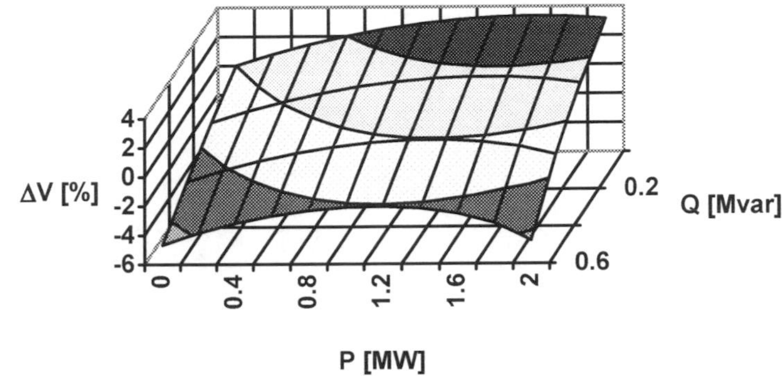


**Fig. 4.10.** Voltage change at node 8 in the network with overhead lines ($V_2 = 1.04$ p.u. = const.; initial values of voltages at point $(P, Q) = (0, 0)$ are $V_1 = 1.0$ p.u., $V_8 = 0.98$ p.u.)

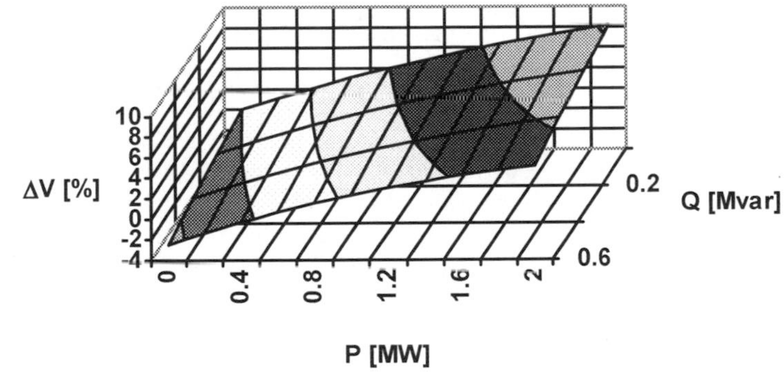


**Fig. 4.11.** Voltage change at node 8 in the network with cables ($V_2 = 1.01$ p.u. = const.; initial values of voltages at point $(P, Q) = (0, 0)$ are $V_1 = 1.0$ p.u., $V_8 = 0.98$ p.u.)

The figures show that an increase of the real power injected into the network causes an increase of the voltage (the case with the high value of the power factor $\tan\varphi$ in the network with overhead lines is an exception). The voltage increase is higher in a network consisting of cables. An increase of the reactive power consumption at a given node causes a decrease of the nodal voltage.

It is also possible to keep the voltage unchanged by controlling the reactive power consumed as a function of the real power generated. For the network with overhead lines, the maximum consumption of reactive power can be found. In the considered case, the maximum reactive power is equal to about $Q = 0.4$ Mvar for a real power injection equal to about $P = 1.2$ MW. For the cabled network, the real power increase needs the reactive power increase to keep the voltage unchanged.

Today, wind turbines equipped with doubly-fed asynchronous machines or synchronous machines operate with small consumption of reactive power (tan$\varphi$ close to zero). The turbines equipped with asynchronous generators with squirrel-cage rotors or equipped with a dynamic slip-control system (variable resistors connected to the rotor) operate with a power factor usually not higher than 0.2–0.25 at rated load (only the no-load reactive power is compensated). Only some of them are equipped with capacitor banks enabling compensation of the reactive power at a higher level.

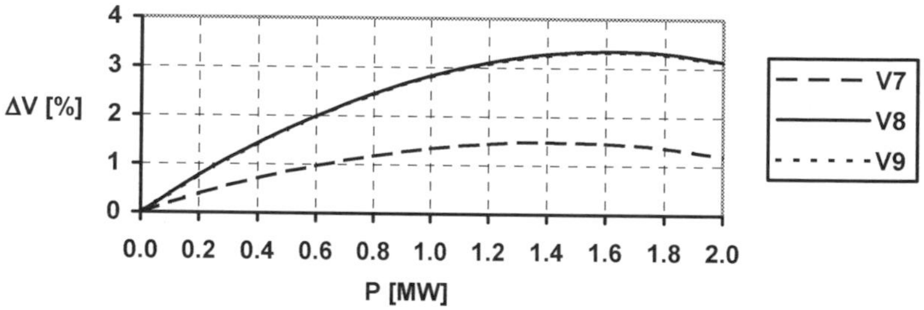


**Fig. 4.12.** Voltage change at LV nodes (network with overhead lines), tan$\varphi$ = 0, (curves $V_8$ and $V_9$ are located close to each other)

In such a case, i.e. full compensation of the reactive power, as a cross-section of the plane presented in Fig. 4.10, it is possible to draw curves showing the voltage change at various nodes. As an example, Fig. 4.12 presents the voltage change in all LV nodes of the considered network. The curves show that the maximum voltage change (in a radial network) takes place at the node of the WTGS connection. The same (very close) voltage change is visible at node(s) located at the point between the wind turbine connection (PCC) and the end of the network (e.g. node 9, voltage $V_9$). The voltages at the nodes located between the PCC (node 5) and the feeding bus (node 2) change to a lower degree (e.g. voltage $V_7$). The voltage change is smaller for the node located far from the PCC (and close to the feeding bus 2).

When the power factor (tan$\varphi$ or cos$\varphi$) as a function of the power generated by the wind turbine is known, it is possible to compute the nodal voltages directly for the given WTGS type. The result of such a computation is presented in Fig. 4.13 for three values of the power factor tan$\varphi$ and for the WTGS with known characteristics $Q = f(P)$ (Fig. 4.13b)[8].

The curves show that for a real power injection (or shedding) at a level less than 1 MW the voltage changes are minimal when the power factor is relatively high (tan$\varphi$ = 0.4). But when the real power change increases, the voltage change increases drastically (the voltage drop is high). This is especially visible for the network with overhead transmission lines. The voltage change in a system with cable transmission lines has a near-linear character.

[8] The characteristic presented in Fig. 6.3 and its extrapolation (above 0.9 MW) is used here.

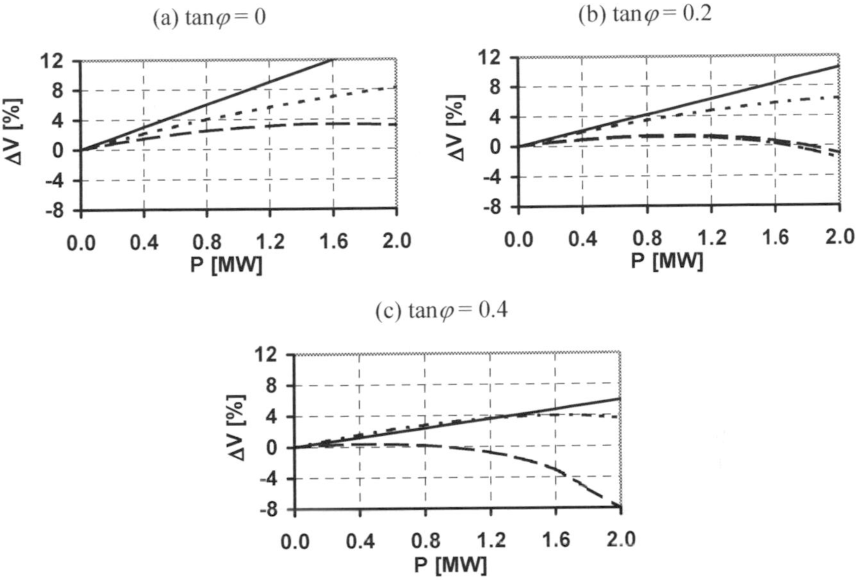


**Fig. 4.13.** Voltage change at node 8 ($V_8$) as a result of power injection at node 1 (dashed line: network with overhead lines; dotted line: network with cables; continuous line: voltage limit according to (4.17)

In the case when the power factor is relatively low, $\tan\varphi \leq 0.2$–$0.25$, the curve showing the maximum change of voltage $\Delta V_{max}$ computed from (4.17) is located above the curves showing the voltage change computed for the network consisting of overhead and cable lines, in Figs. 4.13a and 4.13b. This means that (4.17) can be used for voltage variation evaluation. This equation, on the one hand, gives some margin of safety, but on the other hand it is too rigorous, especially for WTGSs operating with a power factor close to zero.

In the case where the power factor is higher, i.e. $\tan\varphi \geq 0.4$, the maximum change of voltage $\Delta V_{max}$ is located close (above and below) to the curve computed for the network with cable lines[9]. At the same time, for the network with overhead lines the equation gives too rigorous results. In this case, (4.17) can give too optimistic values of voltage variation.

In general, it must be emphasized that using a mathematical model of the network should be common practice in the case of voltage variation computing. In such a case it is possible to obtain more exact values than when using a network model replaced by a single impedance (as in (4.17)). A simplified model can be utilized only for initial step calculations or in the case where a high-capacity network is considered.

And finally, the frequently published requirements linking the maximum rated power of a WTGS and the short-circuit apparent power at the point of common

[9] Because modern WTGSs operate with power factor less than 0.3, this case is not realistic.

coupling (PCC) are worth considering. According to the requirements, the wind turbine rated power should not exceed the short-circuit apparent power divided by 20, or 40, or $20\sqrt{N}$, where $N$ is the number of WTGSs connected to the given PCC. Because the short-circuit apparent power at node 5 of the considered network is equal to $S''_k = 13.6$ MVA, the maximum rated power of the WTGS is about 0.34–0.68 MW, which depends on the condition used and the number of wind turbines considered. In this context we can consider the power change as a result of switching the WTGS (operating at rated power) on or off. Then, according to Fig. 4.13, the voltage change can reach up to 4% in a cabled network and up to 3% in a network with overhead lines when the wind turbine operates with the power factor $\tan\varphi$ equal to zero. When the power factor increases, the voltage change decreases and "stabilizes" at a level of about 3% (for $P = 0.68$ MW) for a cabled network.

Using the more rigorous condition $S_{\text{nWTGS}} \leq S''_k / 40$ – leading to a WTGS with rated power less or equal to $P_{\text{nWTGS}} \leq 0.34$ MW – we achieve lower voltage changes, which in most cases do not exceed 2% (for the considered network).

## 4.4 Flicker

The flicker problem is considered based on the short-term and long-term flicker severity coefficients. At the same time, in the standards related to the power quality and the WTGS operation, the dynamic voltage fluctuations (as a result of switching operations) are considered.

In general, at the point where the wind turbine installation (single turbine or wind farm), is connected to the electric power system, the following conditions should be fulfilled:

$$\begin{aligned} P_{\text{st}} &\leq E_{\text{Pst}i} \\ P_{\text{lt}} &\leq E_{\text{Plt}i} \end{aligned} \tag{4.26}$$

$$d \leq \frac{\Delta V_{\text{dyn}}}{V_{\text{n}}}, \tag{4.27}$$

where $P_{\text{st}}$, $P_{\text{lt}}$ – short-term and long-term flicker emission from the wind turbine installation at the given PCC,

$E_{\text{Pst}i}$, $E_{\text{Plt}i}$ – short-term and long-term flicker emission limits for the relevant PCC,

$d$ – relative voltage change due to a switching operation of the wind turbine installation,

$\Delta V_{\text{dyn}}/V_{\text{n}}$ – maximum permitted dynamic voltage change.

The left-hand sides of (4.26) and (4.27) (emission level) can be computed according to the rules given in the IEC 61000-21 standard[10] "*Wind turbine systems*", while the right sides (limits) can be computed according to the rules given in IEC 61000-3-7 standard "*Assessment of emission limits for fluctuating loads in MV and HV power systems*".

*Flicker emission computation*

According to the IEC 61000-21 standard, flicker emission at the PCC should be computed for the state of the wind turbine(s) in continuous operation and for the state of switching operation. In the second case, the following events should be considered:

- switching operation at start-up at cut-in wind speed,
- switching operation at start-up at rated wind speed,
- the worst case of switching between generators.

During continuous operation of the wind turbine, the flicker emission level at the PCC for the single wind turbine is defined as follows:

$$P_{st} = P_{lt} = c(\psi_k, v_a)\frac{S_n}{S''_k}, \tag{4.28}$$

and for $N$ wind turbines it is defined as:

$$P_{st} = P_{lt} = \frac{1}{S''_k}\sqrt{\sum_{i=1}^{N}(c_i(\psi_k, v_a)S_{ni})^2}, \tag{4.29}$$

where
$c(\psi_k, v_a)$ – wind turbine flicker coefficient,[11]
$S_n$ – rated apparent power of the wind turbine,
$S''_k$ – apparent short-circuit power at the PCC,
$\psi_k$ – network impedance phase angle,
$v_a$ – annual average wind speed.

During the wind turbine switching operation, the flicker emission levels at the PCC for a single WTGS are defined as follows:

$$P_{st} = 18N_{10}^{0.31}k_f(\psi_k)\frac{S_n}{S''_k} \tag{4.30}$$

---

[10] IEC 61000-21 is a draft of the standard, but it is commonly used today by the manufacturers and the electric utilities.

[11] The flicker coefficient $c(\psi_k, v_a)$ and – presented in the following equations – the maximum number of switching operations within 10 and 120 minute periods $N_{10}$, $N_{120}$, the flicker step factor $k_f(\psi_k)$ and the voltage change factor $k_u(\psi_k)$ can be obtained by measurement (measuring method defined in the IEC 61000-21 standard) or from the wind turbine certificate. Approval certificates are prepared by specialized institutions, such as, for example DEWI, (German Wind Energy Institute), Windtest Kaiser-Wilhelm-Koog GmbH, DNV (Det Norske Veritas) or RISO National Laboratory.

$$P_{\mathrm{lt}} = 8N_{120}^{0.31} k_{\mathrm{f}}(\psi_{\mathrm{k}})\frac{S_{\mathrm{n}}}{S_{\mathrm{k}}''}. \tag{4.31}$$

For $N$ wind turbines they are defined as follows:

$$P_{\mathrm{st}} = \frac{18}{S_{\mathrm{k}}''}(\sum_{i=1}^{N} N_{10i}(k_{\mathrm{f}i}(\psi_{\mathrm{k}})S_{\mathrm{n}i})^{3.2})^{0.31} \tag{4.32}$$

$$P_{\mathrm{lt}} = \frac{8}{S_{\mathrm{k}}''}(\sum_{i=1}^{N} N_{120i}(k_{\mathrm{f}i}(\psi_{\mathrm{k}})S_{\mathrm{n}i})^{3.2})^{0.31}, \tag{4.33}$$

where $N_{10}$, $N_{120}$ - maximum number of switching operations within 10 and 120 minute periods (if the values are not available from the WTGS manufacturer then the following will be assumed: $N_{10} = 10$, $N_{120} = 120$),[12]

$k_{\mathrm{f}}(\psi_{\mathrm{k}})$ - flicker step factor.

The relative voltage change (percentage) due to a switching operation of the wind turbine installation according to the IEC 61000-21 standard is defined as

$$d = 100k_{\mathrm{u}}(\psi_{\mathrm{k}})\frac{S_{\mathrm{n}}}{S_{\mathrm{k}}''}, \tag{4.34}$$

where $k_{\mathrm{u}}(\psi_{\mathrm{k}})$ is the voltage change factor.

In the case of an installation with many wind turbines (wind farm), the probability of performing the switching operation at the same time is very low. Then (4.34) can be utilized when wind farms and/or systems with a few (many) wind turbines are considered.

*Flicker limits computation*

According to the IEC 61000-3-7 standard, the flicker emission limits at the PCC can be computed in one of the following three ways (stages):

- by simplified evaluation of disturbance emission,
- by computing the emission limits as proportional to the agreed power of the consumer,
- by acceptance of the higher emission levels on an exceptional and precarious basis.

*Way (stage) 1* - limits for flicker emission levels are not given. It is assumed that when the consumer's agreed power variations[13] $\Delta S$ are small in relation to the apparent short-circuit power $S_{\mathrm{k}}''$ at the PCC then the flicker emission level is small

[12] In real systems the coefficients usually are smaller, e.g. $N_{10} = 1$, $N_{120} = 10$–20.

[13] In the case of WTGS analysis, power variations at maximum can be assumed as equal to the wind turbine rated power $S_{\mathrm{n}}$ (for a single WTGS). In practice, when evaluating flicker, power variations within 95% of the maximum variation band corresponding to a standard deviation are evaluated.

and it should not be necessary to carry out detailed evaluation. The limits of the acceptable relative power variations as a function of the number of variations per minute $r$ for MV and HV networks are presented in Table 4.1.

**Table 4.1.** Limits for relative power variations as a function of the number of variations ($r$)

| $r$ [min$^{-1}$] | MV network $(\Delta S / S''_{\mathrm{k}})_{\max}$ | HV network $(\Delta S / S''_{\mathrm{k}})_{\max}$ |
|---|---|---|
| $r > 200$ | 0.001 | 0.001 |
| $10 \le r \le 200$ | 0.002 | 0.001 |
| $r < 10$ | 0.004 | 0.001 |

*Way (stage) 2* – for loads not fulfilling the above criterion. The flicker emission limits are computed in the following steps.

- Determination of the maximum global contributions $G_{\mathrm{Pst}}$, $G_{\mathrm{Plt}}$ from all the fluctuating loads which are fed by this system to the flicker level. The contributions for an MV system can be computed by using the following equations:

$$G_{\mathrm{PstMV}} = \sqrt[3]{L^3_{\mathrm{PstMV}} - T^3_{\mathrm{PstHM}} L^3_{\mathrm{PstHV}}} \tag{4.35}$$

$$G_{\mathrm{PltMV}} = \sqrt[3]{L^3_{\mathrm{PltMV}} - T^3_{\mathrm{PltHM}} L^3_{\mathrm{PltHV}}}\,, \tag{4.36}$$

where $L_{\mathrm{PstMV}}$, $L_{\mathrm{PltMV}}$ – planning level of the flicker level in the MV system (see Chap. 3),
$L_{\mathrm{PstHV}}$, $L_{\mathrm{PltHV}}$ – planning level of the flicker level in the upstream HV system (see Chap. 3),
$T_{\mathrm{PstHM}}$, $T_{\mathrm{PltHM}}$ – transfer coefficient from the upstream HV system to the MV system (can be assumed as equal to 0.8),

For computing global contributions in the HV system, (4.35) and (4.36) should be also used. In this case, HV and MV should be exchanged (HV→MV and MV→HV). The transfer coefficients from the HV system to the downstream MV system, $T_{\mathrm{PstMH}}$ $T_{\mathrm{PltMH}}$, should be assumed as equal to 0, which is a result of the assumption that the voltage fluctuations from the MV system to the HV system do not pass.

- Determination of individual (for a given load or WTGS) emission limits $E_{\mathrm{Pst}i}$, $E_{\mathrm{Plt}i}$. The limits in an MV system can be computed as follows:

$$E_{\mathrm{Pst}i} = G_{\mathrm{PstMV}} \sqrt[3]{\frac{S_i}{S_{\mathrm{MV}}} \frac{1}{F_{\mathrm{MV}}}} \tag{4.37}$$

$$E_{\mathrm{Plt}i} = G_{\mathrm{PltMV}} \sqrt[3]{\frac{S_i}{S_{\mathrm{MV}}} \frac{1}{F_{\mathrm{MV}}}}\,, \tag{4.38}$$

where $S_i$ – agreed power of $i$th load (e.g. rated power of WTGS),
$S_{\mathrm{MV}}$ – total power of loads directly supplied at MV at saturation of the system,

$F_{MV}$ – coincidence factor (according to the standard, typically equal to 0.2–0.3).

For users with a comparatively low agreed power, the above equations may yield strict limitations. To overcome this problem standard IEC 61000-3-7 proposes that certain minimum emission levels equal to $E_{Pst\,i} = 0.35$ $E_{Plt\,i} = 0.25$ will always be granted for short-term and long-term flicker severity. These values are frequently used as limits when a WTGS flicker problem is considered.

Individual emission limits in the HV system can be computed by using (4.37) and (4.38) as well. Computation for the HV system needs two modifications of the equations. The first consists in neglecting the coincidence factor $F_{HV}$ (the factor by assumption is equal to 1). The second consists in a different definition of the power $S_{HV}$. In the HV system, the power is equal to the total power available for all the users connected to the given bus. This is a part of the total supply capacity of the HV substation, which is devoted to the HV users.

*Way (stage) 3* - the utility can accept higher flicker emission levels of a given load. There are various reasons for this. For example, acceptance of higher flicker emission levels is possible when other loads connected to a given system do not produce significant flicker, e.g. a wind turbine operating in a rural area.

**Table 4.2.** Limits for voltage changes as a function of the number of changes per hour ($r$)

| $r$ | MV network | HV network |
|---|---|---|
| $[h^{-1}]$ | $\Delta V_{dyn}/V_n$ [%] | $\Delta V_{dyn}/V_n$ [%] |
| $r \le 1$ | 4 | 3.0 |
| $1 < r \le 10$ | 3 | 2.5 |
| $10 < r \le 100$ | 2 | 1.5 |
| $100 < r \le 1000$ | 1.25 | 1.0 |

And finally, in the case of voltage fluctuation analysis the IEC 61000-3-7 standard gives emission limits for the rapid voltage change $\Delta V_{dyn}/V_n$, that can be made by the consumer (load, WTGS) during normal operating conditions.[14] The limits are presented in Table 4.2.

The IEC 61000-3-7 standard presents not only the method for computing the flicker limits but also (in an appendix) the method for simplified calculation of the flicker severity. According to this method, the flicker severity level for the known shape, "amplitude" and frequency of the voltage changes can be computed as

$$P_{st} = \sqrt[3]{\frac{\sum t_f}{10 \cdot 60}} = \sqrt[3]{\frac{2.3(100dF)^3}{10 \cdot 60}} \tag{4.39}$$

$$P_{lt} = \sqrt[3]{\frac{\sum t_f}{120 \cdot 60}} = \sqrt[3]{\frac{2.3(100dF)^3}{120 \cdot 60}}, \tag{4.40}$$

[14] The standard does not cover voltage changes less frequent than one per day.

where $F$ – shape factor (function of the voltage change shape),
$d$ – relative voltage change.

The relative voltage change can be computed by using the equations given in the standard:

$$d = \frac{\Delta S_i}{S''_\mathrm{k}}, \qquad d = \frac{R\Delta P_i + X\Delta Q_i}{V_\mathrm{n}^2} \tag{4.41}$$

or by utilizing the model (i.e. mathematical) of the considered network.

It is not necessary to use this method when the voltage step factor $k_\mathrm{u}(\psi_\mathrm{k})$ for the considered turbine is known. When that factor is unknown, (4.39) and (4.40) can be used. But in this case, the voltage change shape during switching operations should be known (or assumed).

Figures containing limits of the short-circuit and rated powers quotient $S''_\mathrm{k} / S_\mathrm{n}$ resulting from various equations presented above (and in standards) are presented as final considerations related to the flicker evaluation in a system with a wind turbine (wind farm). The computation has been made for the Vestas V80-2MW wind turbine. The relevant coefficients – the base for computation – obtained from the DEWI report [21] are presented in Tables 4.3 and 4.4.

**Table 4.3.** Flicker coefficients for V80-2MW wind turbine in normal operation [93]

| $v_\mathrm{a}$ [m/s] | $\psi_\mathrm{k}$ [deg] | | | |
|---|---|---|---|---|
| | 30 | 50 | 70 | 85 |
| | $c(\psi_\mathrm{k}, v_\mathrm{a})$ | | | |
| 6.0 | 4.76 | 4.41 | 5.02 | 5.67 |
| 7.5 | 4.88 | 4.41 | 5.00 | 5.72 |
| 8.5 | 4.90 | 4.41 | 5.00 | 5.72 |
| 10.0 | 4.91 | 4.41 | 4.90 | 5.72 |

Figure 4.14 presents the limits of $S''_\mathrm{k} / S_\mathrm{n}$ quotient computed from (4.28) for two values of the flicker severity coefficients: $P_\mathrm{st} = 0.35$, $P_\mathrm{lt} = 0.25$. These values are often assumed as the flicker limits $E_\mathrm{Pst}$, $E_\mathrm{Plt}$ for WTGS evaluation. The computation has been made for the highest values of the flicker coefficient $c(\psi_\mathrm{k}, v_\mathrm{a})$ presented in Table 4.3.

The curves show that, in general, both conditions (defined by the flicker severity levels during continuous operation) are fulfilled when the short-circuit apparent power at the PCC is higher by at least 20 times than the rated apparent power of the wind turbine. This value $S''_\mathrm{k} / S_\mathrm{n} \geq 20$ often appears as a condition related to the possibilities of WTGS connection to the power system.

Figure 4.15 presents the limits of $S''_\mathrm{k} / S_\mathrm{n}$ quotient computed from (4.30) and (4.31) (switching operations). For the computation the highest values of the flicker step factor $k_\mathrm{f}(\psi_\mathrm{k})$ for various switching operations have been used. The flicker severity coefficients have been assumed as in the previous example, i.e. $P_\mathrm{st} = 0.35$, $P_\mathrm{lt} = 0.25$. The curves show that the conditions related to flicker severity levels during switching operations are fulfilled when the considered quotient is also higher than 20. Contrary to the previous condition, a decrease of the quotient limit can be observed when the network impedance angle increases. The quotient change over an angle change from 30° to 85° is about 10%. It is worth noting that the maximum number of switching operation coefficients $N_{10}$, $N_{120}$ greatly influences the results (increasing the quotient by up to twice). For WTGS operations, $N_{10} = 1$ and $N_{120} = 10–20$ can be used even if the data are not available.[15]

**Table 4.4.** Flicker coefficients in the switching operation of the V80-2MW wind turbine

| Type of switching | | | | | | | | | Maximum number of generator switching in 2 hours ($N_{120}$) | | | |
|---|---|---|---|---|---|---|---|---|---|---|---|---|
| Cut-in of generator/winding set 1 at cut-in wind speed | | | | | | | | | 20 | | | |
| Cut-in of generator/winding set 2 at change over wind speed | | | | | | | | | 10 | | | |
| | Generator 1/winding 1 Cut-in at cut-in wind speed | | | | Generator 2/winding 2 Cut-in at change over wind speed | | | | Generator 1/winding 1 Cut-in at change back wind speed | | | |
| $\psi_k$ [deg] | 30 | 50 | 70 | 85 | 30 | 50 | 70 | 85 | 30 | 50 | 70 | 85 |
| $k_i$[a] | 0.19 | | | | 0.36 | | | | 0.28 | | | |
| $k_u$ | 0.25 | 0.25 | 0.21 | 0.20 | 0.34 | 0.26 | 0.15 | 0.06 | 0.42 | 0.31 | 0.18 | 0.12 |
| $k_f$ | 0.15 | 0.15 | 0.14 | 0.12 | 0.07 | 0.07 | 0.07 | 0.07 | 0.17 | 0.14 | 0.11 | 0.11 |

[a] The current spike factor $k_i$ is calculated as the mean value of the one line period rms current peak of all measured switching operations.

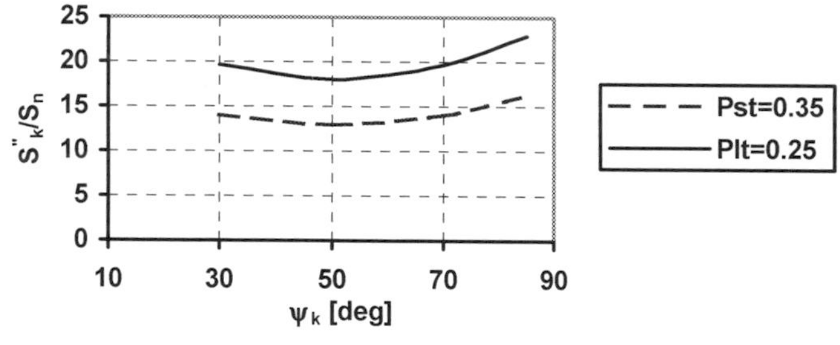


**Fig. 4.14.** $S''_k / S_n$ quotient limit for continuous operation of the wind turbine

Figure 4.16 presents the limits of the $S''_k / S_n$ quotient computed from (4.34), which is related to the voltage change during switching operations. As before, the highest values of the voltage change factor $k_u(\psi_k)$ have been used for the computation. The curves show that when the considered quotient lies between 20 and 40, the voltage change will be less than 1%. It is worth noting that the condition $S''_k / S_n \geq 40$ also appears in publications as a condition related to the possibilities of WTGS connection to the grid. For the voltage change limit 2%, the quotient

[15] IEC 61000-21 standard suggests using values $N_{10} = 10$ and $N_{120} = 120$ when the data are unknown. For high-rated power modern wind turbines the values are too high.

should be less than 20. The curves, and the quotient, highly decrease (by up to about 50%) when the network impedance angle increases.

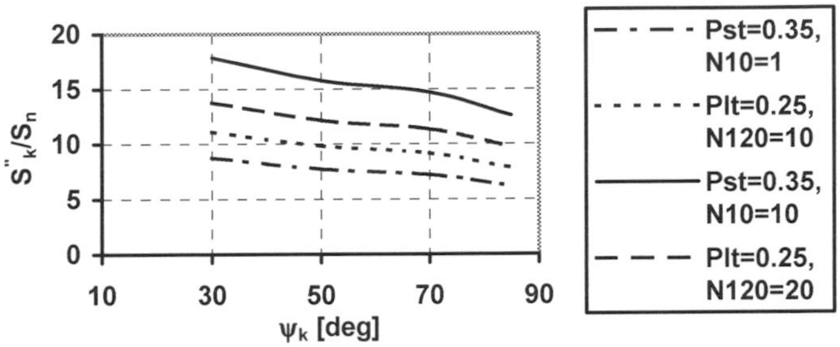


**Fig. 4.15.** $S''_k / S_n$ quotient limit for switching operation of the wind turbine for various flicker severity coefficients and numbers of switching operations

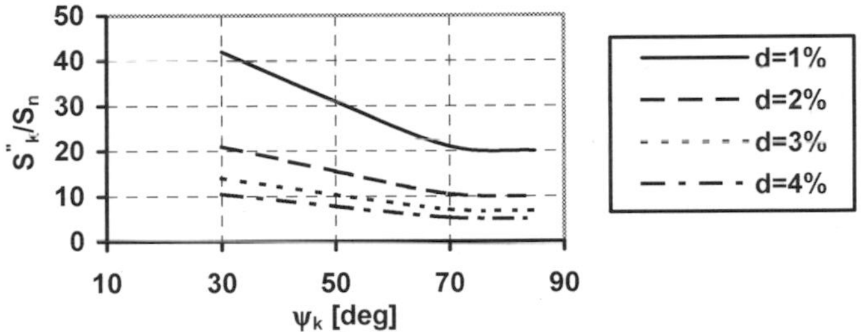


**Fig. 4.16.** $S''_k / S_n$ quotient limit for switching operation of the wind turbine for various voltage change limits

## 4.5 Harmonics

The harmonic voltages and currents should be limited to the acceptable level at the point of the wind turbine (wind farm) connection to the network. The problem of harmonics is considered in detail (e.g. with harmonic distortion limits) in standard IEC 61000-3-6. The standard focuses on harmonic voltages but also gives the relative harmonic current limits. The limits are presented in Table 4.5.

**Table 4.5.** Indicative values for relative harmonic current limits (ranges are given depending on the type of network)

| Harmonic number | 5 | 7 | 11 | 13 | $\sqrt{\sum i_h^2}$ |
|---|---|---|---|---|---|
| Admissible harmonic current $i_h = I_h/I_i$ | 5–6 | 3–4 | 1.5–3 | 1–2.5 | 6–8 |

where $I_h$ – total harmonic current of $h$th order caused by the consumer,
$I_i$ – rms current corresponding to the consumer agreed power (fundamental frequency).

Standard IEC 61000-21 gives guidance for the summation of the harmonic current distortion from various loads. The $h$th order harmonic current $I_{h\Sigma}$ at the PCC in installations with a number of wind turbines can be computed here as

$$I_{h\Sigma} = \sqrt[\beta]{\sum_{i=1}^{N} \left(\frac{I_{hi}}{\upsilon_i}\right)^{\beta}} , \tag{4.42}$$

where $I_{h\Sigma}$ – $h$th order harmonic current distortion at the PCC,
$I_{hi}$ – hth order harmonic current distortion of the $i$th wind turbine,
$N$ – number of wind turbines,
$\upsilon_i$ – ratio of the transformer at the $i$th wind turbine,
$\beta$ – exponent given in Table 4.6.

**Table 4.6.** Specification of exponent $\beta$ according to IEC 61000-3-6

| Harmonic order | Exponent $\beta$ |
|---|---|
| $h < 5$ | 1.0 |
| $5 \leq h \leq 10$ | 1.4 |
| $h > 10$ | 2.0 |

While computing harmonic currents it is necessary to take into account the step-up transformer and the MV/HV transformer (for the wind farm) connection groups. Usually, the MV step-up transformer and the MV/HV transformer at MV side have delta-connected windings. It practically eliminates harmonic currents of the third order and their multiples.

The above considerations should be made for wind turbines equipped with the power electronic converters, i.e. wind turbines equipped with doubly-fed asynchronous generators or synchronous generators connected to the power system through the power electronic converter. In the case of wind turbines equipped with asynchronous generators directly connected to the network, without converters and equipped with directly connected synchronous generators, according to standard IEC 61000-21, the harmonic current assessment is not required.

## 4.6 Short-Circuit Currents

Most of the WTGSs operating all over the world today are equipped with asynchronous machines. The second group, taking into account the number of the wind turbines, is made of WTGSs equipped with synchronous generators connected to the grid through the power electronic converters. The rest (very small) is made of other constructions, e.g. synchronous generators connected to the grid directly. Therefore, when computing the short-circuit currents in systems with wind turbines, it is necessary to take into account the asynchronous machines [47, 75].

According to standard IEC 60909 and other national standards (based on the IEC standard) the asynchronous motors (generators) should be considered in short-circuit analysis when:

1. The total rated current of induction motors is greater than 1% of the initial AC short-circuit current computed without the motors:

$$\sum I_{\mathrm{nM}} \geq \frac{I''_{\mathrm{k}}}{100}. \tag{4.43}$$

The condition (4.43) expressed in terms of apparent power takes the following form:

$$\sum S_{\mathrm{nM}} \geq \frac{S''_{\mathrm{k}}}{100}, \tag{4.44}$$

which for a single asynchronous motor can be described as

$$S_{\mathrm{nM}} \geq \frac{S''_{\mathrm{k}}}{100}, \tag{4.45}$$

where $I''_{\mathrm{k}}$ – initial AC short-circuit current at the point of common coupling (computed without asynchronous machines),
$S''_{\mathrm{k}}$ – initial AC short-circuit apparent current at the point of common coupling (computed without asynchronous machines),
$I_{\mathrm{nM}}$ – rated current of asynchronous machine,
$S_{\mathrm{nM}}$ – rated apparent power of asynchronous machine.

2. The contribution of asynchronous motors in the initial AC short-circuit current exceeds 5%, i.e.

$$I''_{\mathrm{kM}} \geq \frac{I''_{\mathrm{k}}}{20} \quad \text{or} \quad \frac{I''_{\mathrm{k}} + I''_{\mathrm{kM}}}{I''_{\mathrm{k}}} \geq 1.05, \tag{4.46}$$

where $I''_{\mathrm{kM}}$ – contribution of asynchronous motor in the initial AC short-circuit current:

$$I''_{\mathrm{kM}} \geq \frac{cV_N k_{\mathrm{LR}} S_{\mathrm{nM}}}{\sqrt{3}V_{\mathrm{nM}}^2}, \tag{4.47}$$

$c$ – voltage factor representing the difference between the effective voltage and the system voltage – in HV and MV networks $c = 1.1$ for computation of the maximum value of the initial AC short-circuit current,
$V_{\mathrm{nM}}$ – rated voltage of asynchronous machine,
$V_N$ – rated system voltage (line voltage),
$k_{\mathrm{LR}}$ – start-up current factor:

$$k_{LR} = \frac{I_{LR}}{I_{nM}}, \tag{4.48}$$

$I_{\mathrm{LR}}$ – locked-rotor current of motor.

The above, with the assumption $V_N = V_{\text{nM}}$, leads to the condition:

$$S_{\text{nM}} \geq \frac{1}{ck_{\text{LR}}} \frac{S''_{\text{k}}}{20}. \tag{4.49}$$

3. High- and low-voltage motors, connected to the network through two-winding transformers, which fulfill the condition:

$$\frac{\sum P_{\text{nM}}}{\sum S_{\text{nTM}}} \geq \frac{0.8}{\left| \frac{c100\sum S_{\text{nTM}}}{S''_{\text{k}}} - 0.3 \right|}, \tag{4.50}$$

where $S_{\text{nTM}}$ – rated apparent power of transformer,
$P_{\text{nM}}$ – rated real power of asynchronous motor,
$u_{\text{knTM}}$ – nominal short-circuit voltage of transformer [%].
Equation (4.50) is in fact a simplified (as a result of assumption $k_{\text{LR}} = 5$ and $u_{\text{knTM}} = 6\%$[16]) form of the following equation:

$$\frac{k_{\text{LR}} \sum P_{\text{nM}}}{\sum S_{\text{nTM}}} \geq \frac{0.8}{\left| \frac{c20\sum S_{\text{nTM}}}{S''_{\text{k}}} - \frac{u_{\text{knTM}}}{100} \right|}. \tag{4.51}$$

According to the standard, asynchronous motors for short-circuit current computation purposes are modeled as an impedance $\underline{Z}_{\text{M}}$ whose modulus is equal to:

$$Z_{\text{M}} = \left| R_{\text{M}} + jX_{\text{M}} \right| = \frac{1}{k_{\text{LR}}} \frac{V_{\text{nM}}^2}{S_{\text{nM}}}. \tag{4.52}$$

The method of computing the asynchronous motor resistance $R_{\text{M}}$ and reactance $X_{\text{M}}$ from the impedance $Z_{\text{M}}$ is presented in Table 4.7.

**Table 4.7.** Asynchronous motor resistance and reactance coefficients

| $k_R = R_M/X_M$ | $k_X = X_M/Z_M$ | Type of machine to which the data is applicable |
|---|---|---|
| 0.1 | 0.955 | MV motors with power /pole-pair $P_{\text{nM}}/p \geq 1$ MW |
| 0.15 | 0.989 | MV motors with power /pole-pair $P_{\text{nM}}/p < 1$ MW |
| 0.42 | 0.992 | LV motors with cable lines |
| 0.1 | 0.995 | Converter-supplied drives |

Converter-supplied drives are treated in the same way as asynchronous motors. For these drives, the effective impedance $\underline{Z}_{\text{M}}$ may be computed by using the start-up current factor equal to $k_{\text{LR}} = 3$, and by substituting $V_{\text{nM}}$ by the rated voltage of the supply-side of the converter or converter transformer, and by substituting the current $I_{\text{nM}}$ by the rated supply-side current.

[16] These are values typical for present day WTGSs.

The considerations related to the short-circuit current computation in the electric power system with WTGS are conducted here based on a MV network (as an example) which is presented in Fig. 4.17. Assuming that line L1 feeds loads, which consist of a small number of motors, the only sources of the short-circuit currents are the power system Q and the wind turbine generator system (WTGS). Therefore, looking for the initial AC short-circuit current at the point of common coupling (PCC), the transmission lines L1, L2, L3 etc. in the following considerations can be neglected. In other considerations, related to the short-circuit current computation, the lines L1, L2 and L3 will be taken into consideration.

The rated data of the transformers and the MV transmission lines necessary for the analysis are presented in Tables 4.8–4.10.

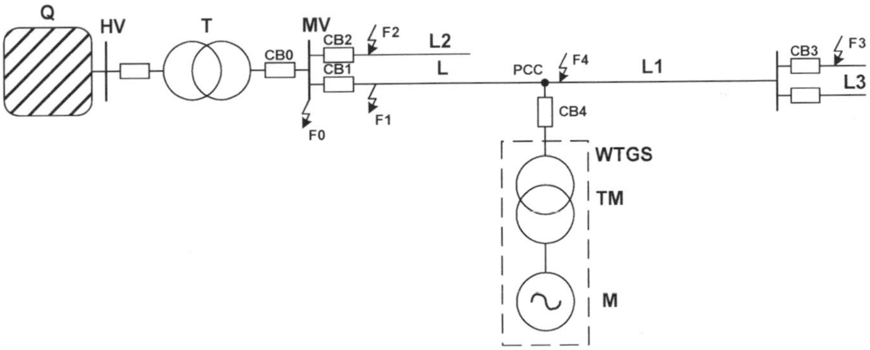


**Fig. 4.17.** A schematic diagram of a typical power system with WTGS

**Table 4.8.** Rated data of feeding transformers (T)[17]

| $S_{nT}$ [MVA] | $\Delta P_{CuT}$ [kW] | $u_{knT}$ [%] | $\vartheta_{nT}$ [kV/kV] | $S_{nT}$ [MVA] | $\Delta P_{CuT}$ [kW] | $u_{knT}$ [%] | $\vartheta_{nT}$ [kV/kV] |
|---|---|---|---|---|---|---|---|
| 6.3 | 44 | 11 | 110/15.75 | 16 | 87 | 11 | 110/15.75 |
| 10 | 69 | 11 | 110/15.75 | 25 | 110 | 11 | 110/15.75 |

**Table 4.9.** Rated data of MV overhead transmission lines (L) for $V_N = 15$ kV

| S [mm$^2$] | $I_{max(pc)}$ [A] | $S_{max(pc)}$ [MVA] | $R'_L$ [Ω/km] | $X'_L$ [Ω/km] | $C'_L$ [μF/km] |
|---|---|---|---|---|---|
| 35 | 145 | 3.7 | 0.840 | 0.40 | $9.9\times10^{-3}$ |
| 50 | 170 | 4.4 | 0.588 | 0.39 | $10.1\times10^{-3}$ |
| 70 | 235 | 6.0 | 0.420 | 0.38 | $10.2\times10^{-3}$ |
| 95 | 290 | 7.5 | 0.309 | 0.37 | $10.3\times10^{-3}$ |
| 120 | 415 | 10.7 | 0.240 | 0.36 | $10.4\times10^{-3}$ |

The considered system, before coupling of the WTGS, can be modeled by serially connected impedances of the system $\underline{Z}_Q$, of the feeding transformer $\underline{Z}_T$, and of

[17] The variables in Tables 4.8 and 4.9 are defined in the following pages.

the transmission line $\underline{Z}_L$. The schematic diagram of the system is presented in Fig. 4.18a. After coupling the WTGS, the system is modeled by adding to the circuit presented in Fig. 4.18a impedance of the wind turbine generator $\underline{Z}_M$ and impedance of the step-up transformer $\underline{Z}_{TM}$.[18] The schematic diagram of the system in such a case is presented in Fig. 4.18b.

**Table 4.10.** Rated data of the WTGS step-up transformers (TM)

| $S_{nTM}$ [MVA] | $\Delta P_{CuTM}$ [KW] | $u_{knTM}$ [%] | $\vartheta_{nT}$ [kV/kV] | $S_{nM}$ [MVA] |
|---|---|---|---|---|
| 0.315 | 5.55 | 6 | 15.75/0.69 | 0.25 |
| 0.63 | 9.45 | 6 | 15.75/0.69 | 0.5 |
| 0.80 | 11.5 | 6 | 15.75/0.69 | 0.75 |
| 1.25 | 16.25 | 6 | 15.75/0.69 | 1.0 |
| 1.60 | 19.5 | 6 | 15.75/0.69 | 1.5 |
| 1.85 | 22.5 | 6 | 15.75/0.69 | 1.75 |
| 2.10 | 27.0 | 6 | 15.75/0.69 | 2.0 |

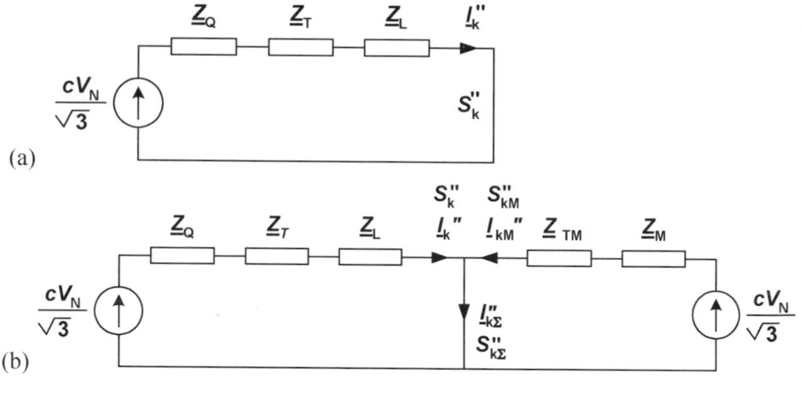


**Fig. 4.18.** Equivalent circuits for the short-circuit current computation: (a) system without WTGS; (b) system with WTGS

The impedances of particular elements of the considered system can be calculated as follows.[19]

1. Power system:

$$Z_Q = \frac{cV_{N,HV}^2}{S''_{kQ}} \vartheta_{nT}^{-2}$$
$$\underline{Z}_Q = R_Q + jX_Q, \tag{4.53}$$

---

[18] The wind turbine should be taken into consideration when (4.43), (4.46) or (4.50) are satisfied.

[19] The impedances in the example are calculated at the MV power system voltage level – the level at the point of common coupling of the wind turbine generator system with the power system.

where $\vartheta_{nT} = V_{nT,HV} / V_{nT,MV}$ – rated transformation ratio of the feeding transformer,
$V_{nT,HV}$ – rated voltage at the HV side of the transformer,
$V_{nT,MV}$ – rated voltage at the MV side of the transformer,
$c$ – voltage factor (explained in (4.47)).

If there is no information about the resistance and reactance of the system, then the following assumption can be made: $X_Q = 0.995Z_Q$, $R_Q = 0.1Z_Q$. In a network with rated voltage greater than 35 kV, it can be assumed that $X_Q = Z_Q$, $R_Q = 0$. But it is worth noticing that in weak HV systems, the $R_Q/X_Q$ quotient can be close to the $R_L/X_L$ quotient of the HV transmission lines feeding the MV subsystem. Then the system resistance should be taken into consideration.

2. Transformer feeding the MV subsystem:

$$R_T = \Delta P_{CuT} \frac{V_{nT,MV}^2}{S_{nT}^2}$$
$$X_T = \sqrt{\left(\frac{u_{knT}}{100} \frac{V_{nT,MV}^2}{S_{nT}}\right)^2 - R_T^2} \tag{4.54}$$
$$\underline{Z}_T = R_T + jX_T ,$$

where $\Delta P_{CuT}$ – copper losses at rated power,
$u_{knT}$ – rated short-circuit voltage.

3. Transmission line:

$$R_L = R'_L l$$
$$X_L = X'_L l \tag{4.55}$$
$$\underline{Z}_L = R_L + jX_L ,$$

where $R'_L$ – transmission line resistance per phase per km,
$X'_L$ – transmission line reactance per phase per km,
$l$ – transmission line length.

4. Transformer of the wind turbine generator system:

$$R_{TM} = \Delta P_{CuTM} \frac{V_{nTM,MV}^2}{S_{nTM}^2}$$
$$X_{TM} = \sqrt{\left(\frac{u_{knTM}}{100} \frac{V_{nTM,MV}^2}{S_{nTM}}\right)^2 - R_{TM}^2} \tag{4.56}$$
$$\underline{Z}_{TM} = R_{TM} + jX_{TM} ,$$

where $V_{nTM,MV}$ is the rated voltage at the MV side of the transformer.

5. Wind turbine generator (asynchronous motor):

$$Z_{\mathrm{M}} = \frac{1}{k_{\mathrm{LR}}} \frac{V_{\mathrm{nM}}^2}{S_{\mathrm{nM}}} \vartheta_{\mathrm{nTM}}^2$$
$$X_{\mathrm{M}} = k_{\mathrm{X}} Z_{\mathrm{M}} \tag{4.57}$$
$$R_{\mathrm{M}} = k_{\mathrm{R}} X_{\mathrm{M}}$$
$$\underline{Z}_{\mathrm{M}} = R_{\mathrm{M}} + \mathrm{j}X_{\mathrm{M}},$$

where $\vartheta_{\mathrm{nTM}} = V_{\mathrm{nT,MV}}/V_{\mathrm{nT,LV}}$ – rated transformation ratio of the transformer,
$V_{\mathrm{nT,LV}}$ – rated voltage at the LV side of the transformer,
$k_{\mathrm{X}}$, $k_{\mathrm{R}}$ – factors from Table 4.7.

The initial AC short-circuit apparent power at the point of the WTGS coupling (but without the WTGS) is equal to

$$S_{\mathrm{k}}'' = \frac{cV_{N,\mathrm{MV}}^2}{Z_{\mathrm{k}}}, \tag{4.58}$$

where $Z_{\mathrm{k}} = |\underline{Z}_{\mathrm{Q}} + \underline{Z}_{\mathrm{T}} + \underline{Z}_{\mathrm{L}}|$ is the effective short-circuit impedance.

After connecting the WTGS to the grid, the total initial AC short-circuit apparent power at the PCC is equal to

$$S_{\mathrm{k}\Sigma}'' = S_{\mathrm{k}}'' + S_{\mathrm{kM}}'' = \frac{cV_{N,\mathrm{MV}}^2}{Z_{\mathrm{k}}} + \frac{cV_{N,\mathrm{MV}}^2}{Z_{\mathrm{kM}}}, \tag{4.59}$$

where $Z_{\mathrm{kM}} = |\underline{Z}_{\mathrm{TM}} + \underline{Z}_{\mathrm{M}}|$ – effective short-circuit impedance of the WTGS,
$S_{\mathrm{kM}}''$ – WTGS contribution to the short-circuit power.

The growth of the initial AC short-circuit apparent power at the PCC after switching the WTGS on is equal to

$$k_{\Delta\mathrm{S}} = \frac{S_{\mathrm{k}\Sigma}''}{S_{\mathrm{k}}''} = \frac{S_{\mathrm{k}}'' + S_{\mathrm{kM}}''}{S_{\mathrm{k}}''} = 1 + \frac{S_{\mathrm{kM}}''}{S_{\mathrm{k}}''} = 1 + \frac{Z_{\mathrm{k}}}{Z_{\mathrm{kM}}}. \tag{4.60}$$

Let us assume that the initial AC short-circuit apparent power at the PCC after switching the WTGS on cannot increase more than $k_{\Delta\mathrm{Smax}}$ times (which, for example, can result from the short-circuit electrical strength of the installed equipment). Then, the requirement related to the contribution of the WTGS to the initial AC short-circuit apparent power (from (4.60)) can be defined as follows:

$$S_{\mathrm{kM}}'' \le (k_{\Delta\mathrm{Smax}} - 1) S_{\mathrm{k}}''. \tag{4.61}$$

Comparing (4.59) and (4.61), the considered requirement for the value of the WTGS effective impedance takes the following form:

$$Z_{\mathrm{kM}} \ge \frac{cV_{N,\mathrm{MV}}^2}{(k_{\Delta\mathrm{Smax}} - 1) S_{\mathrm{k}}''} \tag{4.62}$$

or

$$\left|\underline{Z}_{\text{TM}} + \underline{Z}_{\text{M}}\right| \geq \frac{cV_{N,\text{MV}}^2}{(k_{\Delta\text{Smax}} - 1)S_{\text{k}}''} \,. \tag{4.63}$$

Neglecting the resistances[20] (i.e. assuming that $Z = X$), it is possible to define the rated apparent power of the wind turbine generator system which fulfils the requirements (4.62) and (4.63):

- for the wind turbine generator system connected to the bus through the transformer:

$$S_{\text{nM}} \leq \frac{c}{k_{\text{LR}}\left(\dfrac{c}{(k_{\Delta\text{Smax}} - 1)S_{\text{k}}''} - \dfrac{u_{\text{knTM}}}{100S_{\text{nTM}}}\right)} \,, \tag{4.64}$$

- for the wind turbine generator system connected to the bus directly:

$$S_{\text{nM}} \leq \frac{(k_{\Delta\text{Smax}} - 1)S_{\text{k}}''}{k_{\text{LR}}} \,. \tag{4.65}$$

The literature presents the requirements defining the relation between the initial AC short-circuit apparent power $S_{\text{k}}''$ at the PCC before switching on the WTGS and the rated apparent power $S_{\text{nM}}$ of the WTGS. Some electric utilities define this condition for a single WTGS as[21]:

$$S_{\text{nM}} \leq \frac{S_{\text{k}}''}{20} \,, \tag{4.66}$$

while other utilities use more rigorous condition:

$$S_{\text{nM}} < \frac{S_{\text{k}}''}{40} \,. \tag{4.67}$$

For installations consisting of a number of WTGSs (equipped with asynchronous generators) the requirement often takes the following form:

$$\sum S_{\text{nM}} \leq \frac{\sqrt{N} S_{\text{k}}''}{40} \,, \tag{4.68}$$

where $N$ is the number of asynchronous generators in the wind farm (installation).

For WTGSs equipped with a synchronous generator the following condition is used:

---

20 Estimation of the error related to neglecting the resistance is presented later in this section.

21 It is worth remembering that the origin of the conditions is mainly related to the power quality problem.

$$S_{\text{nG}} \le \frac{S''_{\text{k}}}{250} \tag{4.69}$$

or, which ENERCON proposes for its synchronous generators:

$$S_{\text{nG}} \le \frac{S''_{\text{k}}}{65}\,. \tag{4.70}$$

The conditions (4.66)–(4.70) show that the WTGS equipped with the asynchronous generator with rated power over 1.5 times higher than one equipped with an synchronous generator can be connected at a given PCC. This shows favoritism towards WTGSs equipped with asynchronous generators, especially in weak power systems.

Taking into account the softest conditions, (4.66) and (4.67), and rearranging (4.64) and (4.65), the border values $k_{\Delta S20}$ and $k_{\Delta S40}$ of the factors showing the growth of the initial AC short-circuit apparent power take the following form:

$$k_{\Delta\text{S20}} = 1 + \frac{k_{\text{LR}}}{20}\,, \qquad k_{\Delta\text{S40}} = 1 + \frac{k_{\text{LR}}}{40} \tag{4.71}$$

for the WTGS directly connected to the grid, and

$$k_{\Delta\text{S20}} = 1 + \frac{c}{20}\frac{1}{\dfrac{c}{k_{\text{LR}}} + \dfrac{u_{\text{knTM}}}{100}}\,, \qquad k_{\Delta\text{S40}} = 1 + \frac{c}{40}\frac{1}{\dfrac{c}{k_{\text{LR}}} + \dfrac{u_{\text{knTM}}}{100}} \tag{4.72}$$

for the WTGS connected to the grid through the step-up transformer (with the assumption $S_{\text{nM}} = S_{\text{nTM}}$).

The values of these factors for a typical start-up current factor $k_{\text{LR}} = 5$, transformer rated short-circuit voltage $u_{\text{knTM}} = 6\%$, and factor $c = 1.1$ show the growth limit of the initial AC short-circuit apparent power at PCC equal to 25% and 12.5%, respectively, in the system with the WTGS connected directly, and 20% and 10%, respectively, in the system with the WTGS connected through the step-up transformer. Of course, the real permissible growth of the power $S''_{\text{k}}$ depends on the electrical strength of the equipment installed in the electric power system.

As an example, Fig. 4.19 presents the permissible (in the sense of (4.64) and (4.65)) rated apparent power of the WTGS as a function of the factor $k_{\Delta\text{S}}$. The computation has been made for a system with the feeding transformer rated power equal to $S_{\text{nT}} = 10$ MVA, the transmission line length $l = 10$ km, and the cross-sectional area of conductors $s = 120\ \text{mm}^2$. In the figure, there are marked the border factors $k_{\Delta\text{S20}}$, $k_{\Delta\text{S40}}$ and a rectangle showing the location of conditions (4.43), (4.46) and (4.50).[22]

The location of the points shows that for asynchronous generators with rated power $S_{\text{nM}} < 0.22$–$0.28$ MVA (which depends on $S''_{\text{kQ}}$), the generator can be neglected in the short-circuit currents computation. But for machines with rated

[22] The points fulfilling the conditions are located close to one another.

power greater than one, up to the value defined by the crossing-point of the appropriate curve and the border factor $k_{\Delta S20}$ or $k_{\Delta S40}$ should be taken into consideration.

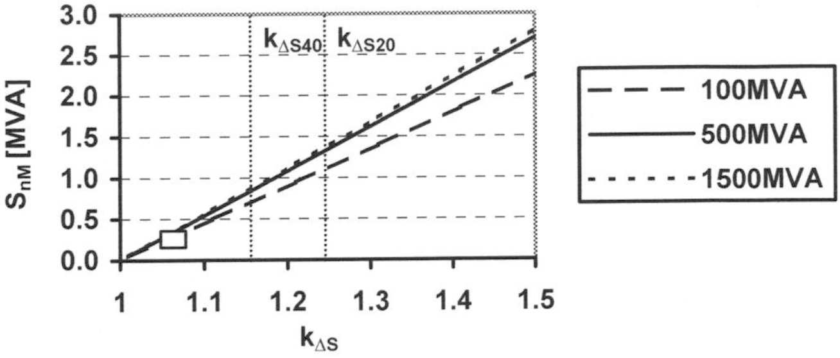


**Fig. 4.19.** Rated apparent power of the WTGS $S_{nM}$ as a function of factor $k_{\Delta S}$ for various values of the initial AC short-circuit apparent power of the system Q. Lines: continuous – $S''_{kQ} = 100$ MVA; dashed – $S''_{kQ} = 500$ MVA; dotted – $S''_{kQ} = 1500$ MVA.

The short-circuit current computation in MV power systems (but also in HV systems, e.g. 110 kV) should be conducted without neglecting resistances (values of resistances, especially of the transmission lines, are comparable here to the values of reactances). Let us check the influence of the resistance on the results of the considered problems.

The initial AC short-circuit apparent power $S''_k$ at the PCC before coupling the WTGS to the grid is equal to (4.73)

$$S''_k = \frac{cV^2_{N,MV}}{\sqrt{\left(\frac{\Delta P_{CuT} V^2_{nT,MV}}{S^2_{nT}} + R'_L l\right)^2 + \left(\frac{cV^2_{N,HV}}{S''_{kWN}}\vartheta^{-2}_{nT} + \sqrt{\left(\frac{u_{knT} V^2_{nT,MV}}{100 S_{nT}}\right)^2 - \left(\frac{\Delta P_{CuT} V^2_{nT,MV}}{S^2_{nT}}\right)^2} + X'_L l\right)^2}} .$$

After neglecting resistances, the initial AC short-circuit apparent power at the PCC can be computed as

$$S''_{kX} = \frac{cV^2_{N,MV}}{\frac{cV^2_{N,HV}}{S''_{kWN}}\vartheta^{-2}_{nT} + \frac{u_{knT} V^2_{nT,MV}}{100 S_{nT}} + X'_L l} . \tag{4.74}$$

The increase of initial AC short-circuit apparent power as a result of neglecting resistance can be defined as:

$$k_{\Delta SX} = \frac{S''_{kX}}{S''_k} . \tag{4.75}$$

The results of computation of the factor $k_{\Delta SX}$ as a function of the initial AC short-circuit apparent power of the feeding system $S''_{kQ}$ are presented in Fig. 4.20. The curves were computed for a system with the feeding transformer rated power $S_{nT}$ equal to 10 and 25 MVA, and the transmission line with length $l$ equal to 10 and 20 km (the cross-sectional area of conductor were equal to $s = 120$ mm$^2$).

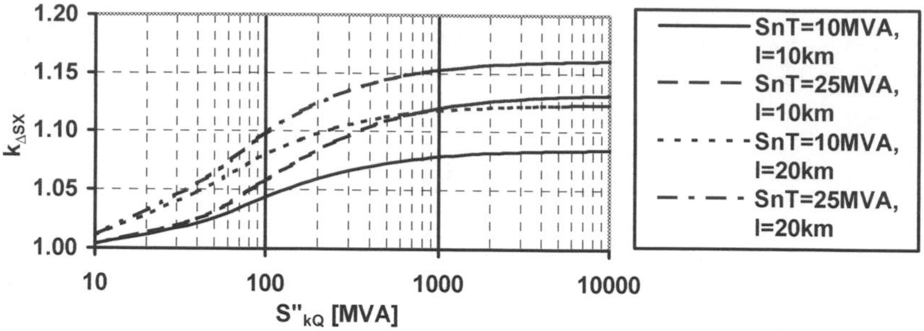


**Fig. 4.20.** Increase of the initial AC short-circuit apparent power $S''_k$ at the PCC as a function of power system short-circuit power $S''_{kQ}$.

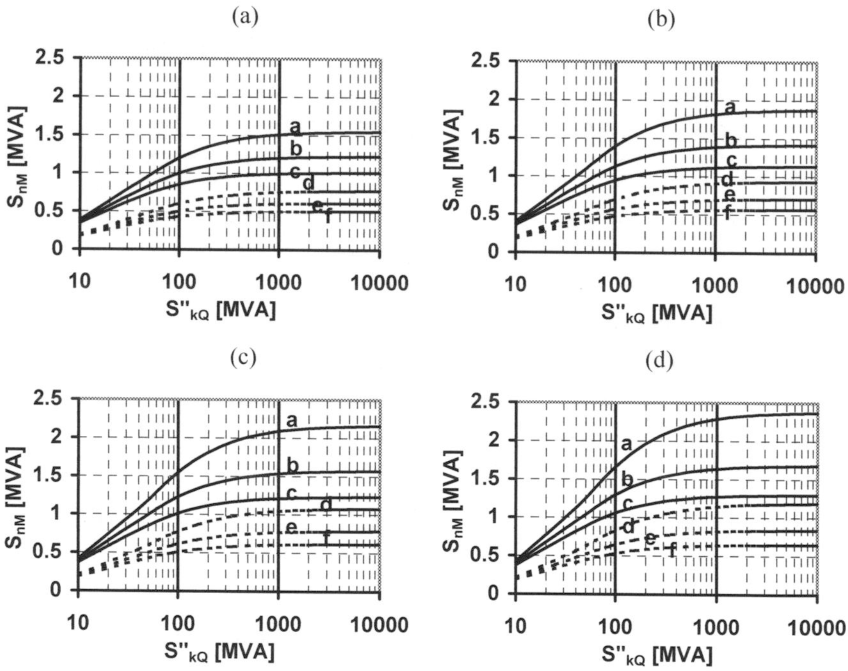


**Fig. 4.21.** Maximum rated apparent power $S_{nM}$ of the WTGS that can be connected to a typical power system as a function of $S''_{kQ}$ for various lengths of transmission line. (a) $S_{nT}$ = 6.3 MVA; (b) $S_{nT}$ = 10 MVA; (c) $S_{nT}$ = 16 MVA; (d) $S_{nT}$ = 25 MVA. Lines: a, d – $l$ = 10 km; b, e – $l$ = 15 km; c, f – $l$ = 20 km; continuous line – $S_{nM} = S''_{kQ}/20$; dotted line – $S_{nM} = S''_{kQ}/40$.

The curves in Fig. 4.20 indicate that the difference between the AC short-circuit power levels as a result of neglecting the resistance in the MV system is relatively high. The difference can reach up to 20%, which depends on the network structure. Therefore, neglecting the resistances in the short-circuit analysis can lead to computing excessively high values of the WTGS rated power, what results not only in the short-circuit current level but also in the voltage (and power) quality in the system after the WTGS connection.

Let us consider the influence of the various elements of a given power system and its capability of connecting various WTGSs (taking into account the rated power). The maximum rated apparent power $S_{nM}$ of the asynchronous generator (or sum of the rated powers $\Sigma S_{nM}$ for the wind farm) is computed here by using (4.66)–(4.68), where $S''_{k}$ is defined by (4.73). The results of the computation are presented in Fig. 4.21. The results show that the increase of the WTGS rated power as a result of the initial AC short-circuit apparent power $S''_{kQ}$ increase has a limited character. For high values of $S''_{kQ}$, the grid impedance ($\underline{Z}_{T}$ + $\underline{Z}_{L}$) limits the value of $S''_{k}$ and at the same time limits the WTGS rated power. In the considered case, an increase of the initial AC short-circuit apparent power $S''_{kQ}$ of the feeding power system above 500 MVA practically does not cause any increase of the WTGS maximum rated power.

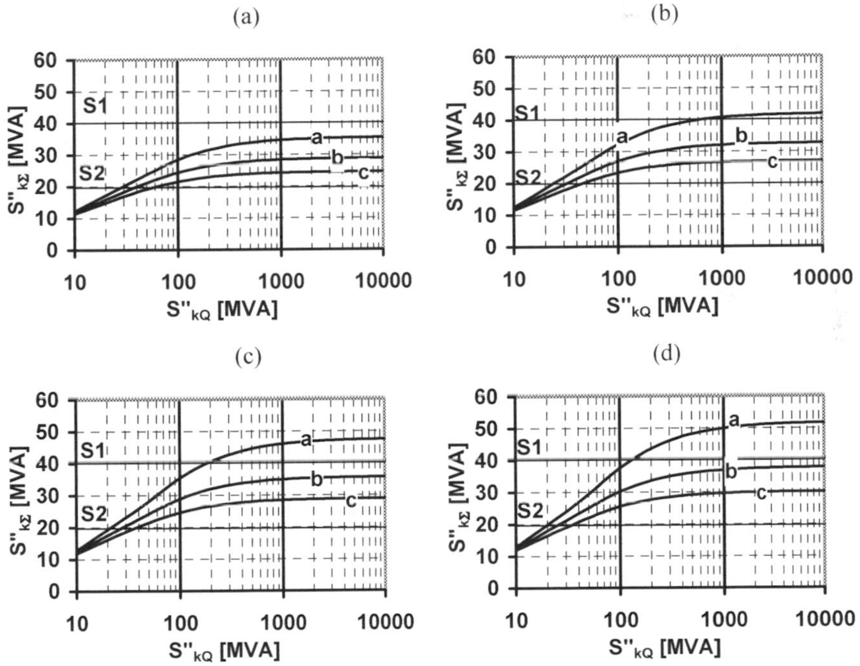


**Fig. 4.22.** Initial AC short-circuit apparent power $S''_{k\Sigma}$ after coupling the WTGS with rated power $S_{nM}$ = 1.0 MVA as a function of $S''_{kQ}$ for various lengths of transmission line. (a) $S_{nT}$ = 6.3 MVA; (b) $S_{nT}$ = 10 MVA; (c) $S_{nT}$ = 16 MVA; (d) $S_{nT}$ = 25 MVA. Lines: a – $l$ = 10 km; b – $l$ = 15 km; c – $l$ = 20 km; S1 – $S''_{k} = 40 S_{nM}$ ; S2 – $S''_{k} = 20 S_{nM}$ .

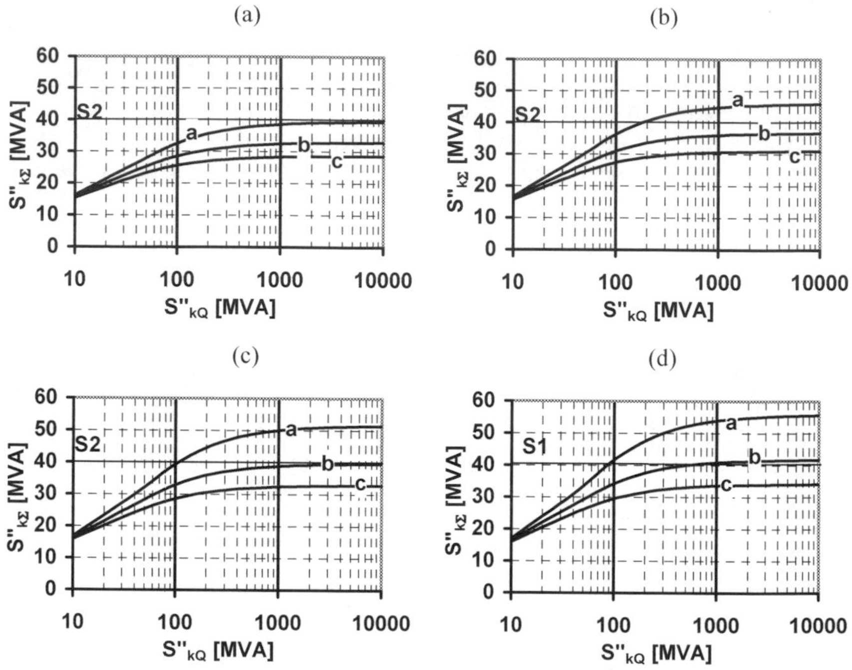


**Fig. 4.23.** Initial AC short-circuit apparent power $S''_{k\Sigma}$ after coupling the WTGS with rated power $S_{nM}$ = 2.0 MVA as a function of $S''_{kQ}$ for various lengths of transmission line. (a) $S_{nT}$ = 6.3 MVA; (b) $S_{nT}$ = 10 MVA; (c) $S_{nT}$ = 16 MVA; (d) $S_{nT}$ = 25 MVA. Lines: a – $l$ = 10 km; b – $l$ = 15 km; c – $l$ = 20 km; S2 – $S''_k = 20S_{nM}$ .

The maximum value of the WTGS rated power depends on the impedance of the transmission system. For the considered network, $S_{nM}$ is directly proportional to the feeding transformer rated power and the cross-sectional area of the transmission line conductor, and is inversely proportional to the transmission line length.

The following tests are related to the evaluation of the influence of power system elements on the short-circuit apparent power. The initial AC short-circuit apparent power $S''_{k\Sigma}$ after the WTGS connection to the grid can be computed as:

$$S''_{k\Sigma} = \frac{cV^2_{N,\mathrm{MV}}}{\left|\dfrac{(\underline{Z}_\mathrm{Q} + \underline{Z}_\mathrm{T} + \underline{Z}_\mathrm{L})(\underline{Z}_\mathrm{M} + \underline{Z}_\mathrm{TM})}{\underline{Z}_\mathrm{Q} + \underline{Z}_\mathrm{T} + \underline{Z}_\mathrm{L} + \underline{Z}_\mathrm{M} + \underline{Z}_\mathrm{TM}}\right|}, \tag{4.76}$$

where the values of impedances are defined by (4.53)–(4.57).

Figures 4.22 and 4.23 present the initial AC short-circuit apparent power $S''_{k\Sigma}$ after the WTGS coupling to the power system as a function of the initial AC short-circuit apparent power $S''_{kQ}$ of the feeding power system Q for various types of feeding transformer, transmission line and two types of WTGS – with rated power equal to 1 MVA and 2 MVA. The figures show that the character of the $S''_{kQ}$ influence on the $S''_{k\Sigma}$ for various considered parameters is similar to that shown in Fig. 4.21, i.e. for high values of $S''_{kQ}$, saturation of the initial AC short-circuit power at the PCC is observed.

Figure 4.24 presents the influence of the current start-up coefficient $k_{LR}$ on the initial AC short-circuit apparent power $S''_{k\Sigma}$ (after WTGS coupling to the power system) as a function of the initial AC short-circuit apparent power $S''_{kQ}$. The lowest curves in the figure show the value of $S''_{k}$ (in a system without WTGS) and therefore enable the influence of the WTGS on the initial AC short-circuit apparent power level to be calculated. The figure shows the increase of the initial AC short-circuit apparent power at the PCC resulting from the WTGS coupling. The increase is higher for smaller values of $S''_{kQ}$. For $S''_{kQ}$ greater than 500 MVA the increase is practically constant, but depends of course on the WTGS rated power. The increase of the coefficient $k_{\Delta S} = S''_{k\Sigma} / S''_{k}$ for high values of $S''_{kQ}$ is equal to 10–20% for a WTGS with rated power above 1 MVA.

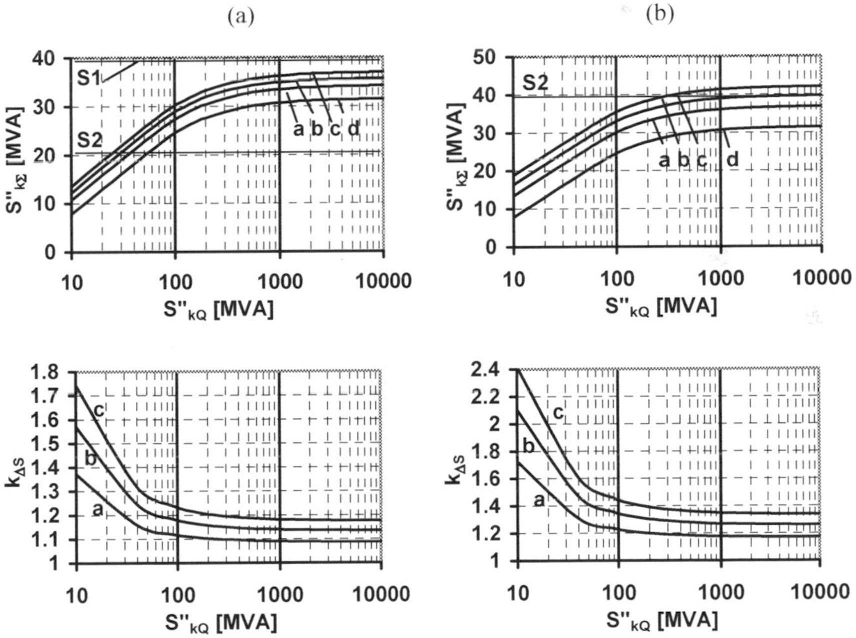


**Fig. 4.24.** Initial AC short-circuit apparent power $S''_{k\Sigma}$ and factor of the $S''_{k}$ increase after coupling WTGS as a function of $S''_{kQ}$ for various $k_{LR}$ coefficients ($S_{nT}$ = 16 MVA, $l$ = 15 km): (a) $S_{nM}$ = 1 MVA; (b) $S_{nM}$ = 2 MVA. Lines: a – $k_{LR}$ = 3; b – $k_{LR}$ = 5; c – $k_{LR}$ = 7; d – without WTGS; S1 – $S''_{k} = 40 S_{nM}$; S2 – $S''_{k} = 20 S_{nM}$.

Figure 4.25 presents the influence of the WTGS rated power $S_{nM}$ on the initial AC short-circuit apparent power $S''_{k\Sigma}$ as a function of the initial AC short-circuit apparent power $S''_{kQ}$. The lowest curves in the figure show the value of $S''_{k}$ at the

PCC for a system without a WTGS. The figures show that the increase of the asynchronous generator rated power with simultaneous increase of the step-up transformer rated power causes an increase of the initial AC short-circuit apparent power at the PCC similar in level and character to that achieved in the previous example (related to the influence of the coefficient $k_{LR}$).

The black squares and triangles located on the curves show the points fulfilling the conditions (4.66) and (4.67), respectively. It means that the points show the minimum value of $S''_{kQ}$ for the given generator in the given network. The condition (4.67) is marked in the figure only for the smallest generator. For other generators, these points are located outside the figure.

The above considerations were related to the computation of the initial AC short-circuit apparent power at the point of common coupling of the system and the WTGS. Considering the protection problem, it is necessary to compute the short-circuit currents as a result of faults at various points of the considered network. This computation is necessary to check that the short-circuit currents, which flow through the elements of the system (lines, transformers, busses, current transformers, etc.) after the WTGS connection, do not exceed the permissible current limits. This is also necessary for selecting adequate electrical equipment. It is also utilized to check and correct, if necessary, the settings (reference values) of the protection systems.

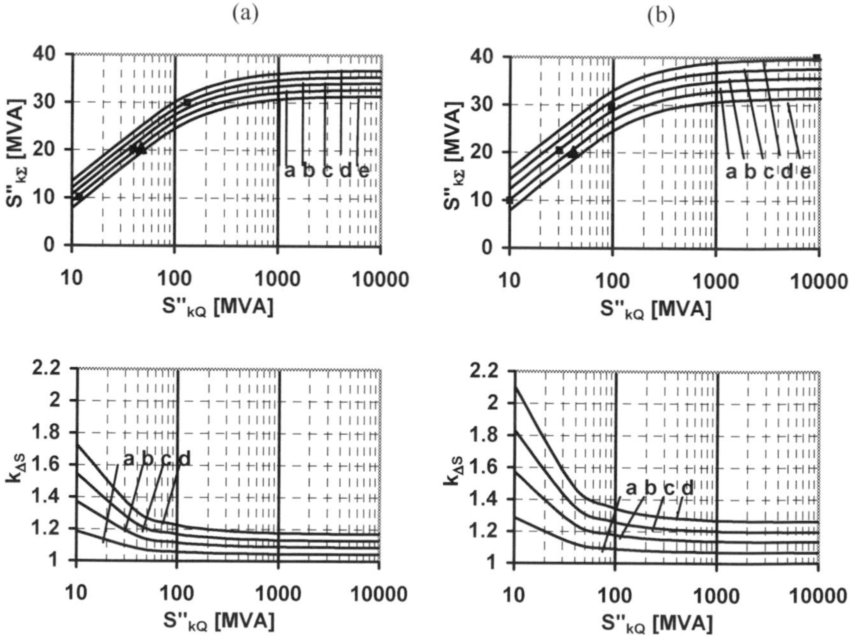


**Fig. 4.25.** Initial AC short-circuit apparent power $S''_{k\Sigma}$ and factor of the $S''_{k}$ increase after coupling WTGS as a function of $S''_{kQ}$ for various rated power levels of the generator ($S_{nT}$ = 16 MVA, $l$ = 15 km): (a) $k_{LR}$ = 3; (b) $k_{LR}$ = 5. Lines: a – $S_{nM}$ = 0.5 MVA; b – $S_{nM}$ = 1.0 MVA; c – $S_{nM}$ = 1.5 MVA; d – $S_{nM}$ = 2.0 MVA; e – without WTGS.

**Table 4.11.** Short-circuit currents in a typical network versus location of the short-circuit

| Short-circuit location | CB0 | CB1 | CB2 | CB3 | CB4 |
|---|---|---|---|---|---|
| F0 | + | ++ | – | – | ++ |
| F1 | + | + | – | – | ++ |
| F2 | ++/— | ++ | ++ | – | ++ |
| F3 | ++/— | ++/— | – | ++ | ++ |
| F4 | + | + | – | – | ++ |

**Table 4.12.** Initial AC short-circuit currents $I_k''$ flowing through the circuit-breakers CB0–CB4 in a typical network

| Short-circuit location | Circuit-breaker | Short-circuit current $I_k''$ |
|---|---|---|
| F0 | CB0 | $\frac{cV_N}{\sqrt{3}\left\lvert\underline{Z}_{\mathrm{Q}}+\underline{Z}_{\mathrm{T}}\right\rvert}$ |
| F1 | CB0 | $\frac{cV_N}{\sqrt{3}\left\lvert\underline{Z}_{\mathrm{Q}}+\underline{Z}_{\mathrm{T}}+\alpha\underline{Z}_{\mathrm{L}}\right\rvert}$ |
| F2 | CB0 | $\frac{cV_N}{\sqrt{3}\left\lvert\alpha\underline{Z}_{\mathrm{L2}}\frac{\underline{Z}_{\mathrm{Q}}+\underline{Z}_{\mathrm{T}}+\underline{Z}_{\mathrm{M}}+\underline{Z}_{\mathrm{TM}}+\underline{Z}_{\mathrm{L}}}{\underline{Z}_{\mathrm{M}}+\underline{Z}_{\mathrm{TM}}+\underline{Z}_{\mathrm{L}}}+\underline{Z}_{\mathrm{Q}}+\underline{Z}_{\mathrm{T}}\right\rvert}$ |
| F3 | CB0 | $\frac{cV_N}{\sqrt{3}\left\lvert(\alpha\underline{Z}_{\mathrm{L3}}+\underline{Z}_{\mathrm{L1}})\frac{\underline{Z}_{\mathrm{Q}}+\underline{Z}_{\mathrm{T}}+\underline{Z}_{\mathrm{M}}+\underline{Z}_{\mathrm{TM}}+\underline{Z}_{\mathrm{L}}}{\underline{Z}_{\mathrm{M}}+\underline{Z}_{\mathrm{TM}}}+\underline{Z}_{\mathrm{Q}}+\underline{Z}_{\mathrm{T}}+\underline{Z}_{\mathrm{L}}\right\rvert}$ |
| F4 | CB0 | $\frac{cV_N}{\sqrt{3}\left\lvert\alpha\underline{Z}_{\mathrm{L1}}\frac{\underline{Z}_{\mathrm{Q}}+\underline{Z}_{\mathrm{T}}+\underline{Z}_{\mathrm{M}}+\underline{Z}_{\mathrm{TM}}+\underline{Z}_{\mathrm{L}}}{\underline{Z}_{\mathrm{M}}+\underline{Z}_{\mathrm{TM}}}+\underline{Z}_{\mathrm{Q}}+\underline{Z}_{\mathrm{T}}+\underline{Z}_{\mathrm{L}}\right\rvert}$ |
| F0 | CB1 | $I_k''=\frac{cV_N}{\sqrt{3}\left\lvert\underline{Z}_{\mathrm{M}}+\underline{Z}_{\mathrm{TM}}+\underline{Z}_{\mathrm{L}}\right\rvert}$ |
| F1 | CB1 | $\frac{cV_N}{\sqrt{3}\left\lvert\underline{Z}_{\mathrm{Q}}+\underline{Z}_{\mathrm{T}}+\alpha\underline{Z}_{\mathrm{L}}\right\rvert}$ |
| F2 | CB1 | $\frac{cV_N}{\sqrt{3}\left\lvert\alpha\underline{Z}_{\mathrm{L2}}\frac{\underline{Z}_{\mathrm{Q}}+\underline{Z}_{\mathrm{T}}+\underline{Z}_{\mathrm{M}}+\underline{Z}_{\mathrm{TM}}+\underline{Z}_{\mathrm{L}}}{\underline{Z}_{\mathrm{Q}}+\underline{Z}_{\mathrm{T}}}+\underline{Z}_{\mathrm{M}}+\underline{Z}_{\mathrm{TM}}+\underline{Z}_{\mathrm{L}}\right\rvert}$ |
| F3 | CB1 | $\frac{cV_N}{\sqrt{3}\left\lvert(\alpha\underline{Z}_{\mathrm{L3}}+\underline{Z}_{\mathrm{L1}})\frac{\underline{Z}_{\mathrm{Q}}+\underline{Z}_{\mathrm{T}}+\underline{Z}_{\mathrm{M}}+\underline{Z}_{\mathrm{TM}}+\underline{Z}_{\mathrm{L}}}{\underline{Z}_{\mathrm{M}}+\underline{Z}_{\mathrm{TM}}}+\underline{Z}_{\mathrm{Q}}+\underline{Z}_{\mathrm{T}}+\underline{Z}_{\mathrm{L}}\right\rvert}$ |
| F4 | CB1 | $\frac{cV_N}{\sqrt{3}\left\lvert\alpha\underline{Z}_{\mathrm{L1}}\frac{\underline{Z}_{\mathrm{Q}}+\underline{Z}_{\mathrm{T}}+\underline{Z}_{\mathrm{M}}+\underline{Z}_{\mathrm{TM}}+\underline{Z}_{\mathrm{L}}}{\underline{Z}_{\mathrm{M}}+\underline{Z}_{\mathrm{TM}}}+\underline{Z}_{\mathrm{Q}}+\underline{Z}_{\mathrm{T}}+\underline{Z}_{\mathrm{L}}\right\rvert}$ |

**Table 4.12.** (cont.)

| Short-circuit location | Circuit-breaker | Short-circuit current $I''_k$ |
|---|---|---|
| F0 | CB2 | – |
| F1 | CB2 | – |
| F2 | CB2 | $\dfrac{cV_N}{\sqrt{3}\left\lvert \alpha \underline{Z}_{L2} + \dfrac{(\underline{Z}_Q + \underline{Z}_T)(\underline{Z}_M + \underline{Z}_{TM} + \underline{Z}_L)}{\underline{Z}_Q + \underline{Z}_T + \underline{Z}_M + \underline{Z}_{TM} + \underline{Z}_L} \right\rvert}$ |
| F3 | CB2 | – |
| F4 | CB2 | – |
| F0 | CB3 | – |
| F1 | CB3 | – |
| F2 | CB3 | – |
| F3 | CB3 | $\dfrac{cV_N}{\sqrt{3}\left\lvert \alpha \underline{Z}_{L3} + \underline{Z}_{L1} + \dfrac{(\underline{Z}_Q + \underline{Z}_T + \underline{Z}_L)(\underline{Z}_M + \underline{Z}_{TM})}{\underline{Z}_Q + \underline{Z}_T + \underline{Z}_M + \underline{Z}_{TM} + \underline{Z}_L} \right\rvert}$ |
| F4 | CB3 | – |
| F0 | CB4 | $\dfrac{cV_N}{\sqrt{3}\left\lvert \underline{Z}_M + \underline{Z}_{TM} + \underline{Z}_L \right\rvert}$ |
| F1 | CB4 | $\dfrac{cV_N}{\sqrt{3}\left\lvert \underline{Z}_M + \underline{Z}_{TM} + (1-\alpha)\underline{Z}_L \right\rvert}$ |
| F2 | CB4 | $\dfrac{cV_N}{\sqrt{3}\left\lvert \alpha \underline{Z}_{L2} \dfrac{\underline{Z}_Q + \underline{Z}_T + \underline{Z}_M + \underline{Z}_{TM} + \underline{Z}_L}{\underline{Z}_Q + \underline{Z}_T} + \underline{Z}_M + \underline{Z}_{TM} + \underline{Z}_L \right\rvert}$ |
| F3 | CB4 | $\dfrac{cV_N}{\sqrt{3}\left\lvert (\alpha \underline{Z}_{L3} + \underline{Z}_{L1}) \dfrac{\underline{Z}_Q + \underline{Z}_T + \underline{Z}_M + \underline{Z}_{TM} + \underline{Z}_L}{\underline{Z}_Q + \underline{Z}_T + \underline{Z}_L} + \underline{Z}_M + \underline{Z}_{TM} \right\rvert}$ |
| F4 | CB4 | $\dfrac{cV_N}{\sqrt{3}\left\lvert \alpha \underline{Z}_{L1} \dfrac{\underline{Z}_Q + \underline{Z}_T + \underline{Z}_M + \underline{Z}_{TM} + \underline{Z}_L}{\underline{Z}_Q + \underline{Z}_T + \underline{Z}_L} + \underline{Z}_M + \underline{Z}_{TM} \right\rvert}$ |

Computation of the short-circuit currents for various locations of fault is the base for the analysis. The possible positions of faults (F0–F4) and possible locations of the circuit-breakers (CB0–CB4), which are controlled by overcurrent protection equipment, are presented in Fig. 4.17. The faults can occur also in lines, and these cases are included in the following considerations as well.

Table 4.11 presents information related to the short-circuit current flowing through the given circuit-breaker as a result of a fault at the given location. The (–) sign indicates that the short-circuit current does not flow through the given circuit-breaker. The (+) sign means that the short-circuit current flows but the WTGS connection to the grid does not increase the current. The (++) sign means that the short-circuit current flows and that after the WTGS connection to the grid the short-circuit current will be higher than in a system without WTGS. And finally, the (++/—) sign means that the short-circuit current flows but that after the WTGS connection to the system it can be higher or lower.

Detailed information related to the values of the short-circuit currents in the form of equations is presented in Table 4.12. The coefficient $\alpha \in <0, 1>$ that appears in the equations defines the position of the short-circuit on the transmission line – counting from the "left-hand" bus in Fig. 4.17.

## 4.7 Real Power Losses

The problem of the real power losses in an electric power system is important, and therefore it is the subject of many investigations and optimization research. To consider the problem of power losses in an electric power network in which a WTGS operates, let us assume a radial network (typical for an MV system) as shown in Fig. 4.26. Let the WTGS be coupled to the network at the end of the last branch (*N*th node).

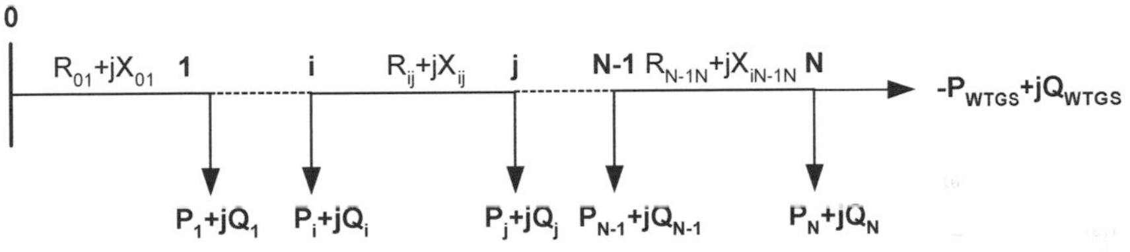


**Fig. 4.26.** Structure of a typical MV power network for power losses calculations

If the following notation defining the apparent power in the *ij*th branch at the *j*th node is used:

$$\underline{S}_{ij}^{(j)} = \sqrt{3}\underline{V}_j \underline{I}_{ij}^* = P_{ij}^{(j)} + jQ_{ij}^{(j)}, \tag{4.77}$$

the *ij*th branch current is defined as

$$\underline{I}_{ij} = \frac{\underline{S}_{ij}^{(j)*}}{\sqrt{3}\underline{V}_j^*}. \tag{4.78}$$

The real power losses in the *ij*th branch are equal to

$$\Delta P_{ij} = \frac{(P_{ij}^{(j)})^2 + (Q_{ij}^{(j)})^2}{V_j^2} R_{ij}. \tag{4.79}$$

For a power network consisting of $N$ serially connected branches with $N$ loads defined by the real $P_j$ and reactive $Q_j$ power, the total power losses are equal to[23]

$$\Delta P = \sum_{i=0}^{N-1} [ ( \sum_{j=i+1}^{N} \frac{P_j}{V_j} )^2 + ( \sum_{j=i+1}^{N} \frac{Q_j}{V_j} )^2 ] R_{i,i+1} . \quad (4.80)$$

After connection of the WTGS, which generates real power $P_{\text{WTGS}}$ and consumes reactive power $Q_{\text{WTGS}}$, which leads to the equation describing the wind turbine apparent power as

$$S_{\text{WTGS}} = -P_{\text{WTGS}} + \text{j}Q_{\text{WTGS}} , \quad (4.81)$$

the total power losses in the considered network are equal to

$$\Delta P_{\Sigma} = \sum_{i=0}^{N-1} [ ( \sum_{j=i+1}^{N} \frac{P_j}{V_j} - \frac{P_{\text{WTGS}}}{V_N} )^2 + ( \sum_{j=i+1}^{N} \frac{Q_j}{V_j} + \frac{Q_{\text{WTGS}}}{V_N} )^2 ] R_{i,i+1} . \quad (4.82)$$

Connection of the WTGS to the electric power network will result in decreasing power losses when:

$$\frac{\Delta P_{\Sigma}}{\Delta P} < 1 , \quad (4.83)$$

which, using (4.80) and (4.82), leads to the following condition:

$$\frac{\sum_{i=0}^{N-1} [ ( \sum_{j=i+1}^{N} \frac{P_j}{V_j} - \frac{P_{\text{WTGS}}}{V_N} )^2 + ( \sum_{j=i+1}^{N} \frac{Q_j}{V_j} + \frac{Q_{\text{WTGS}}}{V_N} )^2 ] R_{i,i+1}}{\sum_{i=0}^{N-1} [ ( \sum_{j=i+1}^{N} \frac{P_j}{V_j} )^2 + ( \sum_{j=i+1}^{N} \frac{Q_j}{V_j} )^2 ] R_{i,i+1}} < 1 . \quad (4.84)$$

Condition (4.84) in the above form is relatively complex and therefore cannot be utilized for analytical considerations without any simplification. Practically speaking, it is possible to carry analytical considerations of the problem only in an extremely simplified system.

Simplifying the network to an extreme degree, let us assume that the network consists of a single transmission line ($N = 1$). Then the power losses in system without WTGS are equal to

$$\Delta P = \frac{P_1^2 + Q_1^2}{V_1^2} R_{01} . \quad (4.85)$$

After coupling the WTGS to the network, the power losses become equal to

[23] Power losses at shunt elements of the transmission line should be added to the loads. For an MV network, the shunt elements of the transmission lines can be neglected.

$$\Delta P = \frac{(P_1 - P_{\text{WTGS}})^2 + (Q_1 + Q_{\text{WTGS}})^2}{V_{1\text{WTGS}}^2} R_{01} . \tag{4.86}$$

Then (4.84) takes the following form:

$$\frac{\Delta P_\Sigma}{\Delta P} = \frac{(P_1 - P_{\text{WTGS}})^2 + (Q_1 + Q_{\text{WTGS}})^2}{P_1^2 + Q_1^2} \frac{V_1^2}{V_{1\text{WTGS}}^2} < 1 . \tag{4.87}$$

Neglecting the difference between node voltage before and after coupling WTGS, (4.87) can be rewritten in the following form:

$$1 + \frac{P_{\text{WTGS}}^2 + Q_{\text{WTGS}}^2 - 2P_1 P_{\text{WTGS}} + 2Q_1 Q_{\text{WTGS}}}{P_1^2 + Q_1^2} < 1 . \tag{4.88}$$

The second part of the left-hand side of (4.88) is a measure of the power loss change as a result of coupling the WTGS. In general, this measure can be negative or positive. The power losses will decrease when this measure becomes negative. Because the component $P_1^2 + Q_1^2$ is always positive, the following formula can be used:

$$f(P_{\text{WTGS}}, Q_{\text{WTGS}}) = P_{\text{WTGS}}^2 - 2P_1 P_{\text{WTGS}} + Q_{\text{WTGS}}^2 + 2Q_1 Q_{\text{WTGS}} < 0 . \tag{4.89}$$

Let us look for values of the WTGS real and reactive power, for a given power consumed by the load, for which condition (4.89) is fulfilled.

For most of the existing WTGSs, the reactive power consumed is controlled (kept equal to zero) or depends – in the known form $Q_{\text{WTGS}} = f(P_{\text{WTGS}})$ – on the real power. Then (4.89) can be treated as a function (quadratic inequality) with $P_{\text{WTGS}}$ as an unknown quantity. The solution of the function is

$$0 < P_{\text{WTGS}} < P_1 + \sqrt{P_1^2 - (Q_{\text{WTGS}}^2 + 2Q_1 Q_{\text{WTGS}})} . \tag{4.90}$$

The left-hand limit results from the assumption that the WTGS generates real power only. The minimum value of the function $f(P_{\text{WTGS}}, Q_{\text{WTGS}})$ – see (4.88) – is achieved for the wind turbine real power equal to the load real power:

$$P_{\text{WTGS}} = P_1 . \tag{4.91}$$

This seems to be obvious, because for the given network state, the real power is produced and consumed in the same place (and does not flow through the transmission line). What is also interesting is that the real power produced by the WTGS for which the minimum power losses are achieved does not depend on the reactive power consumption level.

But, as one can see from (4.90), the reactive power consumed by the WTGS influences the range of the real power $P_{\text{WTGS}}$ for which the decrease of power losses can be achieved. The maximum range of the real power $P_{\text{WTGS}}$ will be achieved when the squared form (right-hand side of (4.90)) becomes maximal. That form becomes maximal for the reactive power consumed by the WTGS equal to

$$Q_{\mathrm{WTGS}} = -Q_1 . \qquad (4.92)$$

Then (including (4.91)) both the real and reactive power are produced and consumed at the point of WTGS common coupling.

Condition (4.92) could be fulfilled when the WTGS produces a reactive power equal to the power consumed by the load or when the WTGS consumes the reactive power produced by the load (capacitive load). Unfortunately, real WTGSs do not produce reactive power. The reactive power is (almost) fully compensated ($Q_{\mathrm{WTGS}} = 0$) or only the no-load reactive power of the asynchronous generator is compensated ($Q_{\mathrm{WTGS}} > 0$). Therefore, the effective possible range of the reactive power change (which influences the range of the real power production, which decreases the power losses) is equal to

$$0 < Q_{\mathrm{WTGS}} < -Q_1 + \sqrt{P_1^2 + Q_1^2} . \qquad (4.93)$$

The right-hand side of condition (4.90), for an inductive load, becomes maximal when the reactive power consumed by the WTGS is equal to zero ($Q_{\mathrm{WTGS}} = 0$). Then the range of the WTGS real power production, for which the real power losses in system are decreased, is equal to

$$0 < P_{\mathrm{WTGS}} < 2P_1 . \qquad (4.94)$$

Condition (4.93) shows that for capacitive loads ($Q_1 < 0$), the right-hand limit of reactive power $Q_{\mathrm{WTGS}}$ becomes higher and simultaneously the range of the real power $P_{\mathrm{WTGS}}$ becomes higher as well.

Unfortunately, in practice, the loads have to compensated to an inductive power factor equal to $0 < \tan\varphi < 0.2$. Then, the maximum range of the real power generated by the WTGS (for which the decrease of the power losses is obtained) can be achieved for the fully compensated WTGS ($Q_{\mathrm{WTGS}} = 0$).

As an example of the above considerations, Fig. 4.27 presents the level of power loss decrease as a function of various values of the load and WTGS power factors $\tan\varphi_1$, $\tan\varphi_{\mathrm{WTGS}}$. The curves show that the real power loss level depends relatively weakly on the load power factor, when the WTGS is fully compensated (Fig. 4.27a). But when the WTGS is not fully compensated, the influence of the load power factor becomes higher (Fig. 4.27b). In that case, the range of WTGS real power generation (in comparison with the load real power) for which the power losses decrease becomes lower.

While utilizing the load and the wind turbine power factors ($\tan\varphi_1$, $\tan\varphi_{\mathrm{WTGS}}$), the condition (4.90) can be rewritten in the following form:

$$0 < P_{\mathrm{WTGS}} < 2P_1 \frac{1 - \tan\varphi_1 \tan\varphi_{\mathrm{WTGS}}}{1 + (\tan\varphi_{\mathrm{WTGS}})^2} . \qquad (4.95)$$

Then the range of the $P_{\mathrm{WTGS}}/P_1$ quotient minimizing the real power losses as a function of the power factors can be limited by the plane presented in Fig. 4.28.

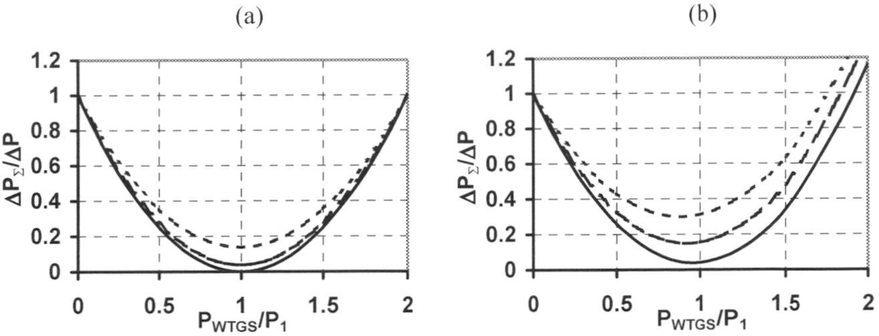


**Fig. 4.27.** Decrease of power losses as a function of $P_{WTGS}/P_1$ for various values of $\tan\varphi_1$ and $\tan\varphi_{WTGS}$: (a) $\tan\varphi_{WTGS} = 0$; (b) $\tan\varphi_{WTGS} = 0.2$. Lines: continuous – $\tan\varphi_1 = 0$; dashed – $\tan\varphi_1 = 0.2$; dotted – $\tan\varphi_1 = 0.4$.

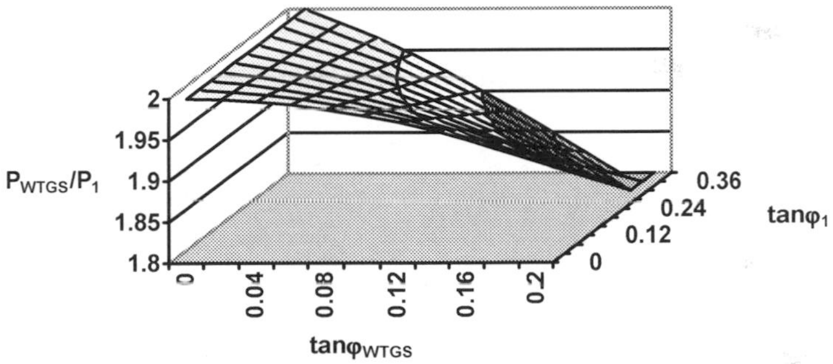


**Fig. 4.28.** Boundary values of $P_{WTGS}/P_1$ quotient minimizing power losses

The above considerations pertain to an extremely simple power network. In general, these considerations can be treated as related to the last branch of the radial network, while the WTGS is connected at the end of one branch. In this case, the conclusions of the above considerations remain valid. Additionally, when, as a result of coupling the WTGS to the network, the power losses in the last branch decrease, then the power losses in the remaining branches of the radial network (when all the loads are inductive) also decrease, which can be treated as a positive feature.

The last question worth asking is how big a decrease in MW can be achieved in the MV power network? Let us consider the overhead MV transmission line with the cross-sectional area of conductor $s = 120\ \text{mm}^2$ ($R' = 0.24\ \Omega/\text{km}$) and the length of the last branch $l = 5$ km. Let the load located at the end of branch be equal to $S_1 = 5$ MVA, $\tan\varphi_1 = 0.2$. For this case, the power losses in the last branch in the case without WTGS are equal to 132 kW. Let us assume that a WTGS that generates real power equal to 1 MW is connected to the network (branch). Therefore, because $P_{WTGS}/P_1 = 0.2$, then, from Fig. 4.27[24], the decrease of the power losses will be equal to about 0.3, which means 40 kW here.[25]

An increase of the power generated by the WTGS, for example to 2 MW, causes a decrease of the power losses to about 0.5, which means 66 kW.

The above values are relatively small but it is worth remembering that the decrease of power losses in other branches of the system can be similar. Then the total decrease of the power losses will be higher. On the other hand, it is worth noting that the power losses in an MV system (and especially in an HV system) are very small in comparison with the power transferred to the loads.

[24] For the assumed value of the WTGS power generation, the real power losses decrease depends only slightly on the WTGS power factor (when $0 \leq \tan\varphi_{WTGS} \leq 0.2$).

[25] Figure 4.27 shows the quotient of the power losses in a network with a WTGS to the power losses in the network without a WTGS. Therefore, the power losses decrease for the given quotient $P_{WTGS}/P_1$ is equal to 1 minus the value read from Fig. 4.27.

# 5 Mathematical Modeling of WTGS Components

## 5.1 Introduction

*Models* of reality (objects, processes, phenomena, etc.) are usually an element of the process presented in Fig. 5.1.

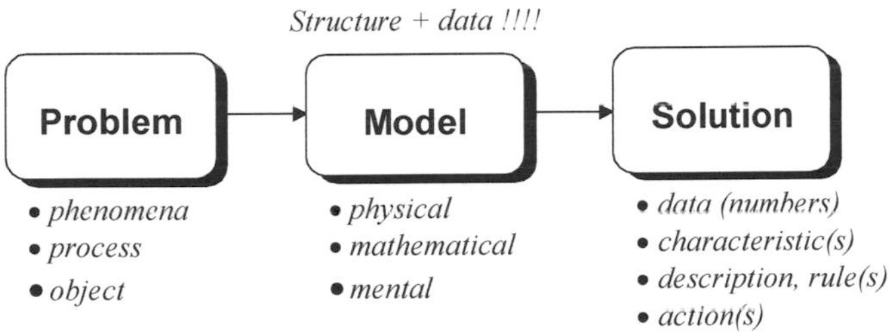


**Fig. 5.1.** Relations: problem, model and solution

And therefore, they should be considered together with other elements of the process, i.e. with *problem* and *solution*. In practice, because of the complexity of reality, general models, which let us solve wide classes of problems, are not elaborated. On the contrary, the *problems* being solved (often defined as individual problems with a narrowly defined level of detail) are the elements that determine the model used. The models differ in the number of introduced simplifications and the domain of applicability. Therefore, the use of a given model depends on the analyzed problem.

Not only the equations creating the model but also the data are related to the *model*, especially when it is considered as a mathematical one. The data, from a model usually describing a wide class of reality (objects, processes, etc.), create a *model* of the defined object (process, phenomena, etc.) that lets us solve the defined *problem* and get an appropriate *solution*.

It is important to realize that all the *solutions* of a *problem* achieved on the basis of a given model are true solutions (e.g. optimal) only for the given *model*. The solutions are not true *solutions* of the *problem* in reality (of a real physical problem).

In the light of the above, one can say that the achieved *solution* of a *problem* is a better *solution* (in relation to reality) if the model better imitates reality, which strongly emphasizes the great significance of *modeling* in solving *problems*.

The above shows how important mathematical models of the analyzed problems are.

Considering WTGS operation in an electric power system it is necessary to utilize two basic types of models: static and dynamic.

The static model of a WTGS is needed for all types of steady-state (and even quasi steady state) analysis, e.g. load flow, power quality assessment (according to the standards), short-circuit calculations, etc. (which were considered in Chap. 4). The static model here is extremely simple and consists of a voltage source ($V$), a voltage and real power source ($P$, $V$), or a real and reactive power source ($P$, $Q$) – which depends on the type of analysis and on the WTGS type [25].

The dynamic model of a WTGS is needed for various types of analysis related to system dynamics, e.g. stability, control system analysis and optimization, etc. [1, 3-7, 13, 14, 18, 56, 65, 68, 70, 74, 81]. Because modern WTGSs are equipped with power electronic systems (converters, soft-switch system), which need a specific approach to their modeling, then at least two types of dynamic models should be used here: a functional model and a mathematical physical model. The models differ in the level of detail of the power electronic system modeling, and therefore can be utilized for various types of analysis – which is described in Table 5.1. Modeling the WTGS operating in an electric power system it is possible and necessary to make decomposition of the model into submodels (components), mainly because of the existence of various types of wind turbines.

In general, WTGSs can be equipped with a synchronous or asynchronous generator. The generator can be connected to the grid directly or through power electronic converter. In the case of asynchronous generators' utilization, the power electronic converter can be switched on between rotor and stator, or a switched resistor can be connected to the machine rotor winding. Finally, the asynchronous generator can also be of the squirrel-cage rotor type.

**Table 5.1.** Model type versus type of analysis

| Model | Type of analysis |
|---|---|
| Steady state – static models – ($V$), ($P$, $V$), ($P$, $Q$) | Analysis of voltage variation<br>Analysis of load flow<br>Analysis of short-circuits |
| Transient state – dynamic models – functional models | Analysis of transient stability<br>Analysis of small-signal stability<br>Analysis of transient response<br>Analysis of steady-state waveforms<br>Synthesis of control<br>Optimization |
| Transient state – dynamic models – mathematical physical models (power electronics) | Analysis of start-up transient effects<br>Analysis of load transient effects<br>Analysis of fault operation<br>Analysis of harmonics and subharmonics<br>Detailed synthesis of control<br>Detailed optimization |

The wind turbine can be equipped with a blade-pitching system or the stall effect can be utilized for the power limiting. The WTGS can be equipped with or without a gearbox. The power system can also be considered (modeled) by the so-called infinite bus (as a single machine system) or as multi-machine one.

The above possibilities of the WTGS modeling, in the form of a general structure of the model, are presented in Fig. 5.2. A description of the submodels (components) is presented in the following sections.

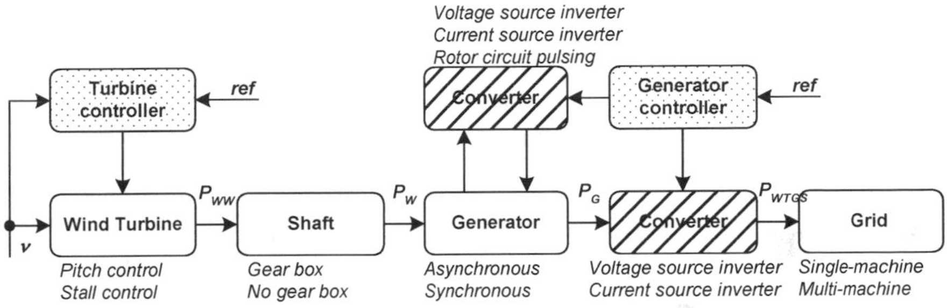


**Fig. 5.2.** Wind turbine generator system model

## 5.2 Wind Turbine Modeling

### 5.2.1 Wind Stream Power

The kinetic energy in air of an object of mass $m$ moving with speed $v$ is equal to:

$$E = \frac{1}{2} m v^2 . \tag{5.1}$$

The power in the moving air (assuming constant wind velocity) is equal to:

$$P_{\text{wind}} = \frac{\mathrm{d}E}{\mathrm{d}t} = \frac{1}{2} \dot{m} v^2 , \tag{5.2}$$

where $\dot{m}$ is the mass flow rate per second. When the air passes across an area $A$ (e.g. the area swept by the rotor blades), the power in the air can be computed as

$$P_{\text{wind}} = \frac{1}{2} \rho A v^3 , \tag{5.3}$$

where $\rho$ is the air density. Air density varies with air pressure and temperature in accordance with the gas law:

$$\rho = \frac{p}{RT} , \tag{5.4}$$

where $p$ – air pressure,
$T$ – temperature,
$R$ – gas constant.

The pressure and temperature varies with the wind turbine location. Some publications (e.g. [72]) give formulas that enable computation of the air density as a function of the turbine elevation above sea level $H$:

$$\rho = \rho_0 - 1.194 \times 10^{-4} H \tag{5.5}$$

where $\rho_0 = 1.225$ kg/m$^3$ is the air density at sea level at temperature $T = 288$ K.

### 5.2.2 Mechanical Power Extracted From the Wind

The air energy described by (5.1) cannot be transferred into another type of energy (e.g. mechanical) with 100% conversion efficiency by any energy converter. In fact, the power extracted from the air stream by any energy converter will be less than the wind power $P_{\mathrm{wind}}$ also because the power achieved by the energy converter $P_{\mathrm{WW}}$ can be computed as the difference between the power in the moving air before and after the converter (the air stream cross-section area before $A_1$ and after $A_2$ converter can be different) [30, 31, 64, 72, 80]:

$$P_{\mathrm{WW}} = P_{\mathrm{wind1}} - P_{\mathrm{wind2}} = \frac{1}{2}\rho(A_1 v_1^3 - A_2 v_2^3)\,. \tag{5.6}$$

Full conversion requires that the air velocity after converter $v_2$ becomes zero, which physically makes no sense, because it constrains the air to be still and further it requires the air velocity before the converter $v_1$ to be equal to zero also.

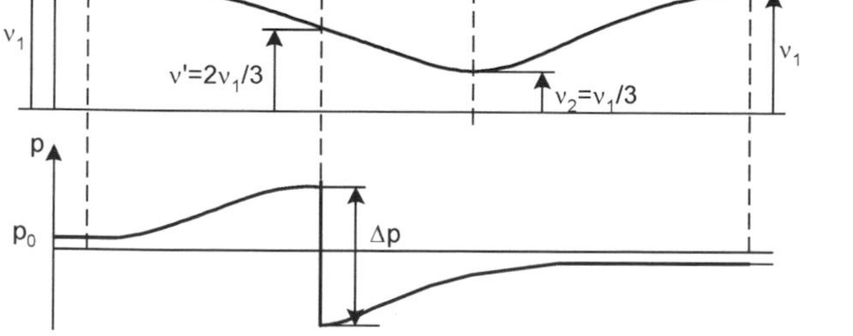

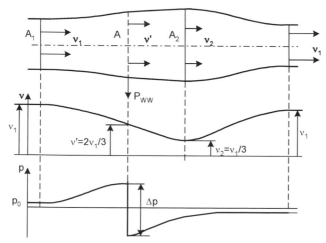

**Fig. 5.3.** Flow conditions due to the extraction of mechanical energy from a free-stream air flow

The real energy converter must be considered as a type of bulkhead. Then the flowing air exerts a force on the converter, the result being that the pressure before the converter increases and simultaneously the air velocity $v'$ decreases (Fig. 5.3).

The force exerted on the converter is equal to

$$F = \dot{m}(v_1 - v_2) \tag{5.7}$$

and the extracted mechanical power is equal to:

$$P_{\mathrm{WW}} = F v' = \dot{m}(v_1 - v_2)v' . \tag{5.8}$$

Comparing (5.6) and (5.8) (assuming that the mass flow rate is constant), we can see that the air velocity through the converter is equal to the average of $v_1$ and $v_2$:

$$v' = \frac{1}{2}(v_1 + v_2) . \tag{5.9}$$

Then the mechanical power extracted from the air stream by the energy converter is equal to

$$P_{\mathrm{WW}} = \frac{1}{4}\rho A(v_1^2 - v_2^2)(v_1 + v_2) \tag{5.10}$$

and less than the power in the air stream before the converter $P_{\mathrm{wind}}$.

Equation (5.10) is usually used in the following form:

$$P_{\mathrm{WW}} = c_{\mathrm{P}} P_{\mathrm{wind}} = c_{\mathrm{P}} \frac{1}{2}\rho A v_1^3 \tag{5.11}$$

where the coefficient $c_{\mathrm{P}} < 1$ – defining the ratio of the mechanical power extracted by the converter to the power in the air stream – is called the power coefficient (Betz's factor). The coefficient is equal to

$$c_{\mathrm{P}} = \frac{P_{\mathrm{WW}}}{P_{\mathrm{wind}}} = \frac{1}{2}\left(1 - \left(\frac{v_2}{v_1}\right)^2\right)\left(1 + \frac{v_2}{v_1}\right). \tag{5.12}$$

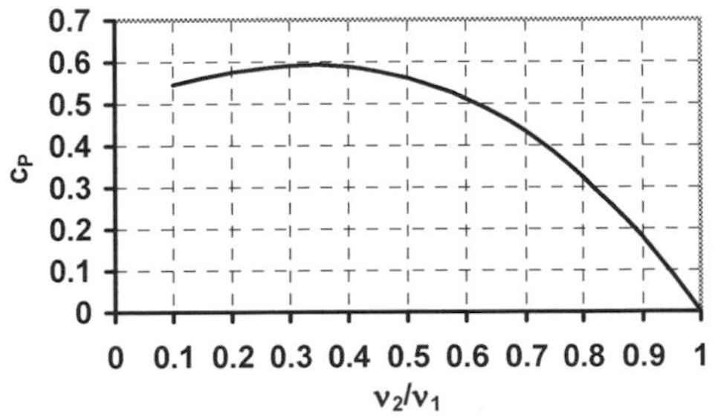


**Fig. 5.4.** Power coefficient as a function of the wind stream speed ratio

The power factor reaches a maximum value (Fig. 5.4) equal to $c_P = 0.593$ when the speed ratio is equal to $v_2/v_1 = 1/3$. Then the air velocity in and behind the energy converter is equal to $v' = 2v_1/3$, $v_2 = v_1/3$, respectively.

The power coefficient of real converters $c_P$ achieves lower values than that computed above because of various aerodynamic losses that depend on the rotor construction (number and shape of blades, weight, stiffness, etc.). The rotor power coefficient is usually given as a function of two parameters: the tip-speed ratio $\lambda$ and the blade pitch angle $\vartheta$. The blade pitch angle is defined as the angle between the plane of rotation and the blade cross-section chord (Fig. 5.7). And the tip-speed ratio $\lambda$ is defined as

$$\lambda = \frac{u}{v_1} = \frac{\omega R}{v_1}, \tag{5.13}$$

where $u$ – tangential velocity of blade tip,
$\omega$ – angular velocity of rotor,
$R$ – rotor radius (≈ blade length).

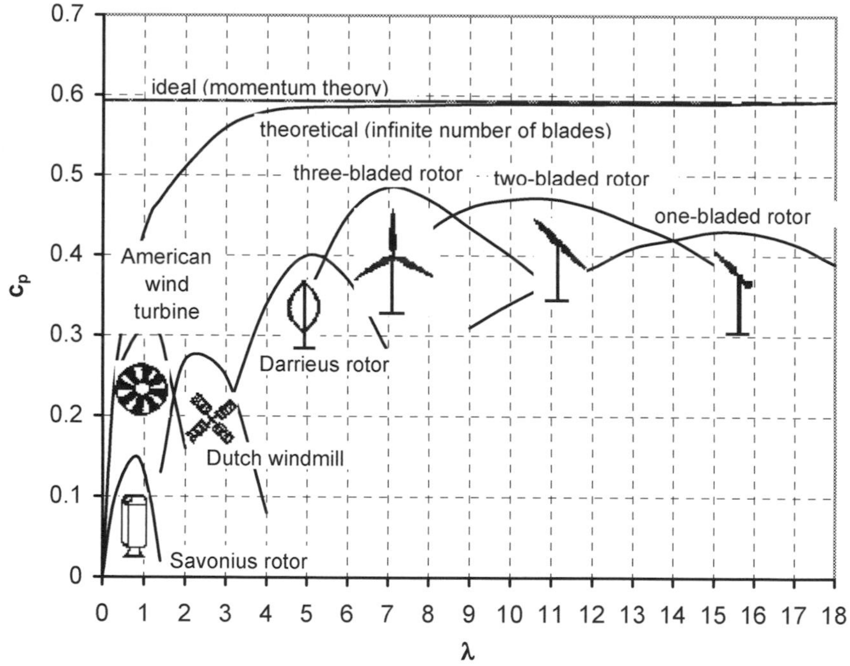


**Fig. 5.5.** Rotor power coefficients of various rotor types [30]

Typical values of the power coefficient $c_P$ and the torque coefficient $c_\tau$ (which is defined analogously to the power coefficient) for various types of wind rotor are presented in Figs. 5.5 and 5.6. The dependence of the rotor power and the torque coefficients for an experimental wind turbine on the blade pitch angle are presented in Figs. 5.7 and 5.8.

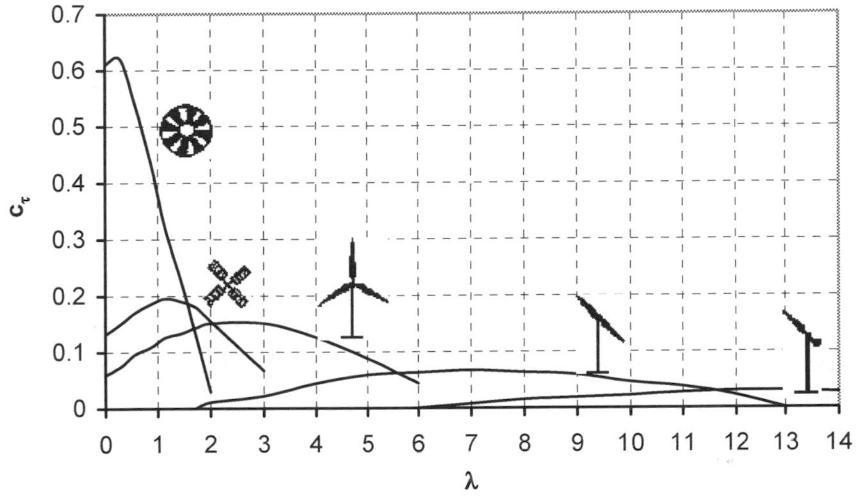


**Fig. 5.6.** Rotor torque coefficients of various rotor types [30]

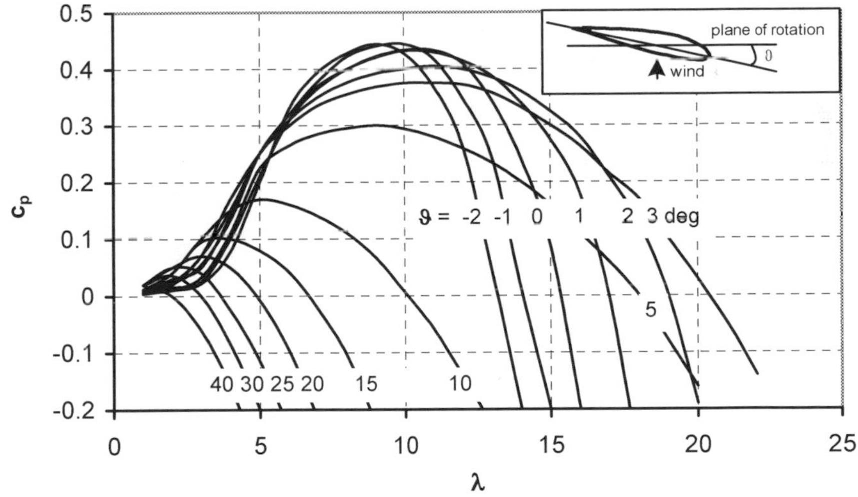


**Fig. 5.7.** Rotor power characteristics of an experimental WKA-60 rotor [30]

Summing up the above, we can say that the mechanical power $P_{WW}$ extracted from the wind converter and the torque $\tau_{WW}$ for the given rotor (described by $c_P$, $c_\tau$ coefficients and the rotor radius $R$) and for the given wind velocity $v_1$, rotor angular velocity $\omega$ and blade pitch angle $\vartheta$ can be computed as

$$P_{\mathrm{WW}} = c_{\mathrm{P}}(\lambda, \vartheta) P_{\mathrm{wind}} = c_{\mathrm{P}}(\lambda, \vartheta) \frac{1}{2} \rho A v_1^3 \tag{5.14}$$

$$\tau_{\mathrm{WW}} = \frac{P_{\mathrm{WW}}}{\omega} = c_\tau(\lambda, \vartheta) \frac{1}{2} \rho A v_1^2 R \,. \tag{5.15}$$

The rotor power and torque coefficients in these models can be utilized in the form of look-up tables or in form of a function. The second approach is presented below, where the general function defining the rotor power coefficient $c_{\mathrm{P}}$ as a function of the tip-speed ratio and the blade pitch angle is defined as follows:

$$c_{\mathrm{P}}(\lambda, \vartheta) = c_1 \left( c_2 \frac{1}{\Lambda} - c_3 \vartheta - c_4 \vartheta^x - c_5 \right) e^{-c_6 \frac{1}{\Lambda}} \,. \tag{5.16}$$

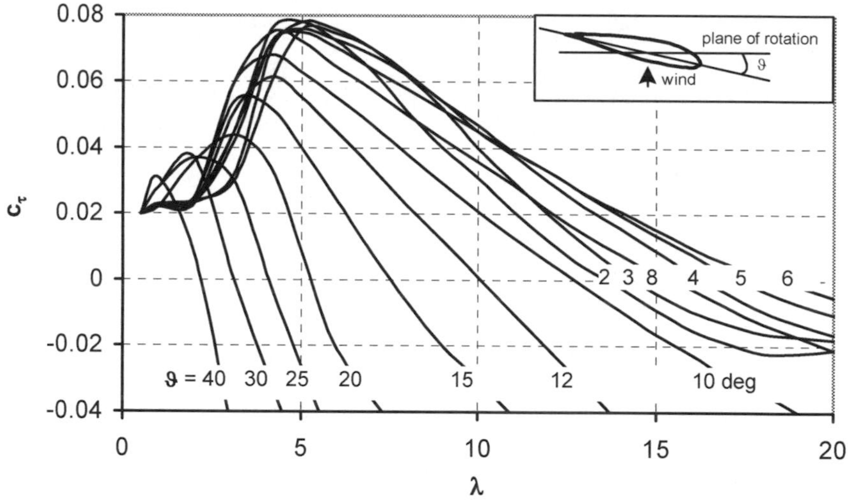


**Fig. 5.8.** Rotor torque characteristics of an experimental WKA-60 rotor [30]

Because the function depends on the WTGS rotor type, the coefficients $c_1$–$c_6$ and $x$ can be different for various turbines. Additionally, the parameter $\Lambda$ is also defined in various ways [8, 9, 30]. For example, the parameter $1/\Lambda$ in [9] is defined as

$$\frac{1}{\Lambda} = \frac{1}{\lambda + 0.08\vartheta} - \frac{0.035}{1 + \vartheta^3} \,, \tag{5.17}$$

while the coefficients $c_1$–$c_6$ are proposed as equal to: $c_1 = 0.5$, $c_2 = 116$, $c_3 = 0.4$, $c_4 = 0$, $c_5 = 5$, $c_6 = 21$ ($x$ is not used here because of $c_4 = 0$).

An example of the $c_{\mathrm{P}} = c_{\mathrm{P}}(\lambda, \vartheta)$ characteristics computed taking into account (5.16) and (5.17) and the above parameters $c_1$–$c_6$, for a given rotor diameter $R$, rotor speed $\omega$ and for various blade pitch angles $\vartheta$ is presented in Fig. 5.9.

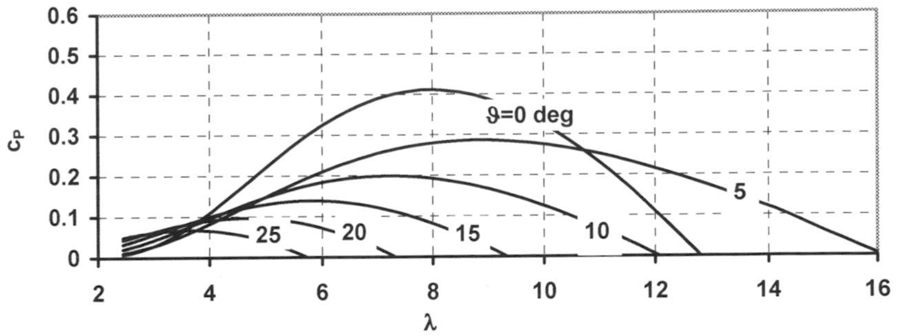


**Fig. 5.9.** Analytical approximation of $c_P(\lambda, \vartheta)$ characteristics ($\omega = 2.09$ rad/s, $R = 35$ m); blade pitch angle in degrees

Often, for a given blade, the characteristics of the drag $c_d$ and the lift $c_l$ ratios as functions of the blade pitch angle are available. In such a case, the rotor power coefficient $c_P$ can be defined as a function of these quantities. Reference [68] gives an example of such a function, which is defined as

$$c_P(\lambda, \vartheta) = \frac{16}{27}\left(1 - \frac{c_1}{N\lambda}\right)^2 \left(e^{\frac{-c_2}{\lambda^{1.29}}} - \frac{c_d}{c_l}\lambda\right) \tag{5.18}$$

where $N$ – number of blades,
$c_d/c_l$ – average drag-to-lift ratio of blade airfoil,
$c_1, c_2$ – coefficients.

### 5.2.3 Blade Pitching System

Blade pitching is a natural way of controlling the mechanical (and further electrical) power extracted from the flowing air stream. The concept of blade pitching is presented in Fig. 5.10.

The mechanical power limitation (reduction) can be achieved here by rotating the blades into the feathered position (minimizing the attack angle $\alpha$) or into the stall position (minimizing the attack angle $\alpha$ above its critical value). In practice, the first method is used.

This type of power control – called pitch control – can be effectively utilized for both constant and variable speed rotors. An example of the influence of the blade pitching on the rotor power is presented in Fig. 5.11. The figure shows that below the rated wind speed the blade pitch angle remains relatively small ($\vartheta < 5°$). And the control system, maximizing the rotor power coefficient $c_P = c_P(\lambda, \vartheta)$, makes only small changes in the pitch angle (in some WTGSs, the pitch angle remains constant here).

For higher wind speed, the blade pitching becomes more active (high changes of pitch angle) because of the necessity for extreme decrease of the power extracted from air.

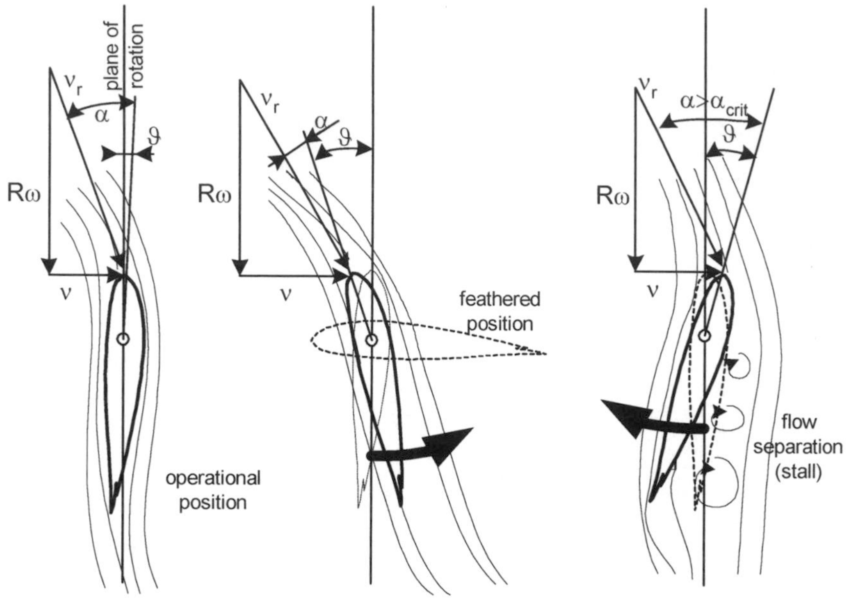


**Fig. 5.10.** Control of power by varying blade pitch angle $\vartheta$ [30] ($v$ – wind velocity, $\omega$ – rotor angular speed, $R$ – wind wheel radius, $v_r$ – relative wind velocity, $\alpha$ – angle of attack)

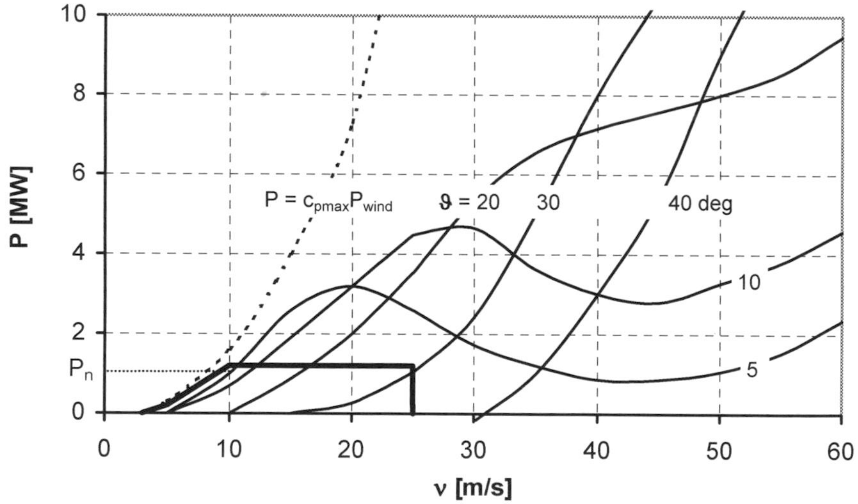


**Fig. 5.11.** Rotor power at a fixed rotor speed for various blade pitch angles (experimental WKA-60 rotor) [30]

### 5.2.4 Fixed Rotor Blades System

As shown in Fig. 5.10, for some values of the blade pitch angle and the attack angle $\alpha > \alpha_{crit}$, the effect of flow separation (stall) at a rotor blade can appear (Fig. 5.12). The effect causes, for a constant speed rotor, a decrease of the rotor power coefficient and simultaneously a decrease of the mechanical power for higher values of wind velocity. Carefully designed rotor blades and properly selected rotor speed allow an appropriate power versus wind speed characteristic to be achieved. The distinctive element of the characteristic of wind turbines with such a type of power "control" is a hummock near the rated wind speed (Fig. 5.13). For higher wind velocity, the power generated by the wind turbine decreases slightly.

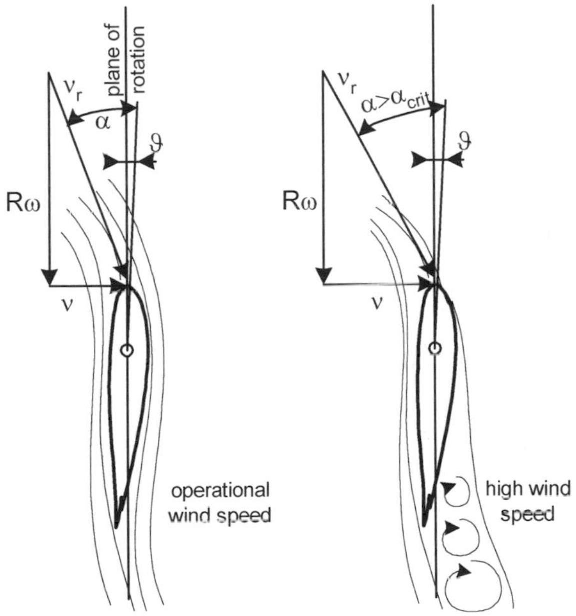


**Fig. 5.12.** Flow separation (stall) at a rotor blade with fixed pitch angle [30]

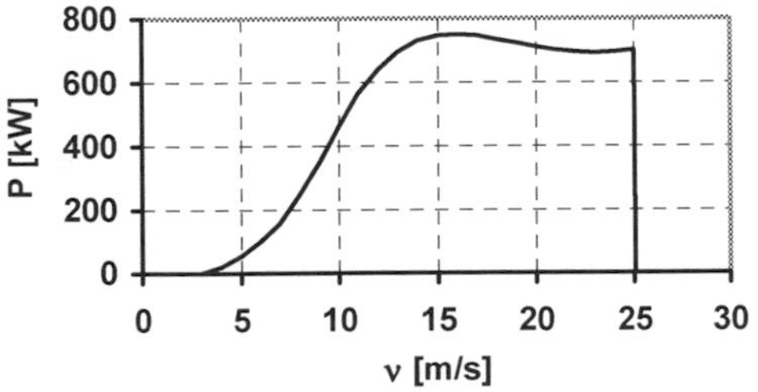


**Fig. 5.13.** Power versus wind speed characteristic for a WTGS with fixed pitch angle – stall control system (NM 750 wind turbine)

For this type of rotor, the stall effect allows the power extracted from air to be limited for high wind velocity. In fact, some type of power control (or power adjustment) is achieved here and therefore this system is usually called passive stall control. The control effect, while the blade pitch angle is constant, is achieved because of the tip-speed ratio change. The rotor power coefficient here depends on the tip-speed ratio $c_P = c_P(\lambda)$ only.

### 5.2.5 Wind Wheel Dynamics

The process of energy conversion from wind to mechanical energy because of the air features and the wind wheel mechanical characteristics has a dynamic nature. This nature is related to the occurrence of a number of phenomena resulting from the aerodynamic behavior of the air at the surface of the blades. The following are the factors that influence the conversion process most:

- building-up of the boundary layer around the blades as the wind speed changes,
- rotational wind wheel speed delaying with respect to the changing wind speed,
- reaching the flapping moment, when the blades have bent to the correct position in accordance with the changing rotational wheel speed.

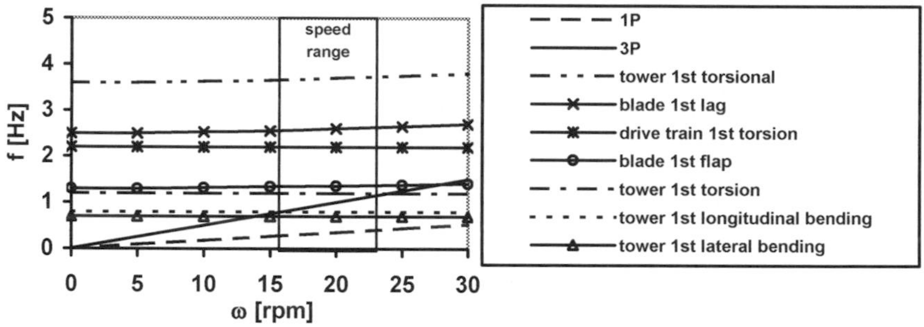


**Fig. 5.14.** Typical resonance diagram of a variable speed wind turbine with 3 blades ($f$ – eigenfrequency, $\omega$ – rotor speed) [28]

The wind turbine construction causes, during its operation, excitation of various forces and torques acting on the blades and the tower, e.g. propeller torque, torque due to lift forces on the blade, torsional resetting torques, torques due to blade deflection, torques due to teetering, frictional torques in the rotor bearing, torques due to air damping and acceleration of the air flow mass, etc. These phenomena cause various types of wind turbine oscillations[1], which for a typical turbine is presented in Fig. 5.14. Some of the oscillations appear in the drive train and further in the power transferred to the grid. These torsional oscillations – phenomena causing the oscillations – should be mainly modeled while considering the WTGS

[1] For a properly designed WTGS, within the wind turbine operating speed range, the curves showing eigenfrequencies as a function of the rotor speed should not cross. This prevents resonance self-excitation.

operation in the electric power system.

In general, modeling of the above processes leads to a set of relatively complex and nonlinear equations [3, 30, 31, 52, 67]. Fortunately, for the purpose of this monograph, the wind-wheel dynamics can be significantly simplified in the way similar to the one applied to steam and water turbine modeling. It is worth noting that in most of today considerations related to analysis of WTGS operation in the electric power system the wind-wheel dynamics is neglected.

The authors of reference [6] propose modeling the process of the energy conversion by the wind wheel as a first-order differential equation, which in the Laplace transfer function is described by the lead–lag function. The function parameters $K_W$, $T_{W1} > T_{W2}$ vary and depend on the wind rotor speed and the blade configuration and type (size, parameters, etc.):

$$\frac{P_R}{P_{WW}} = K_W \frac{1 + sT_{W1}}{1 + sT_{W2}}. \tag{5.19}$$

For medium-size WTGS with asynchronous generator and a stall-type power control system, the parameters can be estimated as $K_W = 1$, $T_{W1} = 3.3$ s, $T_{W2} = 0.9$ s [6].

### 5.2.6 Mechanical Eigenswings

As stated above, during the conversion of wind energy into mechanical energy various forces, e.g. centrifugal, gravity and varying aerodynamic forces acting on blades, gyroscopic forces acting on the tower, etc., operate in the real energy converter (WTGS). These forces produce various mechanical effects, such as tower lateral and longitudinal bending, tower torsion (yawing), rolling, pitching, blade torsion, deflection, etc. All these effects should be considered during the WTGS project design stage.

While analyzing the WTGS operation in the power system and its influence on the electric power system (and power quality), it is necessary to take into consideration the effects that manifest themselves in the rotor as torsional oscillations. The oscillations (mechanical eigenswings) go through the shaft and generator onto the electric part of the WTGS and appear as harmonics in the electric power supplied to the grid.

The type, that is the frequency and magnitude, of the excited mechanical eigenswings depend on the WTGS construction (e.g. number of blades). The main mechanical eigenswings that appear in the electric power are related to the following effects:

- asymmetry in the wind turbine – the 1P frequency,
- vortex tower interaction – the 3P frequency (for wind wheels with 3 blades),
- blade eigenswings – depend on the construction.

The research and experiments that have been carried out show [6, 57] that the above-mentioned eigenswings are the only possible eigenswings to excite themselves in 3-blade wind wheels. Other mechanical swings, which are excited in the WTGS and can be measured in the spectra of electric power (voltage and current) supplied by the WTGS to the grid, are excited in the shaft and gearbox. Therefore the shaft (the drive train in general) should be carefully modeled (see the next section).

The mechanical eigenswings can be considered as systematic. In that case, the mechanical part of the WTGS model can be simplified by modeling the mechanical eigenswings as a set of harmonic signals added to the power $P_R$ extracted from the wind. Therefore the power passed through the drive train can be defined by the following equation:

$$P_W = P_R\left(1+\sum_{k=1}^{3} A_k\left(\sum_{m=1}^{2} a_{km} g_{km}(t)\right)h_k(t)\right)$$
$$g_{km}(t) = \sin\left(\int_0^t m\omega_k(\zeta)\mathrm{d}\zeta + \varphi_{km}\right), \tag{5.20}$$

where $A_k$ – magnitude of $k$th kind of eigenswing,
$\omega_k$ – eigenfrequency of $k$th kind of eigenswing,
$h_k(t)$ – modulation of $k$th kind of eigenswing,
$m$ – harmonics,
$g_{km}$ – distribution of $k$th kind of eigenswing for the $m$th harmonic,
$a_{km}$ – normalised magnitude of $g_{km}$,
$\varphi_{km}$ – phase of $k$th kind of eigenswing for the $m$th harmonic,
$t$, $\zeta$ – time.

Such an approach to mechanical eigenswing modeling is simple but, unfortunately, needs data that can be extracted from experiments or from detailed models of the mechanical part of the WTGS (not considered here).

Typical values of the parameters describing the mechanical eigenswings achieved from the measurements on a real object (medium-size WTGS with asynchronous generator with squirrel-cage rotor) are presented in Table 5.2. The data show relatively high amplitudes of the eigenswings, which is the result of the vortex tower interaction (3P type for a rotor with three blades), and the blade interactions (with frequency depending on the blade, hub and damper construction). It is worth emphasizing that the parameter values depend highly on the WTGS operating point.

**Table 5.2.** Typical parameters of mechanical eigenswings extracted from measurement of a medium-size WTGS [6]

| $k$ | Source of eigenswings | $A_k$ | $\omega_k$ [rad/s] | $h_k(t)$ | $m$ | $a_{km}$ | $\varphi_{km}$ [rad] |
|---|---|---|---|---|---|---|---|
| 1. | Asymmetry | 0.01 | $\omega_W$ (1P) | 1 | 1 | 0.8 | 0 |
| | | | | | 2 | 0.2 | $\pi/2$ |
| 2. | Vortex tower Interaction | 0.08 | $3\omega_W$ (3P) | 1 | 1 | 0.5 | 0 |
| | | | | | 2 | 0.5 | $\pi/2$ |
| 3. | Blades | 0.15 | $2\pi 4.5$ | $0.5(g_{11}(t)+g_{21}(t))$ | 1 | 1.0 | 0 |

### 5.2.7 Drive Train (Shaft) Model

The drive train of a wind turbine generator system in general consists of a blade-pitching mechanism with a spinner, a hub with blades, a rotor shaft (relatively long in WTGS with asynchronous generators), and a gearbox (does not exist in some systems with synchronous generators) with breaker and generator. The moment of inertia of the wind wheel (hub with blades) is about 90% of the drive train total moment, while the generator rotor moment of inertia is equal to about 6–8%. The remaining parts of the drive train comprise the rest (2–4%) of the total moment of inertia.

At the same time, the generator represents the biggest torsional stiffness. The stiffness of the rotor shaft is about 100 times less and the stiffness of the hub with blades is about 50 times less than the generator stiffness. That is why the torsional vibration of the drive train elements are on the one hand inevitable, and on the other hand their character (e.g. frequency, amplitude) can highly influence the WTGS performance. Both factors were proved by tests on models and on real objects. This means that the WTGS drive train cannot be modeled as a single lumped mass.

The acceptable (and common) way to model the WTGS rotor is to treat the rotor as a number of discrete masses connected together by springs defined by damping and stiffness coefficients (Fig. 5.15).

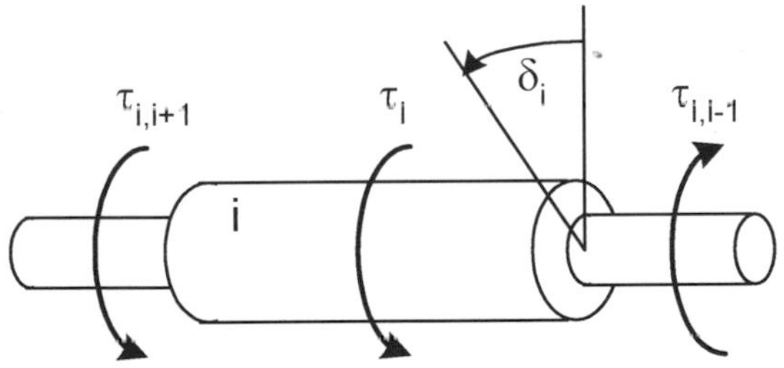


**Fig. 5.15.** Torques acting on the *i*th mass

Therefore the equation of *i*th mass motion – assuming that the torques, when a gear exists, are recalculated into one side – can be described as follows [61]:

$$J_i \frac{\mathrm{d}^2\delta_i}{\mathrm{d}t^2} = \tau_i + \tau_{i,i+1} - \tau_{i,i-1} - D_i \frac{\mathrm{d}\delta_i}{\mathrm{d}t}, \tag{5.21}$$

where $J_i$ – moment of inertia of *i*th mass,
$\delta_i$ – torsion angle of *i*th mass (displacement of the mass),
$t$ – time,
$\tau_i$ – external torque applied to *i*th mass,
$\tau_{i+1}$, $\tau_{i-1}$ – torques applied to *i*th mass (in *i*,(*i*–1)th and *i*,(*i*+1)th shafts),
$D_i$ – damping coefficient representing various damping effects.

The torques $\tau_{i+1}$, $\tau_{i-1}$ of the *i*,(*i*+1)th and *i*,(*i*–1)th shafts are equal to:

$$\tau_{i,i+1} = K_{i,i+1}(\delta_{i+1} - \delta_i) + D_{i,i+1}(\frac{\mathrm{d}\delta_{i+1}}{\mathrm{d}t} - \frac{\mathrm{d}\delta_i}{\mathrm{d}t}) \tag{5.22}$$

$$\tau_{i,i-1} = K_{i,i-1}(\delta_i - \delta_{i-1}) + D_{i,i-1}(\frac{\mathrm{d}\delta_i}{\mathrm{d}t} - \frac{\mathrm{d}\delta_{i-1}}{\mathrm{d}t}) \tag{5.23}$$

where $K_{i,i+1}$, $K_{i,i-1}$ – stiffness coefficients of the shaft sections between mass $i$,($i$+1)th and ($i$–1),$i$th, respectively
$D_{i,i+1}$, $D_{i,i-1}$ – damping coefficients of the shaft sections.

Substituting (5.22) and (5.23) into (5.21), we obtain the equation of motion of the $i$th mass

$$\begin{aligned} J_i \frac{\mathrm{d}^2\delta_i}{\mathrm{d}t^2} &= \tau_i + K_{i,i+1}(\delta_{i+1} - \delta_i) - K_{i,i-1}(\delta_i - \delta_{i-1}) + \\ &D_{i,i+1}(\frac{\mathrm{d}\delta_{i+1}}{\mathrm{d}t} - \frac{\mathrm{d}\delta_i}{\mathrm{d}t}) - D_{i,i-1}(\frac{\mathrm{d}\delta_i}{\mathrm{d}t} - \frac{\mathrm{d}\delta_{i-1}}{\mathrm{d}t}) - D_i \frac{\mathrm{d}\delta_i}{\mathrm{d}t} \end{aligned}. \tag{5.24}$$

For many types of analysis, a set of first-order differential equations is a useful form of equation. Thus, (5.22) takes the form of two equations:

$$\frac{\mathrm{d}\delta_i}{\mathrm{d}t} = \omega_i - \omega_{si} = \Delta\omega_i \tag{5.25}$$

$$\begin{aligned} J_i \frac{\mathrm{d}\Delta\omega_i}{\mathrm{d}t} &= \tau_i + K_{i,i+1}(\delta_{i+1} - \delta_i) - K_{i,i-1}(\delta_i - \delta_{i-1}) + \\ &D_{i,i+1}(\Delta\omega_{i+1} - \Delta\omega_i) - D_{i,i-1}(\Delta\omega_i - \Delta\omega_{i-1}) - D_i\Delta\omega_i \end{aligned}, \tag{5.26}$$

where $\omega_{si}$ is the synchronous speed of the $i$th mass.

Considering the drive train as consisting of $N$ discrete masses, we obtain a set of $2N$ differential equations, which as a state-space matrix system takes the following form:

$$\dot{\boldsymbol{x}} = \boldsymbol{A}\boldsymbol{x} + \boldsymbol{B}\boldsymbol{u}, \tag{5.27}$$

where the state variables vector $\boldsymbol{x}$ and input vector $\boldsymbol{u}$ are equal to, respectively,

$$\boldsymbol{x}^T = [\delta_1 \ \dots \ \delta_i \ \dots \ \delta_N \mid \Delta\omega_1 \ \dots \ \Delta\omega_i \ \dots \ \Delta\omega_N] \tag{5.28}$$

$$\boldsymbol{u}^T = [\tau_1 \ \dots \ \tau_i \ \dots \ \tau_N]. \tag{5.29}$$

The state matrix $\boldsymbol{A}$ and the input matrix $\boldsymbol{B}$ are equal to, respectively,

$$\boldsymbol{A} = \begin{bmatrix} \boldsymbol{0} & \boldsymbol{I} \\ \boldsymbol{K} & \boldsymbol{D} \end{bmatrix} \qquad \boldsymbol{B} = \begin{bmatrix} \boldsymbol{0} \\ \boldsymbol{I} \end{bmatrix}, \tag{5.30}$$

where the non-zero coefficients of stiffness $\boldsymbol{K}$ and damping $\boldsymbol{D}$ matrices are equal to

$$[\boldsymbol{K}]_{i,i-1} = \frac{K_{i-1,i}}{J_i}, \ [\boldsymbol{K}]_{i,i} = \frac{-K_{i-1,i} - K_{i,i+1}}{J_i}, \ [\boldsymbol{K}]_{i,i+1} = \frac{K_{i,i+1}}{J_i} \tag{5.31}$$

$$[\boldsymbol{D}]_{i,i-1} = \frac{D_{i-1,i}}{J_i},\ [\boldsymbol{D}]_{i,i} = \frac{-D_{i-1,i} - D_{i,i} - D_{i,i+1}}{J_i},\ [\boldsymbol{D}]_{i,i+1} = \frac{D_{i,i+1}}{J_i} \tag{5.32}$$

and **0** is the null matrix and ***I*** is the identity matrix.

The state matrix ***A*** can be used to compute the drive-train torsional natural frequencies and associated mode shapes. The natural frequencies of the dynamic system are defined by pairs of complex-conjugate eigenvalues of the form:

$$\lambda_{i,i+1} = -\zeta_i \omega_{ni} \pm j\omega_{di} \tag{5.33}$$

where $\lambda_{i,i+1}$ – eigenvalue of the *i*th mode,
$\zeta_i$ – damping ratio associated with the *i*th mode,
$\omega_{ni}$, $\omega_{di}$ – undamped and damped natural frequency of the *i*th mode, correlated by the dependency $\omega_{di} = \omega_{ni}\sqrt{1-\zeta_i^2}$ .

The eigenvalues can be found as a solution of the equation:

$$\det(\boldsymbol{A} - \lambda \boldsymbol{I}) = 0\ . \tag{5.34}$$

The minimal realization of the drive-train model utilized (as an element of the WTGS model) in the power system operation analysis is based on the assumption of two lumped– masses only: the generator (with gearbox) mass and the hub with blades (wind wheel) mass. The structure of the model is presented in Fig. 5.16.

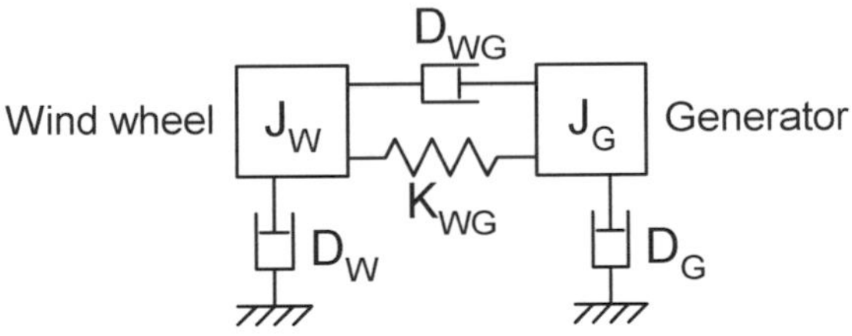


**Fig. 5.16.** WTGS drive train model with two masses

The model is described by the following set of equations extracted from (5.25) and (5.26):

$$\frac{d\delta_W}{dt} = \omega_W - \omega_s = \Delta\omega_W \tag{5.35}$$

$$\frac{d\delta_G}{dt} = \omega_G - \omega_s = \Delta\omega_G \tag{5.36}$$

$$J_W \frac{d\Delta\omega_W}{dt} = \tau_W - K_{WG}(\delta_W - \delta_G) - D_{WG}(\Delta\omega_W - \Delta\omega_G) - D_W\Delta\omega_W \tag{5.37}$$

$$J_G \frac{d\Delta\omega_G}{dt} = \tau_G + K_{WG}(\delta_W - \delta_G) + D_{WG}(\Delta\omega_W - \Delta\omega_G) - D_G\Delta\omega_G\ , \tag{5.38}$$

where all the variables are expressed with respect to the generator or the wind-wheel side. Therefore, set of equations can be applied directly to the WTGS without gearbox modeling (typically with synchronous generators connected to the grid by a power electronic converter). When applying the equations to the WTGS equipped with gearbox modeling, it is necessary to recalculate the relevant data: torque, inertia constant, torsion angle, and/or damping and stiffness coefficients into the generator or the wind wheel side.

The variables can be recalculated from the generator side of the gearbox (high-speed side) into the wind-wheel side (low-speed side), when the transformation ratio of the gearbox is equal to

$$\upsilon = \frac{\omega_{\mathrm{Gn}}}{\omega_{\mathrm{Wn}}} \tag{5.39}$$

where $\omega_{\mathrm{Gn}}$ – rated speed of generator,
$\omega_{\mathrm{Wn}}$ – rated speed of wind wheel,
by using the following equations:

$$\tau_{\mathrm{W}}^{(\mathrm{W})} = \upsilon \tau_{\mathrm{W}}^{(\mathrm{G})} \tag{5.40}$$

$$\omega_{\mathrm{W}}^{(\mathrm{W})} = \frac{\omega_{\mathrm{W}}^{(\mathrm{G})}}{\upsilon} \tag{5.41}$$

$$\delta_{\mathrm{W}}^{(\mathrm{W})} = \frac{\delta_{\mathrm{W}}^{(\mathrm{G})}}{\upsilon} \tag{5.42}$$

$$J_{\mathrm{W}}^{(\mathrm{W})} = \upsilon^2 J_{\mathrm{W}}^{(\mathrm{G})}, \tag{5.43}$$

where the superscript (W) denotes the wind wheel, and the superscript (G) denotes the generator side of the gearbox.

For the sake of completeness, let us consider a drive-train model like the one above (two-mass model) but with an explicitly defined transmission ratio $\upsilon$ of the gearbox. The model, in graphical form, is presented in Fig. 5.17. The rotor-shaft model is defined here by the damping and the stiffness coefficients $D = D_{\mathrm{WG}}$, $K = K_{\mathrm{WG}}$. The rest of the damping coefficients ($D_{\mathrm{W}}$, $D_{\mathrm{G}}$) are neglected because of their relatively small influence on the shaft torsion.

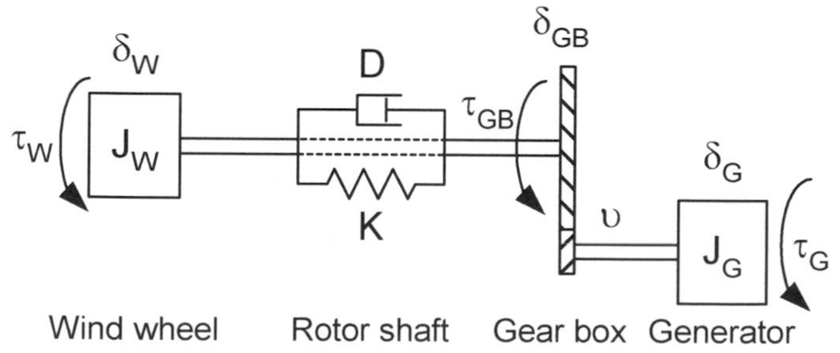


**Fig. 5.17.** Two-masse drive-train model with gearbox

The influence of the generator shaft is neglected here because the generator shaft is usually shorter than the rotor shaft and at the same time the torque acting on the generator shaft is $\upsilon$ times lower than the torque acting on the rotor shaft (gearbox result).

The drive train (Fig. 5.16) can be described by the following set of equations:

$$J_{\mathrm{W}} \frac{\mathrm{d}^2 \delta_{\mathrm{W}}}{\mathrm{d}t^2} = \tau_{\mathrm{W}} - \tau_{\mathrm{GB}} \tag{5.44}$$

$$J_{\mathrm{G}} \frac{\mathrm{d}^2 \delta_{\mathrm{G}}}{\mathrm{d}t^2} = \tau_{\mathrm{G}} + \frac{\tau_{\mathrm{GB}}}{\upsilon} \tag{5.45}$$

$$\tau_{\mathrm{GB}} = K(\delta_{\mathrm{W}} - \delta_{\mathrm{GB}}) + D(\frac{\mathrm{d}\delta_{\mathrm{W}}}{\mathrm{d}t} - \frac{\mathrm{d}\delta_{\mathrm{GB}}}{\mathrm{d}t}) \tag{5.46}$$

$$\begin{aligned} \omega_{\mathrm{G}} &= \upsilon \omega_{\mathrm{GB}} \\ \delta_{\mathrm{G}} &= \upsilon \delta_{\mathrm{GB}} \,. \end{aligned} \tag{5.47}$$

Equations (5.44)–(5.47) can be converted into the following form:

$$\frac{\mathrm{d}\delta_{\mathrm{W}}}{\mathrm{d}t} = \omega_{\mathrm{W}} - \omega_{\mathrm{W0}} = \Delta\omega_{\mathrm{W}} \tag{5.48}$$

$$\frac{\mathrm{d}\delta_{\mathrm{G}}}{\mathrm{d}t} = \omega_{\mathrm{G}} - \omega_{\mathrm{G0}} = \Delta\omega_{\mathrm{G}} \tag{5.49}$$

$$J_{\mathrm{W}} \frac{\mathrm{d}\Delta\omega_{\mathrm{W}}}{\mathrm{d}t} = \tau_{\mathrm{W}} - K(\delta_{\mathrm{W}} - \frac{\delta_{\mathrm{G}}}{\upsilon}) - D(\Delta\omega_{\mathrm{W}} - \frac{\Delta\omega_{\mathrm{G}}}{\upsilon}) \tag{5.50}$$

$$J_{\mathrm{G}} \frac{\mathrm{d}\Delta\omega_{\mathrm{G}}}{\mathrm{d}t} = \tau_{\mathrm{G}} + \frac{K(\delta_{\mathrm{W}} - \frac{\delta_{\mathrm{G}}}{\upsilon}) + D(\Delta\omega_{\mathrm{W}} - \frac{\Delta\omega_{\mathrm{G}}}{\upsilon})}{\upsilon}\,, \tag{5.51}$$

where $\omega_{\mathrm{W0}}$ and $\omega_{\mathrm{G0}}$ are the wind wheel and the generator rotor speed in the steady state, respectively.

It is worth emphasizing that all the quantities here are expressed in SI units, i.e. angular velocity $\omega$ [rad/s], torque $\tau$ [Nm], moment of inertia $J$ [kgm$^2$], torsion angle $\delta$ [rad], stiffness coefficient $K$ [Nm/rad], damping coefficient $D$ [Nms/rad] and time $t$ [s], and that they define the real values of speed, torque, etc. of the given WTGS element. It means, for example, that $\delta_{\mathrm{W}}$ represents the torsion angle of the wind wheel in mechanical radians, while $\delta_{\mathrm{G}}$ represents the generator rotor torsion angle in electrical radians.[2]

The use of models based on per unit quantities is common practice in power system modeling because it has many advantages. Here, the generator synchronous velocity $\omega_{\mathrm{Gs}}$ (in fact the synchronous frequency $\omega_{\mathrm{s}}$) and the rated apparent power of the generator $S_{\mathrm{n}}$ can be used as the basic values. Then, the set of equations (5.49)–(5.52) converted into equations in per unit quantities (with the exception of time $t$ expressed in seconds and torsion angle $\delta$ expressed in radians) takes the following form:

[2] Therefore, these two values cannot be subtracted directly. When computing rotor-shaft twist, it is necessary to divide or multiply the relevant torsion angle.

$$\frac{d\delta_W^*}{dt} = \Delta\overline{\omega}_W \frac{\omega_s}{(\frac{p}{2})} \tag{5.52}$$

$$\frac{d\delta_G}{dt} = \Delta\overline{\omega}_G \frac{\omega_s}{(\frac{p}{2})} \tag{5.53}$$

$$H_W \frac{d\Delta\overline{\omega}_W}{dt} = \overline{\tau}_W - \overline{K}(\delta_W^* - \delta_G) - \overline{D}(\Delta\overline{\omega}_W - \Delta\overline{\omega}_G) \tag{5.54}$$

$$H_G \frac{d\Delta\overline{\omega}_G}{dt} = \overline{\tau}_G + \overline{K}(\delta_W^* - \delta_G) + \overline{D}(\Delta\overline{\omega}_W - \Delta\overline{\omega}_G), \tag{5.55}$$

where $\omega_{Gs} = \frac{\omega_s}{(\frac{p}{2})}$ – synchronous velocity of generator ($\omega_s = 314.15$ rad/s),

$\omega_{Ws} = \frac{\omega_s}{(\frac{p}{2})\upsilon}$ – synchronous velocity of wind wheel,

$p$ – number of generator poles,

$\delta_W^*$ – wind wheel torsion angle expressed in electrical radians (generator side).

The wind wheel and the generator inertia constants $H_W$, $H_G$ and the stiffness and the damping coefficients $\overline{K}$, $\overline{D}$ in (5.52)–(5.55) are given by:

$$H_W = \frac{J_W\omega_s^2}{S_n(\frac{p}{2})^2\upsilon^2} \tag{5.56}$$

$$H_G = \frac{J_G\omega_s^2}{S_n(\frac{p}{2})^2} \tag{5.57}$$

$$\overline{K} = \frac{K\omega_s}{S_n(\frac{p}{2})^2\upsilon^2} \tag{5.58}$$

$$\overline{D} = \frac{D\omega_s^2}{S_n(\frac{p}{2})^2\upsilon^2}. \tag{5.59}$$

### 5.2.8 Wind Turbine Model

The above can be summed up by saying that the mathematical model of the wind turbine for WTGS operation in electric power system analysis can take the form presented in Fig. 5.18. The model consists of the following components, described in previous sections:

- wind-power block – computing the power in the air stream,

- wind-wheel power block – computing the power extracted by the wind wheel from the air on the basis of the rotor power coefficient, the rotor shaft speed and the blade pitch angle (for a WTGS with pitch control system),
- blades-transient block – wind-wheel dynamic model,
- mechanical-eigenswings block – mechanical-eigenswings model,
- drive-train block – drive-train model.

The input for the wind turbine model is the wind velocity $v$, the blade pitch angle $\vartheta$, and the generator torque $\tau_G$, while the model output is the generator shaft speed $\omega_G$. The generator torque is achieved from the generator model, the blade pitch angle is achieved from the turbine control system model, while the wind velocity is the only independent input quantity.

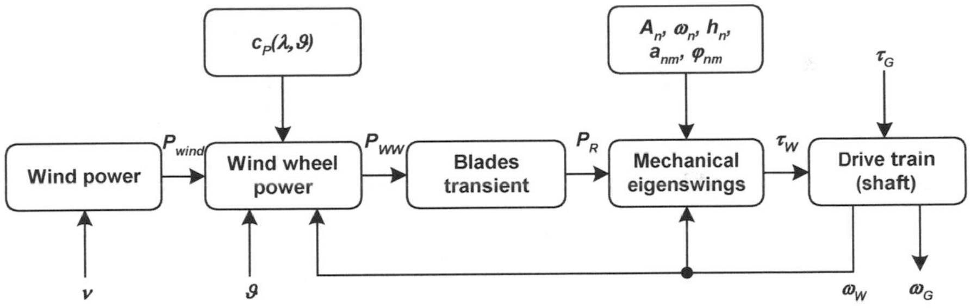


**Fig. 5.18.** Structure of wind turbine model

The wind velocity usually varies considerably, as shown in Fig. 5.19, and has a stochastic character. Therefore, in general, the wind should be modeled as a stochastic process. But for the analysis of WTGS operation in an electric power system, the wind variation can be modeled as a sum of harmonics (see (5.60)) with frequencies in the range 0.1–10 Hz. Wind gusts are usually also included in the wind model.

$$v(t) = v_0(1 + \sum_k A_k \sin(\omega_k t)) + v_g(t) \tag{5.60}$$

where $v_0$ – mean value of wind velocity,[3]
$A_k$ – amplitude of $k$th harmonic,
$\omega_k$ – frequency (pulsation) of $k$th harmonic,
$v_g(t)$ – wind gust.

Wind gusts can be modeled by the following function:

$$v_g(t) = \frac{2v_{g\max}}{1 + e^{-4(\sin(\omega_g t) - 1)}}, \tag{5.61}$$

[3] The mean value of wind velocity at partial load is usually assumed as equal to the value which on the WTGS power characteristic $P=f(v)$ gives the highest slope, e.g. $v_0 = 9$ m/s.

where $v_{gmax}$ – gust amplitude,
$\omega_g$ – gust frequency ($\omega_g = 2\pi/T_g$).

The gust amplitude varies up to 10 m/s and the gust period can be in the range $T_g = 10–50$ s.

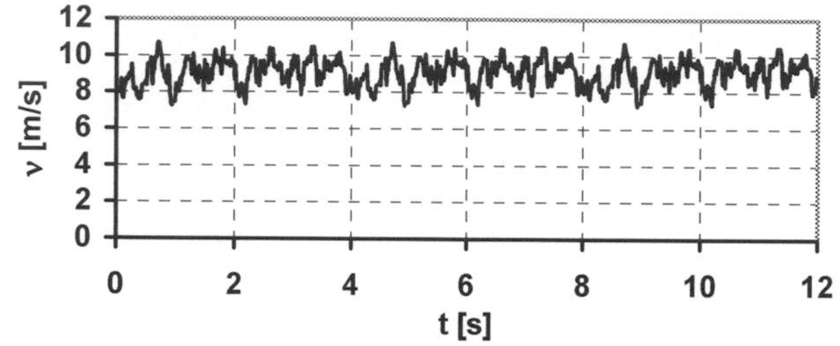


**Fig. 5.19.** Example of wind velocity variation

## 5.3 Asynchronous Generator Modeling

### 5.3.1 Asynchronous Generator Model for Natural Axes

It is assumed that an asynchronous generator (Fig. 5.20) has a three-phase stator armature winding ($A_S$, $B_S$, $C_S$), and a three-phase rotor winding ($A_R$, $B_R$, $C_R$). The mathematical model of an asynchronous generator for power system and converter system analysis is usually based on the following assumptions [77, 85]:

- the stator and rotor windings are placed sinusoidally along the air-gap as far as the mutual effect with the rotor is concerned,
- the stator slots cause no appreciable variations of the rotor inductances with rotor position,
- the rotor slots cause no appreciable variations of the stator inductances with rotor position,
- magnetic hysteresis and saturation effects are negligible,
- the stator and rotor windings are symmetrical,
- the capacitance of all the windings can be neglected.

More detailed modeling usually encounters difficulties in getting appropriate data. Additionally, for machine modeling, such a type of model is adequately precise.

For natural axes, the model can be formulated by the so-called flux equation, the voltage equation and the mechanical system equation.

The flux equation describing the relationship between the flux and current in each winding has the following form:

$$\boldsymbol{\Psi}_N = \boldsymbol{L}_N \boldsymbol{i}_N , \tag{5.62}$$

where the flux vector is equal to

$$\boldsymbol{\Psi}_N = \begin{bmatrix} \boldsymbol{\Psi}_{ABCS} \\ \boldsymbol{\Psi}_{ABCR} \end{bmatrix} = \begin{bmatrix} \Psi_{AS} & \Psi_{BS} & \Psi_{CS} \mid \Psi_{AR} & \Psi_{BR} & \Psi_{CR} \end{bmatrix}^T \tag{5.63}$$

and where $\boldsymbol{\Psi}_{ABCS}$ is the vector of stator fluxes and $\boldsymbol{\Psi}_{ABCR}$ is the vector of rotor fluxes.

The current vector is equal to:

$$\boldsymbol{i}_N = \begin{bmatrix} \boldsymbol{i}_{ABCS} \\ \boldsymbol{i}_{ABCR} \end{bmatrix} = \begin{bmatrix} i_{AS} & i_{BS} & i_{CS} \mid i_{AR} & i_{BR} & i_{CR} \end{bmatrix}^T , \tag{5.64}$$

where $\boldsymbol{i}_{ABCS}$ is the vector of stator currents and $\boldsymbol{i}_{ABCR}$ is the vector of rotor currents, where particular currents are equal to

$$\begin{aligned} i_{AS} &= I_S \cos(\Theta + \varphi_{IS}) \\ i_{BS} &= I_S \cos(\Theta - \tfrac{2}{3}\pi + \varphi_{IS}) \\ i_{CS} &= I_S \cos(\Theta + \tfrac{2}{3}\pi + \varphi_{IS}) \\ i_{AR} &= I_R \cos(\Theta - \Theta_R + \varphi_{IR}) \\ i_{BR} &= I_R \cos(\Theta - \Theta_R - \tfrac{2}{3}\pi + \varphi_{IR}) \\ i_{CR} &= I_R \cos(\Theta - \Theta_R + \tfrac{2}{3}\pi + \varphi_{IR}) . \end{aligned} \tag{5.65}$$

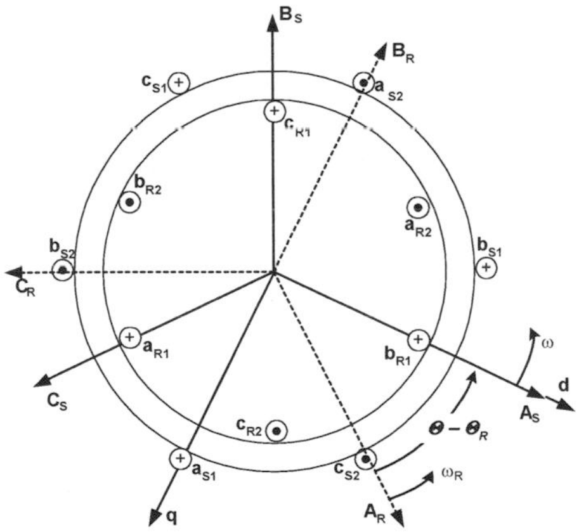


**Fig. 5.20.** The windings in the asynchronous generator

The inductance matrix is equal to:

$$\boldsymbol{L}_N = \begin{bmatrix} \boldsymbol{L}_S & \boldsymbol{L}_{SR} \\ \boldsymbol{L}_{SR}^T & \boldsymbol{L}_R \end{bmatrix} , \tag{5.66}$$

where $\boldsymbol{L}_{\mathrm{S}}$ is a submatrix of the stator's self and mutual inductances, $\boldsymbol{L}_{\mathrm{R}}$ is a submatrix of the rotor's self and mutual inductances and $\boldsymbol{L}_{\mathrm{SR}}$ is a submatrix of the rotor-to-stator mutual inductances. The inductances are given by

$$\boldsymbol{L}_{\mathrm{S}} = \begin{bmatrix} L_{\mathrm{lS}} + L_{\mathrm{mS}} & -\frac{1}{2}L_{\mathrm{mS}} & -\frac{1}{2}L_{\mathrm{mS}} \\ -\frac{1}{2}L_{\mathrm{mS}} & L_{\mathrm{lS}} + L_{\mathrm{mS}} & -\frac{1}{2}L_{\mathrm{mS}} \\ -\frac{1}{2}L_{\mathrm{mS}} & -\frac{1}{2}L_{\mathrm{mS}} & L_{\mathrm{lS}} + L_{\mathrm{mS}} \end{bmatrix}$$

$$\boldsymbol{L}_{\mathrm{R}} = \begin{bmatrix} L_{\mathrm{lR}} + L_{\mathrm{mR}} & -\frac{1}{2}L_{\mathrm{mR}} & -\frac{1}{2}L_{\mathrm{mR}} \\ -\frac{1}{2}L_{\mathrm{mR}} & L_{\mathrm{lR}} + L_{\mathrm{mR}} & -\frac{1}{2}L_{\mathrm{mR}} \\ -\frac{1}{2}L_{\mathrm{mR}} & -\frac{1}{2}L_{\mathrm{mR}} & L_{\mathrm{lR}} + L_{\mathrm{mR}} \end{bmatrix} \tag{5.67}$$

$$\boldsymbol{L}_{\mathrm{SR}} = \begin{bmatrix} L_{\mathrm{SR}}\cos\Theta_{\mathrm{R}} & L_{\mathrm{SR}}\cos(\Theta_{\mathrm{R}} + \frac{2}{3}\pi) & L_{\mathrm{SR}}\cos(\Theta_{\mathrm{R}} - \frac{2}{3}\pi) \\ L_{\mathrm{SR}}\cos(\Theta_{\mathrm{R}} - \frac{2}{3}\pi) & L_{\mathrm{SR}}\cos\Theta_{\mathrm{R}} & L_{\mathrm{SR}}\cos(\Theta_{\mathrm{R}} + \frac{2}{3}\pi) \\ L_{\mathrm{SR}}\cos(\Theta_{\mathrm{R}} + \frac{2}{3}\pi) & L_{\mathrm{SR}}\cos(\Theta_{\mathrm{R}} - \frac{2}{3}\pi) & L_{\mathrm{SR}}\cos\Theta_{\mathrm{R}} \end{bmatrix},$$

where $L_{\mathrm{lS}}$, $L_{\mathrm{lR}}$ are the stator and rotor leakage inductances, $L_{\mathrm{mS}}$, $L_{\mathrm{mR}}$ are the magnetizing inductances of the stator and rotor windings and $L_{\mathrm{SR}}$ is the amplitude of the mutual inductance.

The voltage equation describing the relationship between the voltage, flux and current in each winding (for the generator circuit type shown in Fig. 5.21) has the following form:

$$\boldsymbol{v}_{\mathrm{N}} = -\mathrm{p}\boldsymbol{\Psi}_{\mathrm{N}} - \boldsymbol{R}_{\mathrm{N}}\boldsymbol{i}_{\mathrm{N}}, \tag{5.68}$$

where the voltage vector is equal to

$$\boldsymbol{v}_{\mathrm{N}} = \begin{bmatrix} v_{\mathrm{ABCS}} \\ v_{\mathrm{ABCR}} \end{bmatrix} = \begin{bmatrix} v_{\mathrm{AS}} & v_{\mathrm{BS}} & v_{\mathrm{CS}} \mid v_{\mathrm{AR}} & v_{\mathrm{BR}} & v_{\mathrm{CR}} \end{bmatrix}^{T}, \tag{5.69}$$

where $\boldsymbol{v}_{\mathrm{ABCS}}$ is the vector of stator voltages and $\boldsymbol{v}_{\mathrm{ABCR}}$ is the vector of rotor voltages.

The particular voltages in these vectors are given by

$$\begin{aligned} v_{\mathrm{AS}} &= V_{\mathrm{S}}\cos(\Theta + \varphi_{\mathrm{VS}}) \\ v_{\mathrm{BS}} &= V_{\mathrm{S}}\cos(\Theta - \tfrac{2}{3}\pi + \varphi_{\mathrm{VS}}) \\ v_{\mathrm{CS}} &= V_{\mathrm{S}}\cos(\Theta + \tfrac{2}{3}\pi + \varphi_{\mathrm{VS}}) \\ v_{\mathrm{AR}} &= V_{\mathrm{R}}\cos(\Theta - \Theta_{\mathrm{R}} + \varphi_{\mathrm{VR}}) \\ v_{\mathrm{BR}} &= V_{\mathrm{R}}\cos(\Theta - \Theta_{\mathrm{R}} - \tfrac{2}{3}\pi + \varphi_{\mathrm{VR}}) \\ v_{\mathrm{CR}} &= V_{\mathrm{R}}\cos(\Theta - \Theta_{\mathrm{R}} + \tfrac{2}{3}\pi + \varphi_{\mathrm{VR}}). \end{aligned} \tag{5.70}$$

The resistance matrix is equal to

$$\boldsymbol{R}_{N} = \mathrm{diag}\,[R_{AS} \quad R_{BS} \quad R_{CS} \quad R_{AR} \quad R_{BR} \quad R_{CR}]. \tag{5.71}$$

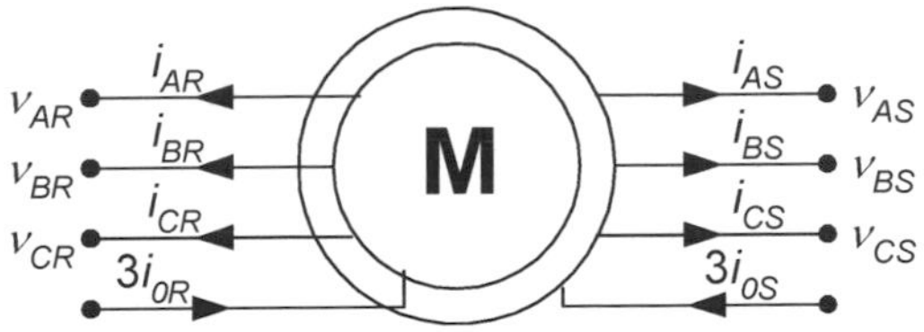


**Fig. 5.21.** Currents and voltages of the asynchronous generator

The mechanical system can be described by the following equation of motion: [4]

$$(\tfrac{2}{p})J\frac{\mathrm{d}\omega_{\mathrm{R}}}{\mathrm{d}t} = \tau_{\mathrm{m}} - \tau_{\mathrm{e}} - D(\tfrac{2}{p})\omega_{\mathrm{R}}\,, \tag{5.72}$$

where the electromagnetic torque $\tau_{\mathrm{e}}$, the angular displacements $\Theta$ and $\Theta_{\mathrm{R}}$, and the angular "mechanical" velocity of the rotor $\omega_{\mathrm{mR}}$ are given by

$$\begin{aligned}
\tau_{\mathrm{e}} &= (\tfrac{p}{2})\frac{\partial W(\boldsymbol{i}_{\mathrm{ABCS}},\boldsymbol{i}_{\mathrm{ABCR}},\Theta)}{\partial\Theta} = \frac{1}{2}(\tfrac{p}{2})\boldsymbol{i}_{\mathrm{ABCS}}^{T}\frac{\partial}{\partial\Theta}\left[\boldsymbol{L}_{\mathrm{SR}}^{T}(\Theta)\right]\boldsymbol{i}_{\mathrm{ABCR}} \\
\Theta &= \int_0^t \omega(\xi)\mathrm{d}\xi + \Theta_0 \\
\Theta_{\mathrm{R}} &= \int_0^t \omega_{\mathrm{R}}(\xi)\mathrm{d}\xi + \Theta_{R0} \\
\omega_{\mathrm{mR}} &= \omega_{\mathrm{R}}(\tfrac{2}{p})\,.
\end{aligned} \tag{5.73}$$

The stator and rotor terminal power are, respectively, given by

$$P_{\mathrm{S}} = \boldsymbol{v}_{\mathrm{ABCS}}^{T}\boldsymbol{i}_{\mathrm{ABCS}} = v_{\mathrm{AS}}i_{\mathrm{AS}} + v_{\mathrm{BS}}i_{\mathrm{BS}} + v_{\mathrm{CS}}i_{\mathrm{CS}} \tag{5.74}$$

$$P_{\mathrm{R}} = \boldsymbol{v}_{\mathrm{ABCR}}^{T}\boldsymbol{i}_{\mathrm{ABCR}} = v_{\mathrm{AR}}i_{\mathrm{AR}} + v_{\mathrm{BR}}i_{\mathrm{BR}} + v_{\mathrm{CR}}i_{\mathrm{CR}}\,. \tag{5.75}$$

### 5.3.2 Asynchronous Generator Model in the *0dq* Reference Frame

The asynchronous generator model for dynamic phenomena in the power system analysis or the converter system modeling and analysis is not usually utilized in the form presented in Sect. 5.3.1. The main disadvantage of the model is the dependence of the angular displacement $\Theta_{\mathrm{R}}$ on the inductance matrix $\boldsymbol{L}_{\mathrm{N}}$. Additionally, the number of quantities, which change with time (AC currents, voltages and fluxes), is relatively large. Therefore, the model (set of equations) is usually converted into a model related to an arbitrarily set reference frame. A special transformation matrix $\boldsymbol{B}_{\mathrm{N}}$ is then used. Here, the machine model is converted into the

---

[4] In general, in WTGS modeling, the drive-train model described in Sect. 5.2 should be used.

so-called *0dq*-reference frame model. Axes *d* and *q* are related to the rotating rotor or stator, as shown in Fig. 5.20. In this case, the transformation of the model in natural axis is realized by using the transformation matrix $\boldsymbol{B}_{\mathrm{N}}$ in the following form:

$$\boldsymbol{B}_{\mathrm{N}} = \begin{bmatrix} \boldsymbol{B}_{\mathrm{S}} & \boldsymbol{0} \\ \boldsymbol{0} & \boldsymbol{B}_{\mathrm{R}} \end{bmatrix}, \tag{5.76}$$

where the submatrices $\boldsymbol{B}_{\mathrm{S}}$ and $\boldsymbol{B}_{\mathrm{R}}$ are equal to

$$\boldsymbol{B}_{\mathrm{S}} = \frac{2}{3}\begin{bmatrix} \frac{1}{2} & \frac{1}{2} & \frac{1}{2} \\ \cos\Theta & \cos(\Theta - \frac{2}{3}\pi) & \cos(\Theta + \frac{2}{3}\pi) \\ \sin\Theta & \sin(\Theta - \frac{2}{3}\pi) & \sin(\Theta + \frac{2}{3}\pi) \end{bmatrix} \tag{5.77}$$

$$\boldsymbol{B}_{\mathrm{R}} = \frac{2}{3}\begin{bmatrix} \frac{1}{2} & \frac{1}{2} & \frac{1}{2} \\ \cos(\Theta - \Theta_{\mathrm{R}}) & \cos(\Theta - \Theta_{\mathrm{R}} - \frac{2}{3}\pi) & \cos(\Theta - \Theta_{\mathrm{R}} + \frac{2}{3}\pi) \\ \sin(\Theta - \Theta_{\mathrm{R}}) & \sin(\Theta - \Theta_{\mathrm{R}} - \frac{2}{3}\pi) & \sin(\Theta - \Theta_{\mathrm{R}} + \frac{2}{3}\pi) \end{bmatrix} \tag{5.78}$$

Multiplying (5.62) by the transformation matrix $\boldsymbol{B}_{\mathrm{N}}$, we achieve the following formula:

$$\boldsymbol{B}_{\mathrm{N}}\boldsymbol{\Psi}_{\mathrm{N}} = \boldsymbol{B}_{\mathrm{N}}\boldsymbol{L}_{\mathrm{N}}\boldsymbol{B}_{\mathrm{N}}^{-1}\boldsymbol{B}_{\mathrm{N}}\boldsymbol{i}_{\mathrm{N}}, \tag{5.79}$$

which, because the flux vector, the current vector and the inductance matrix in the *0dq*-reference frame are given by

$$\begin{aligned} \boldsymbol{B}_{\mathrm{N}}\boldsymbol{\Psi}_{\mathrm{N}} &= \boldsymbol{\Psi}_{\mathrm{M}} \\ \boldsymbol{B}_{\mathrm{N}}\boldsymbol{L}_{\mathrm{N}}\boldsymbol{B}_{\mathrm{N}}^{-1} &= \boldsymbol{L}_{\mathrm{M}} \\ \boldsymbol{B}_{\mathrm{N}}\boldsymbol{i}_{\mathrm{N}} &= \boldsymbol{i}_{\mathrm{M}}, \end{aligned} \tag{5.80}$$

finally gives the flux equation in the *0dq*-reference frame in the following form:

$$\boldsymbol{\Psi}_{\mathrm{M}} = \boldsymbol{L}_{\mathrm{M}}\boldsymbol{i}_{\mathrm{M}}, \tag{5.81}$$

where the inductance matrix, the flux vector and the current vector are, respectively, given by

$$\boldsymbol{L}_{\mathrm{M}} = \begin{bmatrix} L_{0\mathrm{S}} & 0 & 0 & 0 & 0 & 0 \\ 0 & L_{l\mathrm{S}} + \frac{3}{2}L_{\mathrm{mS}} & 0 & 0 & \frac{3}{2}L_{\mathrm{mR}} & 0 \\ 0 & 0 & L_{l\mathrm{S}} + \frac{3}{2}L_{\mathrm{mS}} & 0 & 0 & \frac{3}{2}L_{\mathrm{mR}} \\ 0 & 0 & 0 & L_{0\mathrm{R}} & 0 & 0 \\ 0 & \frac{3}{2}L_{\mathrm{mS}} & 0 & 0 & L_{l\mathrm{R}} + \frac{3}{2}L_{\mathrm{mR}} & 0 \\ 0 & 0 & \frac{3}{2}L_{\mathrm{mS}} & 0 & 0 & L_{l\mathrm{R}} + \frac{3}{2}L_{\mathrm{mR}} \end{bmatrix} \tag{5.82}$$

$$\boldsymbol{\Psi}_{\mathrm{M}} = \begin{bmatrix} \boldsymbol{\Psi}_{0\mathrm{dqS}} \\ \boldsymbol{\Psi}_{0\mathrm{dqR}} \end{bmatrix} = \begin{bmatrix} \Psi_{0\mathrm{S}} & \Psi_{\mathrm{dS}} & \Psi_{\mathrm{qS}} \mid \Psi_{0\mathrm{R}} & \Psi_{\mathrm{dR}} & \Psi_{\mathrm{qR}} \end{bmatrix}^T$$

$$\boldsymbol{i}_{\mathrm{M}} = \begin{bmatrix} \boldsymbol{i}_{0\mathrm{dqS}} \\ \boldsymbol{i}_{0\mathrm{dqR}} \end{bmatrix} = \begin{bmatrix} i_{0\mathrm{S}} & i_{\mathrm{dS}} & i_{\mathrm{qS}} \mid i_{0\mathrm{R}} & i_{\mathrm{dR}} & i_{\mathrm{qR}} \end{bmatrix}^T .$$

Multiplying (5.68) by the transformation matrix $\boldsymbol{B}_{\mathrm{N}}$ we obtain the following formula:

$$\boldsymbol{B}_{\mathrm{N}}\boldsymbol{v}_{\mathrm{N}} = -\boldsymbol{B}_{\mathrm{N}}\frac{\mathrm{d}}{\mathrm{d}t}(\boldsymbol{B}_{\mathrm{N}}^{-1}\boldsymbol{B}_{\mathrm{N}}\boldsymbol{\Psi}_{\mathrm{N}}) - \boldsymbol{B}_{\mathrm{N}}\boldsymbol{R}_{\mathrm{N}}\boldsymbol{B}_{\mathrm{N}}^{-1}\boldsymbol{B}_{\mathrm{N}}\boldsymbol{i}_{\mathrm{N}} , \tag{5.83}$$

where the respective matrices (assuming $R_{\mathrm{AS}} = R_{\mathrm{BS}} = R_{\mathrm{CS}} = R_{\mathrm{S}}$ and $R_{\mathrm{AR}} = R_{\mathrm{BR}} = R_{\mathrm{CR}} = R_{\mathrm{R}}$) and vectors in the *0dq*-reference frame are given by

$$\begin{gathered} \boldsymbol{B}_{\mathrm{N}}\, \boldsymbol{v}_{\mathrm{N}} = \boldsymbol{v}_{\mathrm{M}} \\ \boldsymbol{B}_{\mathrm{N}}\, \boldsymbol{R}_{\mathrm{N}}\, \boldsymbol{B}_{\mathrm{N}}^{-1} = \boldsymbol{R}_{\mathrm{M}} \\ -\boldsymbol{B}_{\mathrm{N}}\frac{\mathrm{d}}{\mathrm{d}t}(\boldsymbol{B}_{\mathrm{N}}^{-1}\boldsymbol{B}_{\mathrm{N}}\boldsymbol{\Psi}_{\mathrm{N}}) = -\boldsymbol{B}_{\mathrm{N}}(\frac{\mathrm{d}\boldsymbol{B}_{\mathrm{N}}^{-1}}{\mathrm{d}t}\boldsymbol{B}_{\mathrm{N}}\boldsymbol{\Psi}_{\mathrm{N}} + \boldsymbol{B}_{\mathrm{N}}^{-1}\frac{\mathrm{d}\boldsymbol{B}_{\mathrm{N}}\boldsymbol{\Psi}_{\mathrm{N}}}{\mathrm{d}t}) = -\boldsymbol{\Omega}\boldsymbol{\Psi}_{\mathrm{M}} - \frac{\mathrm{d}\boldsymbol{\Psi}_{\mathrm{M}}}{\mathrm{d}t} \end{gathered} \tag{5.84}$$

and where the rotation matrix $\boldsymbol{\Omega}$ is given by

$$\boldsymbol{\Omega} = \begin{bmatrix} 0 & 0 & 0 & 0 & 0 & 0 \\ 0 & 0 & \omega & 0 & 0 & 0 \\ 0 & -\omega & 0 & 0 & 0 & 0 \\ 0 & 0 & 0 & 0 & 0 & 0 \\ 0 & 0 & 0 & 0 & 0 & (\omega - \omega_{\mathrm{R}}) \\ 0 & 0 & 0 & 0 & -(\omega - \omega_{\mathrm{R}}) & 0 \end{bmatrix} . \tag{5.85}$$

Assuming the above, the voltage equation in the *0dq*-reference frame finally takes the following form:

$$\boldsymbol{v}_{\mathrm{M}} = -\,\mathrm{p}\,\boldsymbol{\Psi}_{\mathrm{M}} - \boldsymbol{\Omega}\,\boldsymbol{\Psi}_{\mathrm{M}} - \boldsymbol{R}_{\mathrm{M}}\,\boldsymbol{i}_{\mathrm{M}} , \tag{5.86}$$

where the voltage vector and the resistance matrix are, respectively, given by

$$\begin{gathered} \boldsymbol{v}_{\mathrm{M}} = \begin{bmatrix} \boldsymbol{v}_{0\mathrm{dqS}} \\ \boldsymbol{v}_{0\mathrm{dqR}} \end{bmatrix} = \begin{bmatrix} v_{0\mathrm{S}} & v_{\mathrm{dS}} & v_{\mathrm{qS}} \mid v_{0\mathrm{R}} & v_{\mathrm{dR}} & v_{\mathrm{qR}} \end{bmatrix}^T \\ \boldsymbol{R}_{\mathrm{M}} = \mathrm{diag}\,[R_{0\mathrm{S}} \quad R_{\mathrm{S}} \quad R_{\mathrm{S}} \quad R_{0\mathrm{R}} \quad R_{\mathrm{R}} \quad R_{\mathrm{R}}] . \end{gathered} \tag{5.87}$$

The electromagnetic torque $\tau_{\mathrm{e}}$ and the real stator and rotor power $P_{\mathrm{S}}$, $P_{\mathrm{R}}$ and the reactive stator and rotor power $Q_{\mathrm{S}}$, $Q_{\mathrm{R}}$ are, respectively, given by

$$\tau_{\mathrm{e}} = \frac{3}{2}\left(\frac{p}{2}\right)\left(\Psi_{\mathrm{dS}} i_{\mathrm{qS}} - \Psi_{\mathrm{qS}} i_{\mathrm{dS}}\right) \tag{5.88}$$

$$P_{\mathrm{S}} = \frac{3}{2}\boldsymbol{v}_{0\mathrm{dqS}}^{T}\boldsymbol{i}_{0\mathrm{dqS}} = \frac{3}{2}\left(2v_{0\mathrm{S}}i_{0\mathrm{S}} + v_{\mathrm{dS}}i_{\mathrm{dS}} + v_{\mathrm{qS}}i_{\mathrm{qS}}\right) \tag{5.89}$$

$$P_{\mathrm{R}} = \frac{3}{2}\boldsymbol{v}_{0\mathrm{dqR}}^{T}\boldsymbol{i}_{0\mathrm{dqR}} = \frac{3}{2}\left(2v_{0\mathrm{R}}i_{0\mathrm{R}} + v_{\mathrm{dR}}i_{\mathrm{dR}} + v_{\mathrm{qR}}i_{\mathrm{qR}}\right) \tag{5.90}$$

$$Q_{\mathrm{S}} = \frac{3}{2}\left(v_{\mathrm{qS}}i_{\mathrm{dS}} - v_{\mathrm{dS}}i_{\mathrm{qS}}\right) \tag{5.91}$$

$$Q_{\mathrm{R}} = \frac{3}{2}\left(v_{\mathrm{qR}}i_{\mathrm{dR}} - v_{\mathrm{dR}}i_{\mathrm{qR}}\right). \tag{5.92}$$

The complete model of an asynchronous generator in the *0dq*-reference frame consists of equations (5.72), (5.81), (5.86) and (5.88)–(5.92).

### 5.3.3 Asynchronous Generator Model in the $\alpha\beta\gamma$ Reference Frame

When modeling the asynchronous machine operating with the power electronic system (e.g. converter, soft starter modeled in detailed form), the asynchronous generator model using natural axes should be converted into the $\alpha\beta\gamma$-reference frame model [77, 85]. That type of conversion is made to achieve a machine model with the number of inputs relevant to the converter number of inputs (connections), i.e. the 6-pulse rectifier or inverter that has 3 connections (on the AC current side) needs a machine model with 3 inputs (connections).

The aim of the transformation from natural axes (6 inputs) to the $\alpha\beta\gamma$-reference frame, which is 6-inputs model also, is to eliminate the dependence on the machine inductances $\boldsymbol{L}_{\mathrm{N}}$ of the angular displacement $\Theta_{\mathrm{R}}$. The transformation can be realized by using the transformation matrix $\boldsymbol{B}_{\mathrm{N}}$ in the following form:

$$\boldsymbol{B}_{\mathrm{N}} = \begin{bmatrix} \boldsymbol{B}_{\mathrm{S}} & \boldsymbol{0} \\ \boldsymbol{0} & \boldsymbol{B}_{\mathrm{R}} \end{bmatrix}, \tag{5.93}$$

where the submatrices $\boldsymbol{B}_{\mathrm{S}}$ and $\boldsymbol{B}_{\mathrm{R}}$ are given by

$$\boldsymbol{B}_{\mathrm{S}} = \frac{2}{3}\begin{bmatrix} \cos\Theta + \frac{1}{2} & \cos(\Theta - \frac{2}{3}\pi) + \frac{1}{2} & \cos(\Theta + \frac{2}{3}\pi) + \frac{1}{2} \\ \cos(\Theta + \frac{2}{3}\pi) + \frac{1}{2} & \cos\Theta + \frac{1}{2} & \cos(\Theta - \frac{2}{3}\pi) + \frac{1}{2} \\ \cos(\Theta - \frac{2}{3}\pi) + \frac{1}{2} & \cos(\Theta + \frac{2}{3}\pi) + \frac{1}{2} & \cos\Theta + \frac{1}{2} \end{bmatrix} \tag{5.94}$$

$$\boldsymbol{B}_{\mathrm{R}} = \frac{2}{3}\begin{bmatrix} \cos(\Theta - \Theta_{\mathrm{R}}) + \frac{1}{2} & \cos(\Theta - \Theta_{\mathrm{R}} - \frac{2}{3}\pi) + \frac{1}{2} & \cos(\Theta - \Theta_{\mathrm{R}} + \frac{2}{3}\pi) + \frac{1}{2} \\ \cos(\Theta - \Theta_{\mathrm{R}} + \frac{2}{3}\pi) + \frac{1}{2} & \cos(\Theta - \Theta_{\mathrm{R}}) + \frac{1}{2} & \cos(\Theta - \Theta_{\mathrm{R}} - \frac{2}{3}\pi) + \frac{1}{2} \\ \cos(\Theta - \Theta_{\mathrm{R}} - \frac{2}{3}\pi) + \frac{1}{2} & \cos(\Theta - \Theta_{\mathrm{R}} + \frac{2}{3}\pi) + \frac{1}{2} & \cos(\Theta - \Theta_{\mathrm{R}}) + \frac{1}{2} \end{bmatrix}. \tag{5.95}$$

Multiplying (5.62) and (5.68) by $\boldsymbol{B}_{\mathrm{N}}$ (assuming $R_{\mathrm{AS}} = R_{\mathrm{BS}} = R_{\mathrm{CS}} = R_{\mathrm{S}}$ and $R_{\mathrm{AR}} = R_{\mathrm{BR}} = R_{\mathrm{CR}} = R_{\mathrm{R}}$) we obtain

$$\boldsymbol{\Psi}_{\mathrm{M}} = \boldsymbol{L}_{\mathrm{M}}\, \boldsymbol{i}_{\mathrm{M}} \tag{5.96}$$

$$\boldsymbol{v}_{\mathrm{M}} = -\,\mathrm{p}\,\boldsymbol{\Psi}_{\mathrm{M}} - \boldsymbol{\Omega}\,\boldsymbol{\Psi}_{\mathrm{M}} - \boldsymbol{R}_{\mathrm{M}}\,\boldsymbol{i}_{\mathrm{M}}\,, \tag{5.97}$$

where the relevant matrices and vectors are given as follows

the flux vector:

$$\boldsymbol{\Psi}_{\mathrm{M}} = \begin{bmatrix} \boldsymbol{\Psi}_{\alpha\beta\gamma\mathrm{S}} \\ \boldsymbol{\Psi}_{\alpha\beta\gamma\mathrm{R}} \end{bmatrix} = \begin{bmatrix} \Psi_{\alpha\mathrm{S}} & \Psi_{\beta\mathrm{S}} & \Psi_{\gamma\mathrm{S}} & | & \Psi_{\alpha\mathrm{R}} & \Psi_{\beta\mathrm{R}} & \Psi_{\gamma\mathrm{R}} \end{bmatrix}^{T} \tag{5.98}$$

the current vector:

$$\boldsymbol{i}_{\mathrm{M}} = \begin{bmatrix} \boldsymbol{i}_{\alpha\beta\gamma\mathrm{S}} \\ \boldsymbol{i}_{\alpha\beta\gamma\mathrm{R}} \end{bmatrix} = \begin{bmatrix} i_{\alpha\mathrm{S}} & i_{\beta\mathrm{S}} & i_{\gamma\mathrm{S}} & | & i_{\alpha\mathrm{R}} & i_{\beta\mathrm{R}} & i_{\gamma\mathrm{R}} \end{bmatrix}^{T} \tag{5.99}$$

the inductance matrix:

$$\boldsymbol{L}_{\mathrm{M}} = \left[\begin{array}{ccc|ccc} L_{\mathrm{lS}} + L_{\mathrm{mS}} & -\frac{1}{2}L_{\mathrm{mS}} & -\frac{1}{2}L_{\mathrm{mS}} & L_{\mathrm{SR}} & -\frac{1}{2}L_{\mathrm{SR}} & -\frac{1}{2}L_{\mathrm{SR}} \\ -\frac{1}{2}L_{\mathrm{mS}} & L_{\mathrm{lS}} + L_{\mathrm{mS}} & -\frac{1}{2}L_{\mathrm{mS}} & -\frac{1}{2}L_{\mathrm{SR}} & L_{\mathrm{SR}} & -\frac{1}{2}L_{\mathrm{SR}} \\ -\frac{1}{2}L_{\mathrm{mS}} & -\frac{1}{2}L_{\mathrm{mS}} & L_{\mathrm{lS}} + L_{\mathrm{mS}} & -\frac{1}{2}L_{\mathrm{SR}} & -\frac{1}{2}L_{\mathrm{SR}} & L_{\mathrm{SR}} \\ \hline L_{\mathrm{SR}} & -\frac{1}{2}L_{\mathrm{SR}} & -\frac{1}{2}L_{\mathrm{SR}} & L_{\mathrm{lR}} + L_{\mathrm{mR}} & -\frac{1}{2}L_{\mathrm{mR}} & -\frac{1}{2}L_{\mathrm{mR}} \\ -\frac{1}{2}L_{\mathrm{SR}} & L_{\mathrm{SR}} & -\frac{1}{2}L_{\mathrm{SR}} & -\frac{1}{2}L_{\mathrm{mR}} & L_{\mathrm{lR}} + L_{\mathrm{mR}} & -\frac{1}{2}L_{\mathrm{mR}} \\ -\frac{1}{2}L_{\mathrm{SR}} & -\frac{1}{2}L_{\mathrm{SR}} & L_{\mathrm{SR}} & -\frac{1}{2}L_{\mathrm{mR}} & -\frac{1}{2}L_{\mathrm{mR}} & L_{\mathrm{lR}} + L_{\mathrm{mR}} \end{array}\right] \tag{5.100}$$

the rotation matrix $\boldsymbol{\Omega}$:

$$\boldsymbol{\Omega} = \frac{1}{\sqrt{3}} \left[\begin{array}{ccc|ccc} 0 & -\omega & \omega & 0 & 0 & 0 \\ \omega & 0 & -\omega & 0 & 0 & 0 \\ -\omega & \omega & 0 & 0 & 0 & 0 \\ \hline 0 & 0 & 0 & 0 & -(\omega-\omega_{\mathrm{R}}) & (\omega-\omega_{\mathrm{R}}) \\ 0 & 0 & 0 & (\omega-\omega_{\mathrm{R}}) & 0 & -(\omega-\omega_{\mathrm{R}}) \\ 0 & 0 & 0 & -(\omega-\omega_{\mathrm{R}}) & (\omega-\omega_{\mathrm{R}}) & 0 \end{array}\right] \tag{5.101}$$

the voltage vector:

$$\boldsymbol{v}_{\mathrm{M}} = \begin{bmatrix} \boldsymbol{v}_{\alpha\beta\gamma\mathrm{S}} \\ \boldsymbol{v}_{\alpha\beta\gamma\mathrm{R}} \end{bmatrix} = \begin{bmatrix} v_{\alpha\mathrm{S}} & v_{\beta\mathrm{S}} & v_{\gamma\mathrm{S}} & | & v_{\alpha\mathrm{R}} & v_{\beta\mathrm{R}} & v_{\gamma\mathrm{R}} \end{bmatrix}^{T} \tag{5.102}$$

and the resistance matrix:

$$\boldsymbol{R}_{\mathrm{M}} = \mathrm{diag}\,[R_{\mathrm{S}} \quad R_{\mathrm{S}} \quad R_{\mathrm{S}} \quad R_{\mathrm{R}} \quad R_{\mathrm{R}} \quad R_{\mathrm{R}}]\,. \tag{5.103}$$

The electromagnetic torque $\tau_{\mathrm{e}}$ and the real stator and rotor power $P_{\mathrm{S}}$, $P_{\mathrm{R}}$ are given by

$$\tau_e = \frac{1}{\sqrt{3}}\left(\tfrac{p}{2}\right)\left(i_{\alpha R}\left(\Psi_{\gamma R} - \Psi_{\beta R}\right) + i_{\beta R}\left(\Psi_{\alpha R} - \Psi_{\gamma R}\right) + i_{\gamma R}\left(\Psi_{\beta R} - \Psi_{\alpha R}\right)\right) \tag{5.104}$$

or, for a magnetically linear system,

$$\tau_e = \frac{3}{2\sqrt{3}}\left(\tfrac{p}{2}\right)L_{SR}\left(i_{\alpha S}\left(i_{\beta R} - i_{\gamma R}\right) + i_{\beta S}\left(i_{\gamma R} - i_{\alpha R}\right) + i_{\gamma S}\left(i_{\alpha R} - i_{\beta R}\right)\right) \tag{5.105}$$

$$P_S = \boldsymbol{v}_{\alpha\beta\gamma S}^T \boldsymbol{i}_{\alpha\beta\gamma S} = v_{\alpha S} i_{\alpha S} + v_{\beta S} i_{\beta S} + v_{\gamma S} i_{\gamma S} \tag{5.106}$$

$$P_R = \boldsymbol{v}_{\alpha\beta\gamma R}^T \boldsymbol{i}_{\alpha\beta\gamma R} = v_{\alpha R} i_{\alpha R} + v_{\beta R} i_{\beta R} + v_{\gamma R} i_{\gamma R} \,. \tag{5.107}$$

The complete model of an asynchronous generator consists of equations (5.72), (5.96), (5.97) and (5.104)–(5.107).

Usually, in the development of an asynchronous machine models, all the quantities are referred to the stator windings because the machine parameters are measured with respect to the stator windings. Additionally, the assumption is made that both 3-axis windings are Y-connected 3-wire systems. Therefore the stator and rotor currents are given by

$$i_{\alpha S} + i_{\beta S} + i_{\gamma S} = 0 \tag{5.108}$$

$$i_{\alpha R} + i_{\beta R} + i_{\gamma R} = 0 \,. \tag{5.109}$$

With the assumptions mentioned above, the asynchronous generator model (Fig. 5.22) with the rotor quantities referred to stator windings (primed quantities) is described as follows:

$$\boldsymbol{\Psi}_M = \boldsymbol{L}_M \, \boldsymbol{i}_M \tag{5.110}$$

$$\boldsymbol{v}_M = -\,\mathrm{p}\,\boldsymbol{\Psi}_M - \boldsymbol{\Omega}\,\boldsymbol{\Psi}_M - \boldsymbol{R}_M \, \boldsymbol{i}_M \,, \tag{5.111}$$

where the appropriate matrices and vectors are as follows:
the voltage vector:

$$\boldsymbol{v}_M = \begin{bmatrix} \boldsymbol{v}_{\alpha\beta\gamma S} \\ \boldsymbol{v}'_{\alpha\beta\gamma R} \end{bmatrix} = \begin{bmatrix} v_{\alpha S} & v_{\beta S} & v_{\gamma S} & | & v'_{\alpha R} & v'_{\beta R} & v'_{\gamma R} \end{bmatrix}^T \tag{5.112}$$

and the resistance matrix:

$$\boldsymbol{R}_M = \mathrm{diag}\,[R_S \quad R_S \quad R_S \quad R'_R \quad R'_R \quad R'_R] \tag{5.113}$$

the flux vector:

$$\boldsymbol{\Psi}_M = \begin{bmatrix} \boldsymbol{\Psi}_{\alpha\beta\gamma S} \\ \boldsymbol{\Psi}'_{\alpha\beta\gamma R} \end{bmatrix} = \begin{bmatrix} \Psi_{\alpha S} & \Psi_{\beta S} & \Psi_{\gamma S} & | & \Psi'_{\alpha R} & \Psi'_{\beta R} & \Psi'_{\gamma R} \end{bmatrix}^T \tag{5.114}$$

the current vector:

$$\boldsymbol{i}_{\mathrm{M}} = \begin{bmatrix} \boldsymbol{i}_{\alpha\beta\gamma\mathrm{S}} \\ \boldsymbol{i}'_{\alpha\beta\gamma\mathrm{R}} \end{bmatrix} = \begin{bmatrix} i_{\alpha\mathrm{S}} & i_{\beta\mathrm{S}} & i_{\gamma\mathrm{S}} & | & i'_{\alpha\mathrm{R}} & i'_{\beta\mathrm{R}} & i'_{\gamma\mathrm{R}} \end{bmatrix}^T \tag{5.115}$$

the inductance matrix:

$$\boldsymbol{L}_{\mathrm{M}} = \left[\begin{array}{ccc|ccc} L_{\mathrm{S}} & 0 & 0 & L_{\mathrm{m}} & 0 & 0 \\ 0 & L_{\mathrm{S}} & 0 & 0 & L_{\mathrm{m}} & 0 \\ 0 & 0 & L_{\mathrm{S}} & 0 & 0 & L_{\mathrm{m}} \\ \hline L_{\mathrm{m}} & 0 & 0 & L'_{\mathrm{R}} & 0 & 0 \\ 0 & L_{\mathrm{m}} & 0 & 0 & L'_{\mathrm{R}} & 0 \\ 0 & 0 & L_{\mathrm{m}} & 0 & 0 & L'_{\mathrm{R}} \end{array}\right], \tag{5.116}$$

where $L_{\mathrm{S}} = L_{\mathrm{lS}} + L_{\mathrm{m}}$, $L'_{\mathrm{R}} = L'_{\mathrm{lR}} + L_{\mathrm{m}}$ and $L_{\mathrm{m}} = (3/2)L_{\mathrm{mS}} = (3/2)L'_{\mathrm{mR}}$.

The electromagnetic torque $\tau_{\mathrm{e}}$ is given by

$$\tau_{\mathrm{e}} = \sqrt{3}\left(\tfrac{p}{2}\right)\left(\Psi'_{\gamma\mathrm{R}} i'_{\alpha\mathrm{R}} - \Psi'_{\alpha\mathrm{R}} i'_{\gamma\mathrm{R}}\right) \tag{5.117}$$

or, for an electromagnetically linear system, is given by

$$\tau_{\mathrm{e}} = \sqrt{3}\left(\tfrac{p}{2}\right)L_{\mathrm{m}}\left(i_{\gamma\mathrm{S}} i'_{\alpha\mathrm{R}} - i_{\alpha\mathrm{S}} i'_{\gamma\mathrm{R}}\right). \tag{5.118}$$

The rotation matrix $\boldsymbol{\Omega}$ is defined by (5.101), the real power of the stator and rotor side is defined by (5.106) and (5.107), and the mechanical system is described by (5.72) and (5.73).

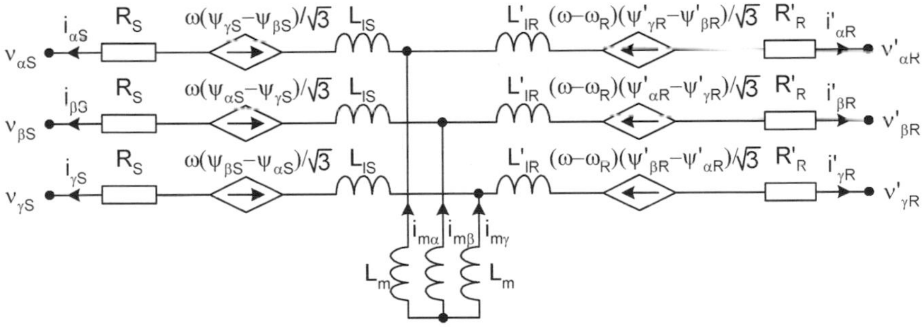


**Fig. 5.22.** Mathematical model of an asynchronous generator in the $\alpha\beta\gamma$-reference frame (for a model with the reference frame fixed in the stator, substitute $\omega = 0$; for a model with the reference frame fixed in the rotor, substitute $\omega = \omega_{\mathrm{R}}$)

# 5.4 Synchronous Generator Modeling

## 5.4.1 Synchronous Generator Model for Natural Axes

It is assumed that the synchronous generator (Fig. 5.23) has a three-phase armature winding (*A, B, C*), a rotor field winding (*f*) and two rotor damper windings – one in the d-axis (*D*) and one in the q-axis (*Q*).

The mathematical model of a synchronous generator for power system and converter system analysis is usually based on the following assumptions [15, 19, 29, 51, 55, 60–63, 71, 77, 82, 85]:

- the stator windings are positioned sinusoidally along the air-gap as far as the mutual effect with the rotor is concerned,
- the stator slots cause no appreciable variations of the rotor inductances with rotor position,
- magnetic hysteresis and saturation effects are negligible,
- the stator winding is symmetrical,
- damping and solid elements are replaced by two amortize circuits,
- the capacitance of all the windings can be neglected,
- the resistances are constant.

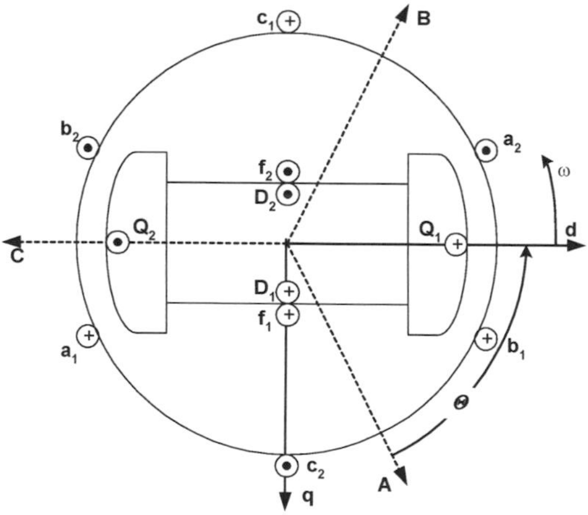


**Fig. 5.23.** The windings of a synchronous generator

A more detailed modeling usually encounters difficulties in getting the data of such generators (for such models).

Using natural axes, the model of synchronous generator can be formulated by the so called flux equation, voltage equation and mechanical system equation.

The flux equation describing the relationship between the flux and current in each winding has the form:

$$\boldsymbol{\Psi}_{\mathrm{N}} = \boldsymbol{L}_{\mathrm{N}}\, \boldsymbol{i}_{\mathrm{N}}\,, \tag{5.119}$$

where the flux vector is given by

$$\boldsymbol{\Psi}_{\mathrm{N}} = \begin{bmatrix} \boldsymbol{\Psi}_{\mathrm{ABC}} \\ \boldsymbol{\Psi}_{\mathrm{fDQ}} \end{bmatrix} = \begin{bmatrix} \Psi_{\mathrm{A}} & \Psi_{\mathrm{B}} & \Psi_{\mathrm{C}} & | & \Psi_{\mathrm{f}} & \Psi_{\mathrm{D}} & \Psi_{\mathrm{Q}} \end{bmatrix}^{T}, \tag{5.120}$$

where $\boldsymbol{\Psi}_{\mathrm{ABC}}$ is the vector of stator fluxes and $\boldsymbol{\Psi}_{\mathrm{fDQ}}$ is the vector of rotor fluxes.

The current vector is equal to:

$$\boldsymbol{i}_{\mathrm{N}} = \begin{bmatrix} \boldsymbol{i}_{\mathrm{ABC}} \\ \boldsymbol{i}_{\mathrm{fDQ}} \end{bmatrix} = \begin{bmatrix} i_{\mathrm{A}} & i_{\mathrm{B}} & i_{\mathrm{C}} & | & i_{\mathrm{f}} & i_{\mathrm{D}} & i_{\mathrm{Q}} \end{bmatrix}^{T}, \tag{5.121}$$

where $\boldsymbol{i}_{\mathrm{ABC}}$ is the vector of stator currents and $\boldsymbol{i}_{\mathrm{fDQ}}$ is the vector of rotor currents.

The inductance matrix is given by

$$\boldsymbol{L}_{\mathrm{N}} = \begin{bmatrix} \boldsymbol{L}_{\mathrm{S}} & \boldsymbol{L}_{\mathrm{SR}} \\ \boldsymbol{L}_{\mathrm{SR}}^{T} & \boldsymbol{L}_{\mathrm{R}} \end{bmatrix} = \left[\begin{array}{ccc|ccc} L_{\mathrm{AA}} & L_{\mathrm{AB}} & L_{\mathrm{AC}} & L_{\mathrm{Af}} & L_{\mathrm{AD}} & L_{\mathrm{AQ}} \\ L_{\mathrm{BA}} & L_{\mathrm{BB}} & L_{\mathrm{AC}} & L_{\mathrm{Bf}} & L_{\mathrm{BD}} & L_{\mathrm{BQ}} \\ L_{\mathrm{CA}} & L_{\mathrm{CB}} & L_{\mathrm{CC}} & L_{\mathrm{Cf}} & L_{\mathrm{CD}} & L_{\mathrm{CQ}} \\ \hline L_{\mathrm{fA}} & L_{\mathrm{fB}} & L_{\mathrm{fC}} & L_{\mathrm{ff}} & L_{\mathrm{fD}} & L_{\mathrm{fQ}} \\ L_{\mathrm{DA}} & L_{\mathrm{DB}} & L_{\mathrm{DC}} & L_{\mathrm{Df}} & L_{\mathrm{DD}} & L_{\mathrm{DQ}} \\ L_{\mathrm{QA}} & L_{\mathrm{QB}} & L_{\mathrm{QC}} & L_{\mathrm{Qf}} & L_{\mathrm{QD}} & L_{\mathrm{QQ}} \end{array}\right], \tag{5.122}$$

where $\boldsymbol{L}_{\mathrm{S}}$ is a submatrix of the stator's self and mutual inductances, $\boldsymbol{L}_{\mathrm{R}}$ is a submatrix of rotor's self and mutual inductances and $\boldsymbol{L}_{\mathrm{SR}}$ is a submatrix of the rotor to stator mutual inductances.

The self and mutual inductances vary, and their values depend on the rotor position relative to the stator winding (electrical rotor angular displacement $\Theta$ in Fig. 5.23). The stator's self inductances are defined as

$$\begin{aligned} L_{\mathrm{AA}} &= L_{\mathrm{S}} + \Delta L_{\mathrm{S}} \cos 2\Theta \\ L_{\mathrm{BB}} &= L_{\mathrm{S}} + \Delta L_{\mathrm{S}} \cos(2\Theta - \tfrac{2}{3}\pi) \\ L_{\mathrm{CC}} &= L_{\mathrm{S}} + \Delta L_{\mathrm{S}} \cos(2\Theta + \tfrac{2}{3}\pi)\,, \end{aligned} \tag{5.123}$$

where both inductances $L_{\mathrm{S}}$ and $\Delta L_{\mathrm{S}}$ are constant and $L_{\mathrm{S}} > \Delta L_{\mathrm{S}}$.

The mutual inductances between the stator windings are defined as

$$\begin{aligned} L_{\mathrm{AB}} = L_{\mathrm{BA}} &= -M_{\mathrm{S}} - \Delta L_{\mathrm{S}} \cos 2(\Theta + \tfrac{1}{6}\pi) \\ L_{\mathrm{BC}} = L_{\mathrm{CB}} &= -M_{\mathrm{S}} - \Delta L_{\mathrm{S}} \cos 2(\Theta - \tfrac{1}{2}\pi) \\ L_{\mathrm{CA}} = L_{\mathrm{AC}} &= -M_{\mathrm{S}} - \Delta L_{\mathrm{S}} \cos 2(\Theta + \tfrac{5}{6}\pi)\,, \end{aligned} \tag{5.124}$$

where $M_{\mathrm{S}} > \Delta L_{\mathrm{S}}$.

The mutual inductances between the stator and rotor windings are defined as

$$\begin{aligned}
L_{\mathrm{Af}} &= L_{\mathrm{fA}} = M_{\mathrm{f}} \cos\Theta \\
L_{\mathrm{AD}} &= L_{\mathrm{DA}} = M_{\mathrm{D}} \cos\Theta \\
L_{\mathrm{AQ}} &= L_{\mathrm{QA}} = M_{\mathrm{Q}} \sin\Theta \\
L_{\mathrm{Bf}} &= L_{\mathrm{fB}} = M_{\mathrm{f}} \cos(\Theta - \tfrac{2}{3}\pi) \\
L_{\mathrm{BD}} &= L_{\mathrm{DB}} = M_{\mathrm{D}} \cos(\Theta - \tfrac{2}{3}\pi) \\
L_{\mathrm{BQ}} &= L_{\mathrm{QB}} = M_{\mathrm{Q}} \sin(\Theta - \tfrac{2}{3}\pi) \\
L_{\mathrm{Cf}} &= L_{\mathrm{fC}} = M_{\mathrm{f}} \cos(\Theta + \tfrac{2}{3}\pi) \\
L_{\mathrm{CD}} &= L_{\mathrm{DC}} = M_{\mathrm{D}} \cos(\Theta + \tfrac{2}{3}\pi) \\
L_{\mathrm{CQ}} &= L_{\mathrm{QC}} = M_{\mathrm{Q}} \sin(\Theta + \tfrac{2}{3}\pi) \,.
\end{aligned} \tag{5.125}$$

And because the $d$- and $q$-axis are perpendicular to each other, their mutual inductances are equal to zero:

$$\begin{aligned}
L_{\mathrm{fQ}} &= L_{\mathrm{Qf}} = 0 \\
L_{\mathrm{DQ}} &= L_{\mathrm{QD}} = 0 \,.
\end{aligned} \tag{5.126}$$

The voltage equation describing the relationship between the voltage, flux and current in each winding has the following form (Fig. 5.24):

$$\boldsymbol{v}_{\mathrm{N}} = -\mathrm{p}\boldsymbol{\Psi}_{\mathrm{N}} - \boldsymbol{R}_{\mathrm{N}}\,\boldsymbol{i}_{\mathrm{N}} \,, \tag{5.127}$$

where the voltage vector is given by

$$\boldsymbol{v}_{\mathrm{N}} = \begin{bmatrix} v_{\mathrm{ABC}} \\ v_{\mathrm{fDQ}} \end{bmatrix} = \begin{bmatrix} v_{\mathrm{A}} & v_{\mathrm{B}} & v_{\mathrm{C}} & | & -v_{\mathrm{f}} & 0 & 0 \end{bmatrix}^T , \tag{5.128}$$

where $\boldsymbol{v}_{\mathrm{ABC}}$ is the vector of stator voltages and $\boldsymbol{v}_{\mathrm{fDQ}}$ is the vector of rotor voltages.

The resistance matrix is given by

$$\boldsymbol{R}_{\mathrm{N}} = \mathrm{diag}\,[R_{\mathrm{A}} \quad R_{\mathrm{B}} \quad R_{\mathrm{C}} \quad R_{\mathrm{f}} \quad R_{\mathrm{D}} \quad R_{\mathrm{Q}}]\,. \tag{5.129}$$

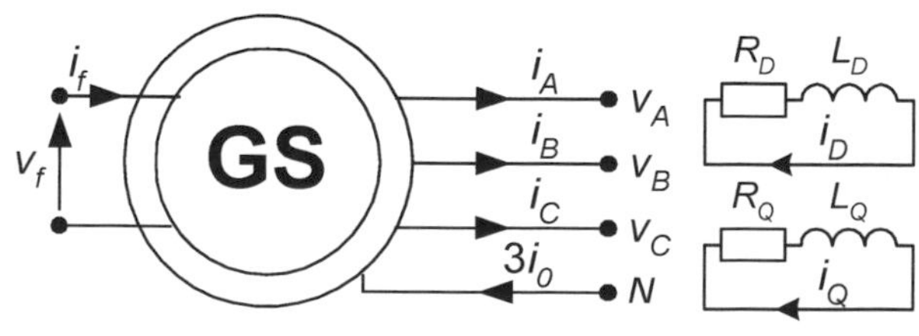


**Fig. 5.24.** Currents and voltages of the synchronous generator

The mechanical system of the WTGS can be described by the following equation of motion[5]:

$$(\tfrac{2}{p})J\frac{\mathrm{d}\omega}{\mathrm{d}t} = \tau_\mathrm{m} - \tau_\mathrm{e} - D(\tfrac{2}{p})\omega \tag{5.130}$$

where, the electromagnetic torque $\tau_\mathrm{e}$, the angular displacement $\Theta$ and the "mechanical" angular velocity of the rotor $\omega_\mathrm{m}$ are given by

$$\begin{aligned}\tau_\mathrm{e} &= (\tfrac{p}{2})\frac{\partial W(\boldsymbol{i}_\mathrm{ABC},\boldsymbol{i}_\mathrm{fDQ},\Theta)}{\partial\Theta} = \frac{1}{2}(\tfrac{p}{2})\left[\boldsymbol{i}_\mathrm{ABC}\boldsymbol{i}_\mathrm{fDQ}\right]^T \frac{\partial}{\partial\Theta}\begin{bmatrix}\boldsymbol{L}_\mathrm{S}(\Theta) & \boldsymbol{L}_\mathrm{SR}(\Theta)\\ \boldsymbol{L}_\mathrm{SR}^T(\Theta) & \boldsymbol{L}_\mathrm{R}\end{bmatrix}\begin{bmatrix}\boldsymbol{i}_\mathrm{ABC}\\ \boldsymbol{i}_\mathrm{fDQ}\end{bmatrix}\\ \Theta &= \int_0^t \omega(\xi)\mathrm{d}\xi + \Theta_0\\ \omega_\mathrm{m} &= \omega(\tfrac{2}{p})\,.\end{aligned} \tag{5.131}$$

The terminal power is given by

$$P = \boldsymbol{v}_\mathrm{ABC}^T\boldsymbol{i}_\mathrm{ABC} = v_\mathrm{A}i_\mathrm{A} + v_\mathrm{B}i_\mathrm{B} + v_\mathrm{C}i_\mathrm{C}\,. \tag{5.132}$$

### 5.4.2 Synchronous Generator Model in the *0dq* Reference Frame

The synchronous generator model for the dynamic phenomena in the power system analysis or converter system (with synchronous generator) modeling and analysis is not usually utilized in the form presented above. The main disadvantage of the model is the same as in the case of asynchronous machine modeling, i.e. the dependence of the inductance matrix $\boldsymbol{L}_\mathrm{N}$ on the angular displacement $\Theta$. Therefore, the model (set of equations) is usually converted into a model related to arbitrarily set rotating axes (reference frame) by utilizing the transformation matrix $\boldsymbol{B}$.

The synchronous generator model is usually converted into the so-called *0dq*-reference frame model. Axes $d$ and $q$ are related to the rotating rotor here, as shown in Fig. 5.23. The transformation is realized by using the transformation matrix $\boldsymbol{B}_\mathrm{N}$ in the form[6]:

$$\boldsymbol{B}_\mathrm{N} = \begin{bmatrix}\boldsymbol{B} & \boldsymbol{0}\\ \boldsymbol{0} & \boldsymbol{1}\end{bmatrix}, \tag{5.133}$$

where the submatrix $\boldsymbol{B}$ is given by

$$\boldsymbol{B} = \frac{2}{3}\begin{bmatrix}\frac{1}{2} & \frac{1}{2} & \frac{1}{2}\\ \cos\Theta & \cos(\Theta-\frac{2}{3}\pi) & \cos(\Theta+\frac{2}{3}\pi)\\ \sin\Theta & \sin(\Theta-\frac{2}{3}\pi) & \sin(\Theta+\frac{2}{3}\pi)\end{bmatrix} \tag{5.134}$$

[5] In general, in WTGS modeling, the drive-train model described in Sect. 5.2 should be used.

[6] Other types of transformation are also utilized.

and submatrix $\boldsymbol{1}$ is the diagonal unity matrix.

Multiplying (5.119) by the transformation matrix $\boldsymbol{B}_\mathrm{N}$ we obtain

$$\boldsymbol{B}_\mathrm{N}\boldsymbol{\Psi}_\mathrm{N} = \boldsymbol{B}_\mathrm{N}\boldsymbol{L}_\mathrm{N}\boldsymbol{B}_\mathrm{N}^{-1}\boldsymbol{B}_\mathrm{N}\boldsymbol{i}_\mathrm{N}\,. \tag{5.135}$$

Because the flux vector, the current vector and the inductance matrix in the *0dq*-reference frame satisfy the following relations:

$$\begin{aligned}\boldsymbol{B}_\mathrm{N}\boldsymbol{\Psi}_\mathrm{N} &= \boldsymbol{\Psi}_\mathrm{G}\\ \boldsymbol{B}_\mathrm{N}\boldsymbol{L}_\mathrm{N}\boldsymbol{B}_\mathrm{N}^{-1} &= \boldsymbol{L}_\mathrm{G}\\ \boldsymbol{B}_\mathrm{N}\boldsymbol{i}_\mathrm{N} &= \boldsymbol{i}_\mathrm{G}\,,\end{aligned} \tag{5.136}$$

the flux equation in the *0dq*-reference frame takes the following form:

$$\boldsymbol{\Psi}_\mathrm{G} = \boldsymbol{L}_\mathrm{G}\boldsymbol{i}_\mathrm{G}\,, \tag{5.137}$$

where the flux vector, the current vector and the inductance matrix are, respectively, given by

$$\boldsymbol{\Psi}_\mathrm{G} = \begin{bmatrix}\boldsymbol{\Psi}_\mathrm{0dq}\\ \boldsymbol{\Psi}_\mathrm{fDQ}\end{bmatrix} = \begin{bmatrix}\Psi_0 & \Psi_\mathrm{d} & \Psi_\mathrm{q} & | & \Psi_\mathrm{f} & \Psi_\mathrm{D} & \Psi_\mathrm{Q}\end{bmatrix}^T$$

$$\boldsymbol{i}_\mathrm{G} = \begin{bmatrix}\boldsymbol{i}_\mathrm{0dq}\\ \boldsymbol{i}_\mathrm{fDQ}\end{bmatrix} = \begin{bmatrix}i_0 & i_\mathrm{d} & i_\mathrm{q} & | & i_\mathrm{f} & i_\mathrm{D} & i_\mathrm{Q}\end{bmatrix}^T \tag{5.138}$$

$$\boldsymbol{L}_\mathrm{G} = \begin{bmatrix} L_0 & 0 & 0 & 0 & 0 & 0\\ 0 & L_\mathrm{d} & 0 & \frac{3}{2}M_\mathrm{f} & \frac{3}{2}M_\mathrm{D} & 0\\ 0 & 0 & L_\mathrm{q} & 0 & 0 & \frac{3}{2}M_\mathrm{Q}\\ 0 & \frac{3}{2}M_\mathrm{f} & 0 & L_\mathrm{f} & L_\mathrm{fD} & 0\\ 0 & \frac{3}{2}M_\mathrm{D} & 0 & L_\mathrm{fD} & L_\mathrm{D} & 0\\ 0 & 0 & \frac{3}{2}M_\mathrm{Q} & 0 & 0 & L_\mathrm{Q}\end{bmatrix}.$$

Multiplying (5.127) by the transformation matrix $\boldsymbol{B}_\mathrm{N}$ we obtain the following formula:

$$\boldsymbol{B}_\mathrm{N}\boldsymbol{v}_\mathrm{N} = -\boldsymbol{B}_\mathrm{N}\frac{\mathrm{d}}{\mathrm{d}t}(\boldsymbol{B}_\mathrm{N}^{-1}\boldsymbol{B}_\mathrm{N}\boldsymbol{\Psi}_\mathrm{N}) - \boldsymbol{B}_\mathrm{N}\boldsymbol{R}_\mathrm{N}\boldsymbol{B}_\mathrm{N}^{-1}\boldsymbol{B}_\mathrm{N}\boldsymbol{i}_\mathrm{N}\,. \tag{5.139}$$

Because the respective matrices (assuming $R_\mathrm{A} = R_\mathrm{B} = R_\mathrm{C} = R_\mathrm{S}$) and vectors in the *0dq*- reference frame satisfy the following relations:

$$\begin{aligned}\boldsymbol{B}_\mathrm{N}\boldsymbol{v}_\mathrm{N} &= \boldsymbol{v}_\mathrm{G}\\ \boldsymbol{B}_\mathrm{N}\boldsymbol{R}_\mathrm{N}\boldsymbol{B}_\mathrm{N}^{-1} &= \boldsymbol{R}_\mathrm{G}\\ -\boldsymbol{B}_\mathrm{N}\frac{\mathrm{d}}{\mathrm{d}t}(\boldsymbol{B}_\mathrm{N}^{-1}\boldsymbol{B}_\mathrm{N}\boldsymbol{\Psi}_\mathrm{N}) &= -\boldsymbol{B}_\mathrm{N}\left(\frac{\mathrm{d}\boldsymbol{B}_\mathrm{N}^{-1}}{\mathrm{d}t}\boldsymbol{B}_\mathrm{N}\boldsymbol{\Psi}_\mathrm{N} + \boldsymbol{B}_\mathrm{N}^{-1}\frac{\mathrm{d}\boldsymbol{B}_\mathrm{N}\boldsymbol{\Psi}_\mathrm{N}}{\mathrm{d}t}\right) = -\boldsymbol{\Omega}\boldsymbol{\Psi}_\mathrm{G} - \frac{\mathrm{d}\boldsymbol{\Psi}_\mathrm{G}}{\mathrm{d}t}\,,\end{aligned} \tag{5.140}$$

where the rotation matrix $\boldsymbol{\Omega}$ is given by

$$\boldsymbol{\Omega} = \omega\begin{bmatrix} 0 & 0 & 0 & 0 & 0 & 0 \\ 0 & 0 & 1 & 0 & 0 & 0 \\ 0 & -1 & 0 & 0 & 0 & 0 \\ 0 & 0 & 0 & 0 & 0 & 0 \\ 0 & 0 & 0 & 0 & 0 & 0 \\ 0 & 0 & 0 & 0 & 0 & 0 \end{bmatrix}, \tag{5.141}$$

the voltage equation in the *0dq*-reference frame takes the following form (Fig. 5.25):

$$\boldsymbol{v}_{\mathrm{G}} = -\,\mathrm{p}\,\boldsymbol{\Psi}_{\mathrm{G}} - \boldsymbol{\Omega}_{\mathrm{G}}\,\boldsymbol{\Psi}_{\mathrm{G}} - \boldsymbol{R}_{\mathrm{G}}\,\boldsymbol{i}_{\mathrm{G}}\,, \tag{5.142}$$

where the voltage vector and the resistance matrix are, respectively, given by

$$\begin{aligned} \boldsymbol{v}_{\mathrm{G}} &= \begin{bmatrix} \boldsymbol{v}_{\mathrm{0dq}} \\ \boldsymbol{v}_{\mathrm{fDQ}} \end{bmatrix} = \begin{bmatrix} v_0 & v_{\mathrm{d}} & v_{\mathrm{q}} & | & -v_{\mathrm{f}} & 0 & 0 \end{bmatrix}^T \\ \boldsymbol{R}_{\mathrm{G}} &= \mathrm{diag}\,[R_0 \quad R_{\mathrm{S}} \quad R_{\mathrm{S}} \quad R_{\mathrm{f}} \quad R_{\mathrm{D}} \quad R_{\mathrm{Q}}]. \end{aligned} \tag{5.143}$$

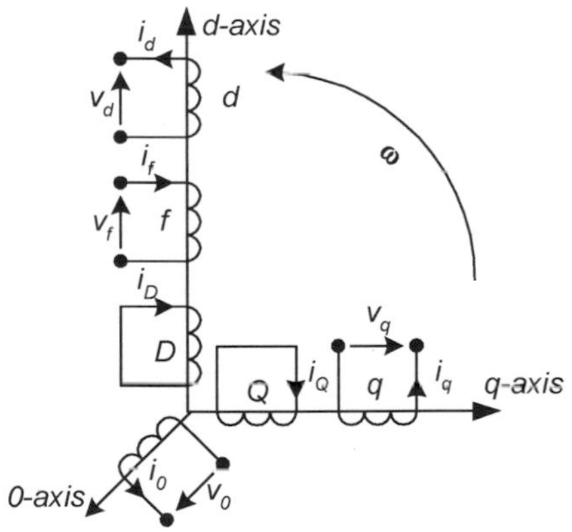


**Fig. 5.25.** Windings representing of the *0dq*-reference frame model of the synchronous generator

The electromagnetic torque $\tau_{\mathrm{e}}$ and the real and reactive terminal power *P, Q* are given by

$$\tau_{\mathrm{e}} = \frac{3}{2}\left(\tfrac{p}{2}\right)\left(\Psi_{\mathrm{d}} i_{\mathrm{q}} - \Psi_{\mathrm{q}} i_{\mathrm{d}}\right) \tag{5.144}$$

$$P = \frac{3}{2}\boldsymbol{v}_{\mathrm{0dq}}^T \boldsymbol{i}_{\mathrm{odq}} = \frac{3}{2}\left(2 v_0 i_0 + v_{\mathrm{d}} i_{\mathrm{d}} + v_{\mathrm{q}} i_{\mathrm{q}}\right) \tag{5.145}$$

$$Q = \frac{3}{2}\left(v_{\mathrm{q}} i_{\mathrm{d}} - v_{\mathrm{d}} i_{\mathrm{q}}\right). \tag{5.146}$$

The complete model of a synchronous generator in the *0dq*-reference frame consists of equations (5.130), (5.137), (5.142) and (5.144)–(5.146).

In many applications, a model whose equations are defined in per unit can be useful. In such a model, the variables expressed per unit are related to the: peak value of the rated line current $I_n$, the peak value of the rated line-to-neutral voltage $V_n$, the rated apparent power of machine $S_n$, the rated frequency $\omega_n$, the rated no-load field current $I_{fn0}$, and the rated no-load field voltage $V_{fn0}$. Usually, only the power angle $\delta$ and the shaft angular velocity $\omega$ are expressed in their natural units, which are radians and radians per second, respectively.

Then the model that represents a synchronous generator connected to the power system (infinite bus with voltage $V_S$) through a transformer (impedance $\underline{Z}_T = R_T + jX_T$) and transmission line (impedance $\underline{Z}_L = R_L + jX_L$) can be described by the following set of equations:

Flux linkage equation:

$$\boldsymbol{\Psi}_G = \boldsymbol{L}_G \, \boldsymbol{I}_G \,, \tag{5.147}$$

Voltage equation:

$$\mathrm{p}\,\boldsymbol{\Psi}_G = \omega_n \left(-\boldsymbol{V}_G - \boldsymbol{\Omega}\,\boldsymbol{\Psi}_G - \boldsymbol{R}_G \boldsymbol{I}_G\right), \tag{5.148}$$

which can be used in the following form to eliminate the flux elimination from the set of differential equations:

$$\mathrm{p}\boldsymbol{I}_G = \omega_n \boldsymbol{L}_G^{-1}\left[-\boldsymbol{V}_G - (\boldsymbol{\Omega}\,\boldsymbol{L}_G + \boldsymbol{R}_G)\,\boldsymbol{I}_G\right], \tag{5.149}$$

Motion equations:

$$\begin{aligned} \mathrm{p}s &= (\tau_m - \tau_e - Ds)(2H_j)^{-1} \\ \mathrm{p}\delta &= \omega_n \left(\tfrac{p}{2}\right) s \,, \end{aligned} \tag{5.150}$$

where the shaft slip is defined as

$$s = (\omega - \omega_n)\,\omega_n^{-1} \,. \tag{5.151}$$

The electromagnetic torque and the terminal real and reactive power equations are

$$\begin{aligned} \tau_e &= \Psi_d I_q - \Psi_q I_d \\ P &= V_d I_d + V_q I_q \\ Q &= V_q I_d - V_d I_q \,. \end{aligned} \tag{5.152}$$

The infinite bus and generator terminal voltage equations are

$$\begin{aligned} V_{dS} &= -V_S \sin\delta \\ V_{qS} &= V_S \cos\delta \\ V_d &= V_{dS} + (R_T + R_L)I_d + (X_T + X_L)(\omega_n^{-1}\mathrm{p}I_d + (1+s)I_q) \end{aligned} \tag{5.153}$$

$$V_{\mathrm{q}} = V_{\mathrm{qS}} + (R_{\mathrm{T}} + R_{\mathrm{L}})I_{\mathrm{q}} + (X_{\mathrm{T}} + X_{\mathrm{L}})(\omega_{\mathrm{n}}^{-1}\mathrm{p}I_{\mathrm{q}} - (1+s)I_{\mathrm{d}})$$

$$V = \sqrt{V_{\mathrm{d}}^2 + V_{\mathrm{q}}^2}\,,$$

because of the existence of external elements, i.e. transformer and transmission line, the vectors and matrices occurring in the above equations should be modified as follows:

$$\boldsymbol{I}_{\mathrm{G}} = \begin{bmatrix} I_0 & I_{\mathrm{d}} & I_{\mathrm{q}} & I_{\mathrm{f}}/L_{\mathrm{ad}} & I_{\mathrm{D}} & I_{\mathrm{Q}} \end{bmatrix}^T$$

$$\boldsymbol{V}_{\mathrm{G}} = \begin{bmatrix} V_0 & V_{\mathrm{dS}} & V_{\mathrm{qS}} & -V_{\mathrm{f}}R_{\mathrm{f}}/L_{\mathrm{ad}} & 0 & 0 \end{bmatrix}^T$$

$$\boldsymbol{R}_{\mathrm{G}} = \mathrm{diag}\,[R_0 \quad R_{\mathrm{S}} + R_{\mathrm{T}} + R_{\mathrm{L}} \quad R_{\mathrm{S}} + R_{\mathrm{T}} + R_{\mathrm{L}} \quad R_{\mathrm{f}} \quad R_{\mathrm{D}} \quad R_{\mathrm{Q}}] \tag{5.154}$$

$$\boldsymbol{L}_{\mathrm{G}} = \begin{bmatrix} L_0 & 0 & 0 & 0 & 0 & 0 \\ 0 & L_{\mathrm{d}} + L_{\mathrm{T}} + L_{\mathrm{L}} & 0 & L_{\mathrm{ad}} & L_{\mathrm{ad}} & 0 \\ 0 & 0 & L_{\mathrm{q}} + L_{\mathrm{T}} + L_{\mathrm{L}} & 0 & 0 & L_{\mathrm{aq}} \\ 0 & L_{\mathrm{ad}} & 0 & L_{\mathrm{f}} & L_{\mathrm{ad}} & 0 \\ 0 & L_{\mathrm{ad}} & 0 & L_{\mathrm{ad}} & L_{\mathrm{D}} & 0 \\ 0 & 0 & L_{\mathrm{aq}} & 0 & 0 & L_{\mathrm{Q}} \end{bmatrix}.$$

The values of resistances and reactances appearing in the above equations can be computed from the synchronous generator rated data as follows:

$$L_{\mathrm{ad}} = L_{\mathrm{d}} - L_{\mathrm{l}} \qquad L_{\mathrm{aq}} = L_{\mathrm{q}} - L_{\mathrm{l}}$$

$$L_{\mathrm{D}} = L_{\mathrm{ad}}\left(1 + \frac{(L'_{\mathrm{d}} - L_{\mathrm{l}})(L''_{\mathrm{d}} - L_{\mathrm{l}})}{L'_{\mathrm{d}} - L''_{\mathrm{d}}}\right) \qquad L_{\mathrm{Q}} = L_{\mathrm{aq}}\left(1 + \frac{L''_{\mathrm{q}} - L_{\mathrm{l}}}{L_{\mathrm{q}} - L''_{\mathrm{q}}}\right)$$

$$R_{\mathrm{D}} = \frac{1}{\omega_{\mathrm{n}}T''_{\mathrm{d}}}\frac{L''_{\mathrm{d}}}{L'_{\mathrm{d}}}\frac{(L'_{\mathrm{d}} - L_{\mathrm{l}})^2}{L'_{\mathrm{d}} - L''_{\mathrm{d}}} \qquad R_{\mathrm{Q}} = \frac{1}{\omega_{\mathrm{n}}T''_{\mathrm{q}}}\frac{L''_{\mathrm{q}}}{L_{\mathrm{q}}}\frac{L_{\mathrm{aq}}^2}{L_{\mathrm{q}} - L''_{\mathrm{q}}} \tag{5.155}$$

$$L_{\mathrm{f}} = L_{\mathrm{ad}}\left(1 + \frac{L'_{\mathrm{d}} - L_{\mathrm{l}}}{L_{\mathrm{d}} - L'_{\mathrm{d}}}\right) \qquad R_{\mathrm{f}} = \frac{1}{\omega_{\mathrm{n}}T''_{\mathrm{d}}}\frac{L'_{\mathrm{d}}}{L_{\mathrm{d}}}\frac{L_{\mathrm{ad}}^2}{L_{\mathrm{d}} - L'_{\mathrm{d}}}$$

$$H_{\mathrm{j}} = \frac{J\omega_{\mathrm{n}}^2}{2S_{\mathrm{n}}}$$

where $H_{\mathrm{j}}$ – inertia constant,
$J$ – moment of inertia,
$L_{\mathrm{ad}}$, $L_{\mathrm{aq}}$ – mutual inductance in *d*- and *q*-axis,
$L_{\mathrm{d}}$, $L_{\mathrm{q}}$ – synchronous inductance in *d*- and *q*-axis,
$L_{\mathrm{l}}$ – stator leakage inductance,
$L_{\mathrm{f}}$ – field inductance,
$L_{\mathrm{D}}$, $L_{\mathrm{Q}}$ – damping circuit inductance in *d*- and *q*-axis,
$L'_{\mathrm{d}}$, $L'_{\mathrm{q}}$, $L''_{\mathrm{d}}$, $L''_{\mathrm{q}}$ – *d*- and *q*-axis transient and subtransient inductances,
$R_{\mathrm{D}}$, $R_{\mathrm{Q}}$ – damping circuit resistance in *d*- and *q*-axis,
$R_{\mathrm{f}}$ – field-winding resistance,

$T'_d, T'_q, T''_d, T''_q$ – $d$- and $q$-axis transient and subtransient time constants,
$V$ – synchronous generator terminal voltage,
$V_d, V_q$ – $d$- and $q$-axis synchronous generator voltages,
$V_{dS}, V_{qS}$ – $d$- and $q$-axis infinite bus (power system) voltages.

### 5.4.3 Circuit-Oriented Model of the Synchronous Generator

The circuit-oriented model of the synchronous generator is designed for the analysis of the machine operation with the power electronic converter, in which the power electronic system is modeled in a detailed form. Unfortunately, because of the synchronous machine construction (field winding), we do not know the transformation matrix for transforming the natural-axis model into the $\alpha\beta\gamma$-reference frame model (in the same way as in the case of an asynchronous machine). Then the *0dq*-reference frame model of the synchronous generator remains as the only one and is the basic model for the analysis considered.

The power electronic 3-phase system (with 3 connections) needs the 3-connections machine model. To achieve this, the synchronous generator model in the *0dq*-reference frame (described above) has to be supplemented by the *ABC/0dq* ideal transformer. The ideal transformer described by the transformation matrix $\boldsymbol{B}$ (see (5.134)) and by its inverse $\boldsymbol{B}^{-1}$ is presented in Fig. 5.26. Then the circuit-oriented model of the synchronous generator in the $\alpha\beta\gamma$-reference frame takes the form presented in Fig. 5.27.

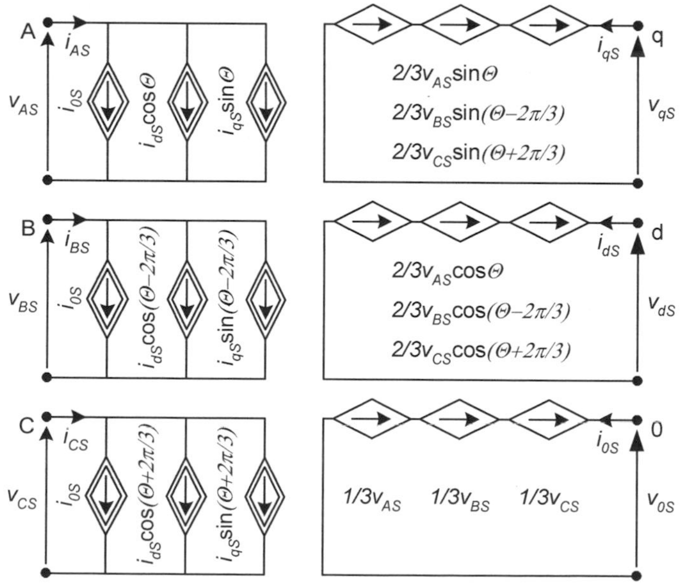


**Fig. 5.26.** Circuit-oriented model of ideal *ABC/0dq* transformer [77, 85] (symbols: parallelogram – controlled voltage source; double line parallelogram – controlled current source)

A method for transforming the field-winding parameters and quantities (voltage and current) to the stator side is defined by the following equations:

$$L'_{\mathrm{f}} = \frac{3}{2}\upsilon^2 L_{\mathrm{f}} \qquad R'_{\mathrm{f}} = \frac{3}{2}\upsilon^2 R_{\mathrm{f}} \tag{5.156}$$

$$i'_{\mathrm{f}} = \frac{2}{3}\frac{i_{\mathrm{f}}}{\upsilon} \qquad v'_{\mathrm{f}} = \upsilon v_{\mathrm{f}} \qquad \upsilon = \frac{N_{\mathrm{S}}}{N_{\mathrm{f}}}\frac{k_{\mathrm{vS}}}{k_{\mathrm{vf}}}, \tag{5.157}$$

where $N_{\mathrm{S}}$ – number of turns on the stator windings,
$N_{\mathrm{f}}$ – number of turns per 2-pole of the field winding,
$k_{\mathrm{vS}}$ – coefficient of the stator winding,
$k_{\mathrm{vf}}$ – coefficient of the field winding.

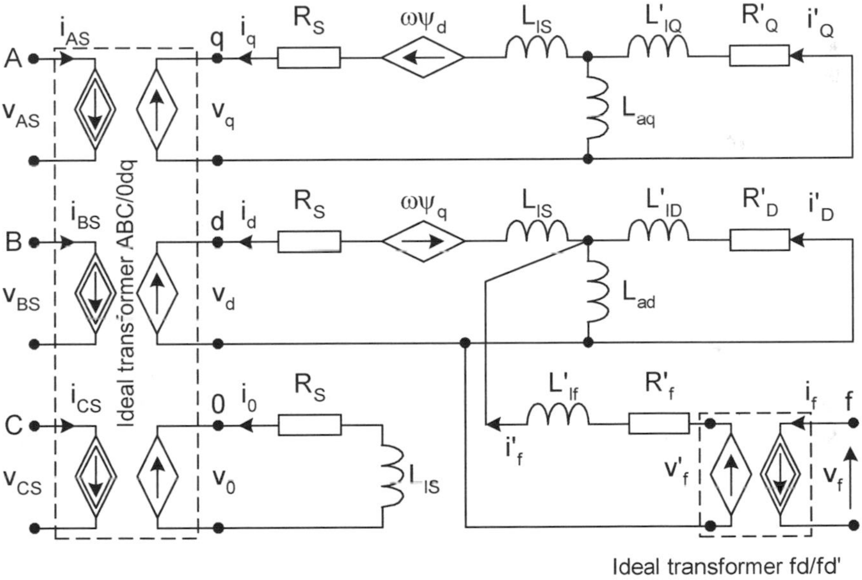


**Fig. 5.27.** Synchronous generator model in the $\alpha\beta\gamma$-reference frame

## 5.5 Converter Modeling

### 5.5.1 Converter Analysis Techniques

The modeling of real objects, phenomena, etc. is nowadays an extremely popular technique of system analysis. The need to use modeling for system analysis arises when:

- there are no possibilities of testing the system (object), i.e. during the design process,
- tests on a real object are expensive or can be dangerous for people or for the object,
- tests cannot be made because of the size of the system (process, object), i.e. when the system is too big or too small,
- the process taking place in the system is too fast or too slow to make the tests in real time,
- there is a need to test the system under various operating conditions, e.g. to simulate accidents which can damage the real system.

The first and the last reasons imply, in particular, the necessity for converter modeling. The modeling usually takes the form of mathematical modeling.

Mathematical modeling of converter systems is realized by using various types of models, which can be broadly divided into two groups: *mathematical functional models* and *mathematical physical models*.

The *functional models* describe the dependence between the input and output signals of the modeled system in the form of a mathematical function. The system is treated as the black box. In fact, the elements making up the system are not modeled separately. The information about the existence of the elements in the system or about the interdependence between them does not exist in an explicit form (this is not necessary). Such types of models are relatively simple, which can be treated as a disadvantage, but they are "fast" when utilized in computation (e.g. time-domain simulation), which is desirable in many applications.

The *physical models* are based on the description of the elements making up the system. Here, the elements exist in an explicit form, and the interdependence between them is described. The level of precision of the element modeling varies and depends on the particular needs. Such types of models are relatively complex. When utilized in computation (e.g. time-domain simulation), the models need the utilization of special mathematical methods.

The models mentioned above (mathematical physical models) can be utilized to make various types of analysis, e.g. time-domain analysis, frequency-domain analysis, etc. The type of analysis usually determines the type and/or structure of the model.

For frequency-domain analysis, the form of the state equation is a useful form of the mathematical model of a converter.

For time-domain analysis, various converter system models can be used, and unlike the models utilized in frequency analysis, it is difficult to point out the best type of modeling. The most popular types of converter system modeling (in conjunction with the simulation method) are the following (Fig. 5.28) [77, 84, 85]:

- Switching events schedule. This is a sequential method where the sequence of states occurring in the converter operation is defined a priori, usually on the basis of quasi-steady state analysis. The model consists of sets of differential equations (or a set of equivalent circuits) that are defined for various states. The moments of the state change, related to the switch state change, are defined.

This method ensures relatively fast simulation, but unfortunately it is not possible to analyze the states and events that have not been considered during the model implementation to simulation.

- Switching events mapping. This is a sequential method where the converter topology is also described as a set of differential equations (or a set of equivalent circuits) that are defined a priori according to specified switching events mapping. The model consists, in comparison to the model/method described above, of sets of equations which describe all possible states – excluding states that are not permissible. Therefore, this type of model permits the analysis of various types of events. The analysis consists of switching between states according to the map of defined states.
- Constant topology. This is a global method where the topology of the converter system is described by a single set of differential equations (or equivalent circuit) that is modified during simulation according to the current state of the switches. In this method, only the coefficients of the equations are modified as a result of changing the switch model, which is here usually defined as a resistance that takes on two values $R_{on}$ (close to zero[7]) and $R_{off}$ (very high). More complex models are also used. Because they introduce a variable resistance (with high change of value), these types of model introduce numerical problems, which require fluency in numerical techniques for solution. These problems can be partly eliminated by modeling the switch as a serial RL element (the reactance here varies from $L_{on}$, which is low, to $L_{off}$, which is high). Moreover, these models are time-consuming because of the need to solve the full set of equations in each integration step. But, on the other hand, these models are easy to implement and allow us to simulate (analyze) various states and events.

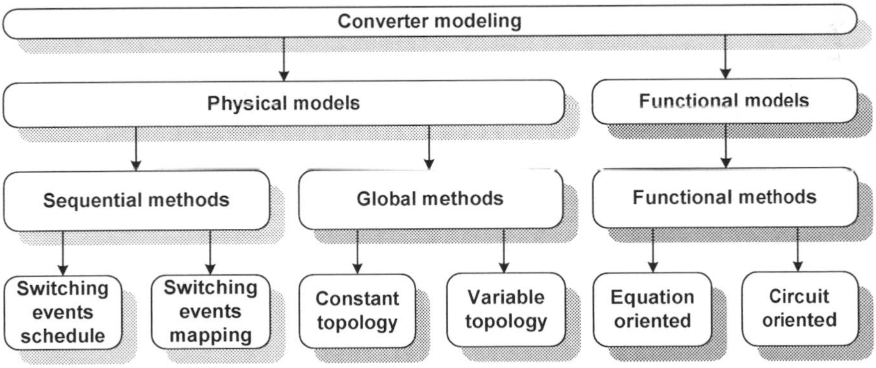


**Fig. 5.28.** Converter system modeling for time-domain analysis

- Variable topology. This is a global method where the topology of the converter system is described by a set of differential equations (or equivalent circuit) that is modified during simulation according to the current state of the switches. In

[7] The subscript *on* defines the on-state of the switch, e.g. thyristor, while the subscript *off* defines the off-state of the switch.

this method, the differential equations are modified as a result of changing the switch model, which is here defined as a resistance that takes on two values $R_{on} = 0$ and $R_{off} = \infty$. Variable topology models are faster than constant topology models because fewer branches and nodes are considered in a given computation step. Unfortunately, the models can use only one idealized type of switch model, which limits the domain of the models/methods utilization.

### 5.5.2 Mathematical Functional Model of Converter

Converters in the wind turbine generator system are utilized for connecting (soft start) and control purposes. In the WTGS with a synchronous generator, the converter realizes both aims while in the WTGS with the a doubly-fed asynchronous machine, the converter is utilized for control purposes only. But in both types of WTGS, the structure of the converter is similar, although the physical realization can vary. A typical structure of the converter is presented in Fig. 5.29. The system consists of a rectifier, an intermediate element and an inverter.

The converter operates connected between the power system element and the machine (synchronous or asynchronous) terminal. In the WTGS with synchronous machine, the rectifier is connected to the machine terminal while the inverter is connected to the step-up transformer. In the WTGS with a doubly-fed asynchronous machine – where the power transferred through the converter can flow in both directions – one side of the converter is connected to the machine rotor terminal and the other side of the converter is connected to the converter's transformer (which connects the converter with the machine terminal).

The intermediate element, which in general is a serial inductor or shunt capacitor, defines the operational characteristics of the converter.

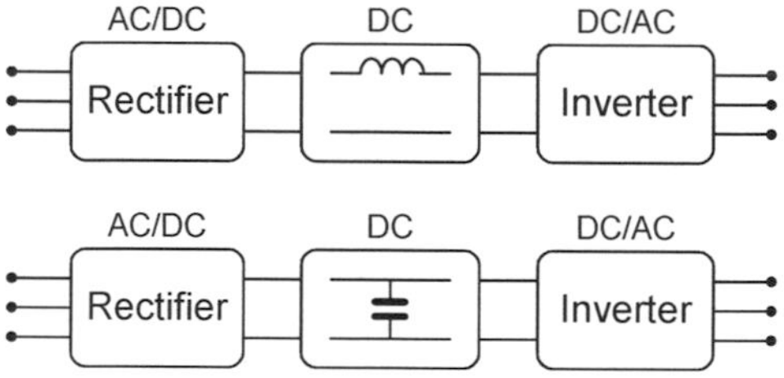


**Fig. 5.29.** Structure of converter system (upper – system with current-source inverter; lower – system with voltage-source inverter)

In general, the rectifier and inverter can have various structures and can be built based on various types of switches, e.g. diodes, thyristors, e.g. classical, GTO (gate turn-off thyristor), MCT (MOS-controlled thyristor), SITh (static induction thyristor), and of transistors, e.g. MOS, IGBT (isolated-gate bipolar transistor), SIT (static induction transistor), etc.

The utilization of separate models of the converter elements, i.e. the rectifier, intermediate element and inverter, is common practice in WTGS converter modeling.

The intermediate element is usually modeled by first-order differential equations. In the case a reactor used as the intermediate element (Fig. 5.30a), the model is defined as

$$V_{\mathrm{drec}} = L_{\mathrm{d}} \frac{\mathrm{d}I_{\mathrm{d}}}{\mathrm{d}t} + R_{\mathrm{d}} I_{\mathrm{d}} + V_{\mathrm{dinv}}, \tag{5.158}$$

where $V_{\mathrm{drec}}$ – rectifier DC-side voltage,
$V_{\mathrm{dinv}}$ – inverter DC-side voltage,
$I_{\mathrm{d}}$ – intermediate-circuit current,
$L_{\mathrm{d}}$, $R_{\mathrm{d}}$ – intermediate-circuit inductance and resistance.

In the case a capacitor used as the intermediate element (Fig. 5.30b), the model is defined as:

$$C_{\mathrm{d}} \frac{\mathrm{d}V_{\mathrm{d}}}{\mathrm{d}t} = I_{\mathrm{drec}} - I_{\mathrm{dinv}}, \tag{5.159}$$

where $V_{\mathrm{d}}$ – rectifier and inverter DC-side voltage,
$I_{\mathrm{drec}}$ – rectifier DC-side current,
$I_{\mathrm{dinv}}$ – inverter DC-side current,
$C_{\mathrm{d}}$ – intermediate-circuit capacitance.

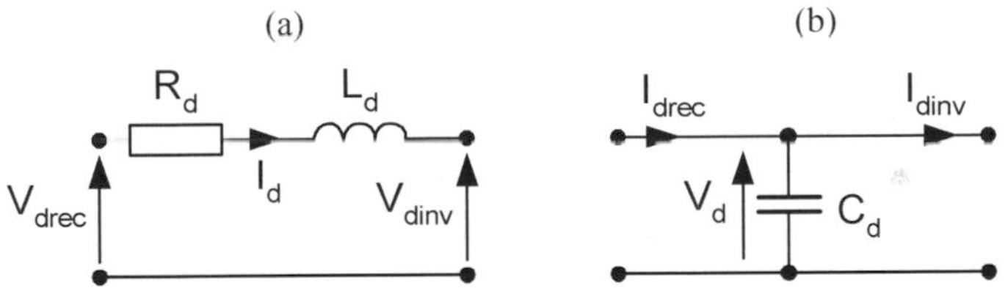


**Fig. 5.30.** Intermediate circuit model

The rectifier and the converter, because of their similar (in general) structure, are modeled in a similar way. For the basic type of thyristor-controlled rectifier (3-phase, 6-pulses) – presented in Fig. 5.31 – (externally commutated), the relationship between input and output is defined as [10, 69, 90]

$$V_{\mathrm{d}} = \frac{3\sqrt{2}}{\pi} V_2 \cos\alpha - \frac{3}{\pi} \omega L_{\mathrm{k}} I_{\mathrm{d}}, \tag{5.160}$$

where $V_{\mathrm{d}}$ – mean value of DC voltage,
$V_2$ – phase-to-phase rms AC feeding voltage,
$\alpha$ – thyristor firing angle,
$I_{\mathrm{d}}$ – mean value of DC current,
$\omega$ – AC voltage frequency in rad/s,
$L_{\mathrm{k}}$ – commutation inductance.

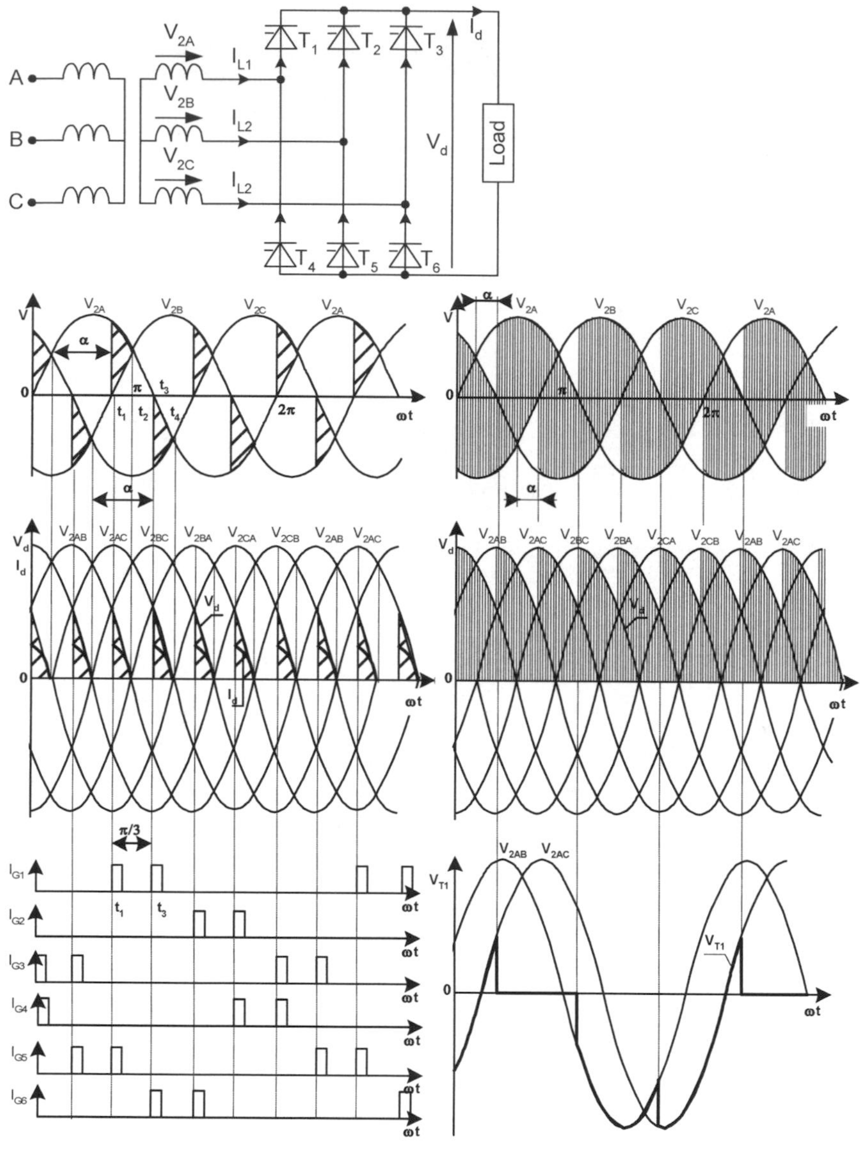


Fig. 5.31. 6-pulse thyristor-controlled rectifier – the structure and the voltages and currents during the rectifier operation

The commutation inductance $L_k$ consists mainly of inductance of the feeding element of the system, i.e. transformer, machine or additional inductor connected to the AC-side of the inverter. The commutation inductance is often approximated by these external inductances. When feeding the rectifier form the transformer or the round-rotor machine, the inductance $L_k$ is equal to the leakage inductance:

transformer or machine, respectively. When feeding the rectifier from the salient-pole machine terminal the inductance is computed as $(X''_d + X''_q)/2\omega$.

Sometimes, the commutation process is neglected and then the voltage drop on the commutation reactance is neglected as well. Then the DC mean voltage of the rectifier is given by

$$V_d = \frac{3\sqrt{2}}{\pi} V_2 \cos\alpha = V_{d0} \cos\alpha \,. \tag{5.161}$$

The amplitude of the basic (first) harmonics of the AC current feeding the rectifier depends on the mean value of the DC current as follows:

$$I_{Lmax} = \frac{2\sqrt{3}}{\pi} I_d \,, \tag{5.162}$$

but the rms current (which is not sinusoidal) in the AC side of the rectifier is equal to:

$$I_{L(rms)} = \sqrt{\frac{2}{3}} I_d \,. \tag{5.163}$$

The first harmonics of AC feeding current lags the feeding voltage $V_1$, and the angle between the voltage and current (neglecting commutation) is proportional to the firing angle:

$$\varphi_1 = \alpha \,. \tag{5.164}$$

In general, in a functional model of a converter, the relation between the AC and DC side currents (neglecting harmonics) can be defined as:

$$I_L = K I_d \,, \tag{5.165}$$

where $K$ is a constant computed by assuming the algebraic identity of the real power in both (AC and DC) sides of the converter.

In the thyristor-controlled type of rectifier, the DC current flows only in one direction, but the power can flow in both directions. When the DC voltage is positive (continuous line in Fig. 5.32), the power flows from the AC side to the DC side of the rectifier. When the DC voltage becomes negative (dashed line), the power flows from the DC side to the AC side of the rectifier – the rectifier operates as an inverter.

The inverter characteristic (dotted and dash-dotted lines) is symmetrical to the rectifier one. The characteristics show that when the DC voltage is positive (dotted line) the converter operates as an inverter. For a negative DC voltage, the converter operates as a rectifier.

In a system consisting of a rectifier, a capacitor and an inverter (voltage inverter in Fig. 5.29), the characteristics presented in Fig. 5.32 for a given $V_d$ enable the rectifier and inverter firing angles to be determined. For example, when $V_d/V_{d0} = 0.5$ (arrows in Fig. 5.32) and when $V_{d0rec} = V_{d0inv}$, the firing angles are equal to $\alpha_{rec} = 60°$, $\alpha_{inv} = 120°$.

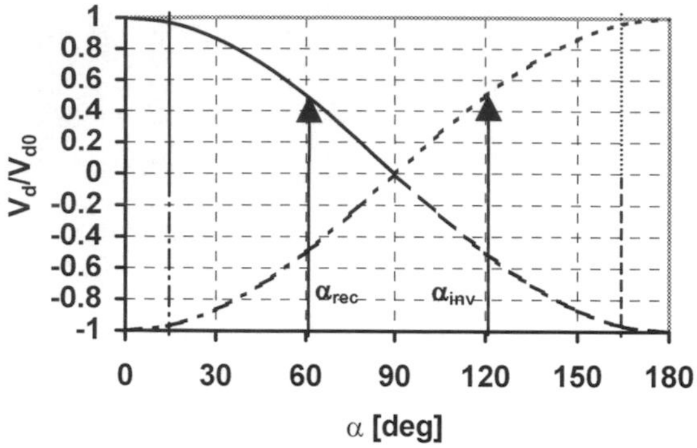


**Fig. 5.32.** Control characteristic of a 6-pulse controlled rectifier and inverter (lines: continuous and dashed – rectifier; dotted and dash-dotted – inverter; perpendicular lines define minimum and maximum values of the rectifier and inverter firing angles)

As shown above, the rectifier and inverter can be modeled in the same way i.e. by utilizing the same equations. The only difference appears when the voltage drop on the commutation inductance is not neglected. In such a case, the relevant sign of the voltage drop (see (5.160)) should be applied.

### 5.5.3 Mathematical Physical Model of Converter

For converter modeling in the form of a mathematical physical model, the constant topology global method is utilized. The model in the form of a circuit for the energy storage element (inductor or capacitor) utilizes the so-called associated discrete circuit model. The nodal-voltage method associated with an integration algorithm is then used for time-domain simulation.

In general, the energy storage elements – inductors L or capacitors C – are described by differential equations, but this is not of course the only way of describing them. The elements L, C here are approximated by the one-port (two-connection) discretized resistive circuit associated with the integration algorithm. Then the model, presented in Fig. 5.33, is referred to as an associated discrete circuit model [84, 85].

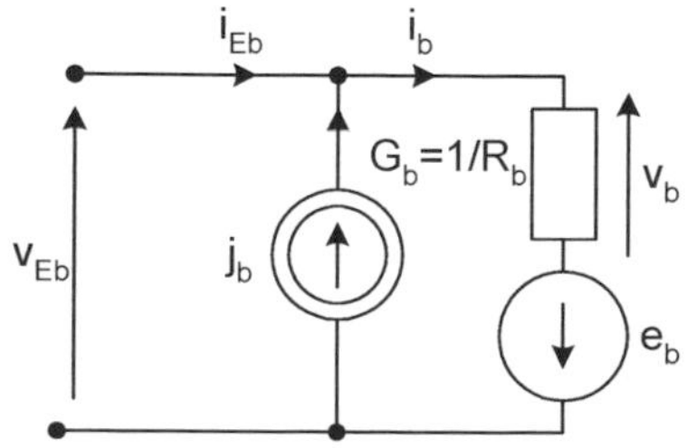


**Fig. 5.33.** Generalized associated discrete circuit model for network energy storage element

The model consists of voltage $e_b$ and current $j_b$ sources and conductance $G_b$ (dependent on inductance L or capacitance C and constant during the integration step $k$). The symbols $v_{Eb}$ and $i_{Eb}$ denote the port voltage and current across an inductor or a capacitor. And the symbols $i_b$ and $v_b$ denote the current across the conductance $G_b$ and voltage $v_b$ on the conductance.

Replacing the energy storage elements L, C in the considered network, we achieve a network with energy sources (voltage or current) and resistances only. Then the time-domain analysis of the dynamic network (RLC) can be replaced by an analysis of the DC network, which is repeated in steps related to the integration steps. This DC network analysis can be made by utilization of various methods. The nodal-voltage method is presented here.

Let us consider a network with $N$ nodes (without the reference node) and with $M < N$ associated branches consisting of associated discrete circuit models (inductors or capacitors). For such a network, the following vectors can be defined:

port (branch) voltages vector:

$$\boldsymbol{v}_{\mathrm{E}}^{T} = \begin{bmatrix} v_{\mathrm{E}1} & v_{\mathrm{E}2} & \dots & v_{\mathrm{E}b} & \dots & v_{\mathrm{E}M} \end{bmatrix} \tag{5.166}$$

conductance voltages vector:

$$\boldsymbol{v}^{T} = \begin{bmatrix} v_1 & v_2 & \dots & v_b & \dots & v_M \end{bmatrix} \tag{5.167}$$

source voltages vector:

$$\boldsymbol{e}^{T} = \begin{bmatrix} e_1 & e_2 & \dots & e_b & \dots & e_M \end{bmatrix} \tag{5.168}$$

port currents vector:

$$\boldsymbol{i}_{\mathrm{E}}^{T} = \begin{bmatrix} i_{\mathrm{E}1} & i_{\mathrm{E}2} & \dots & i_{\mathrm{E}b} & \dots & i_{\mathrm{E}M} \end{bmatrix} \tag{5.169}$$

conductance currents vector:

$$\boldsymbol{i}^{T} = \begin{bmatrix} i_1 & i_2 & \dots & i_b & \dots & i_M \end{bmatrix} \tag{5.170}$$

current sources vector:

$$\boldsymbol{j}^{T} = \begin{bmatrix} j_1 & j_2 & \dots & j_b & \dots & j_M \end{bmatrix}. \tag{5.171}$$

For the branches consisting of resistors, the currents in the associated branches $j_b$ are equal to zero. Then the port currents and voltages are equal to $i_{\mathrm{E}b} = i_b$ and $v_{\mathrm{E}b} = v_b$.

For a set of associated branches, the following equations are true:

$$\boldsymbol{v}_{\mathrm{E}} = \boldsymbol{v} - \boldsymbol{e} \tag{5.172}$$

$$\boldsymbol{i}_{\mathrm{E}} = \boldsymbol{i} - \boldsymbol{j}\,. \tag{5.173}$$

Additionally, the first Kirchhoff law, in the following form, can be introduced for the network:

$$\boldsymbol{A}_{\mathrm{I}}\boldsymbol{i}_{\mathrm{E}} = \boldsymbol{0}\,, \tag{5.174}$$

where $\boldsymbol{A}_{\mathrm{I}} = [a_{nb}]$ is the reduced incidence matrix, where $n = 1...N$ is the node number and $b = 1...M$ is the branch number. The matrix $\boldsymbol{A}_{\mathrm{I}}$ is created from the incidence matrix by removing the line related to the reference node 0. The incidence matrix elements are defined as follows:

$$a_{nb} = \begin{cases} 1 & \text{when current in the } b\text{th branch flows from the } n\text{th node} \\ -1 & \text{when current in the } b\text{th branch flows to the } n\text{th node} \\ 0 & \text{when the } b\text{th branch is not connected to the } n\text{th node} \end{cases} \tag{5.175}$$

Substituting (5.174) in (5.173) we obtain

$$\boldsymbol{A}_{\mathrm{I}}\boldsymbol{i} = \boldsymbol{A}_{\mathrm{I}}\boldsymbol{j}\,. \tag{5.176}$$

Defining the square matrix $\boldsymbol{Y}_{\mathrm{E}}$ of branch conductances, whose elements are equal to

$$y_{ij} = \begin{cases} 0 & \text{when } i \neq j \\ G_{ij} = G_b = 1/R_b & \text{when } i = j \end{cases}, \tag{5.177}$$

we get the relationship between the vector of conductance currents $\boldsymbol{i}$ and the vector of conductance voltages $\boldsymbol{v}$ in the following form:

$$\boldsymbol{i} = \boldsymbol{Y}_{\mathrm{E}}\boldsymbol{v}\,. \tag{5.178}$$

From (5.176) and (5.178) we get:

$$\boldsymbol{A}_{\mathrm{I}}\boldsymbol{Y}_{\mathrm{E}}\boldsymbol{v} = \boldsymbol{A}_{\mathrm{I}}\boldsymbol{j}\,. \tag{5.179}$$

And, after substituting (5.172) in (5.179) we obtain

$$\boldsymbol{A}_{\mathrm{I}}\boldsymbol{Y}_{\mathrm{E}}\boldsymbol{v}_{\mathrm{E}} = \boldsymbol{A}_{\mathrm{I}}(\boldsymbol{j} - \boldsymbol{Y}_{\mathrm{E}}\boldsymbol{e})\,. \tag{5.180}$$

Because the vector of nodal voltages $\boldsymbol{v}_{\mathrm{n}}$ is correlated with the port voltages vector $\boldsymbol{v}_{\mathrm{E}}$ by the reduced incidence matrix:

$$\boldsymbol{v}_{\mathrm{E}} = \boldsymbol{A}_{\mathrm{I}}^{T}\boldsymbol{v}_{\mathrm{n}}\,, \tag{5.181}$$

then, after substituting (5.181) in (5.180) we get

$$\boldsymbol{A}_{\mathrm{I}}\boldsymbol{Y}_{\mathrm{E}}\boldsymbol{A}_{\mathrm{I}}^{T}\boldsymbol{v}_{\mathrm{n}} = \boldsymbol{A}_{\mathrm{I}}(\boldsymbol{j} - \boldsymbol{Y}_{\mathrm{E}}\boldsymbol{e})\,. \tag{5.182}$$

Equation (5.182) is the basic nodal equation, which is usually shown in the following form:

$$\boldsymbol{Y}_{\mathrm{n}}\boldsymbol{v}_{\mathrm{n}} = \boldsymbol{j}_{\mathrm{n}}\,, \tag{5.183}$$

where the vector of nodal currents is given by

$$\boldsymbol{j}_{\mathrm{n}} = \boldsymbol{A}_{\mathrm{I}}(\boldsymbol{j} - \boldsymbol{Y}_{\mathrm{E}}\boldsymbol{e}) \tag{5.184}$$

and the matrix of admittances is given by

$$\boldsymbol{Y}_{\mathrm{n}} = \boldsymbol{A}_{\mathrm{I}}\boldsymbol{Y}_{\mathrm{E}}\boldsymbol{A}_{\mathrm{I}}^{T} . \tag{5.185}$$

The admittance matrix[8] $\boldsymbol{Y}_{\mathrm{n}}$ is usually computed directly (without utilizing the incidence matrix $\boldsymbol{A}_{\mathrm{I}}$) and its elements are defined as follows:

$$\begin{aligned} Y_{ij} &= -G_{ij} \\ Y_{ii} &= \sum_{i=0}^{N} G_{ij} , \end{aligned} \tag{5.186}$$

where $G_{ij}$ is the conductance of the branch between nodes $i$ and $j$, and $G_{ii}$ is the conductance of the branch between node $i$ and the reference node 0.

The final equations that are utilized in the considered nodal-voltage method, are (5.183) in the following form:

$$\boldsymbol{v}_{\mathrm{n}} = \boldsymbol{Y}_{\mathrm{n}}^{-1}\boldsymbol{j}_{\mathrm{n}} \tag{5.187}$$

and (5.184), where the elements of the admittance matrix are computed by using (5.186).

The nodal voltage method combined with the integration algorithm enables us to solve the dynamic network (system) in the time domain as a DC network in time steps that are related to the assumed step of the integration method.

The procedure for implementing the integration algorithm in the associated model of the energy storage element is – as follows assuming the utilization of Euler's interpolation algorithm.

For a linear inductance L, the voltage as a function of current is described by

$$v_{\mathrm{L}}(t) = L\frac{\mathrm{d}i_{\mathrm{L}}(t)}{\mathrm{d}t} \tag{5.188}$$

or

$$\frac{\mathrm{d}i_{\mathrm{L}}(t)}{\mathrm{d}t} = \frac{1}{L}v_{\mathrm{L}}(t) . \tag{5.189}$$

Utilizing Euler's interpolating algorithm, defined as

$$x_{k+1} = x_k + hf(x_{k+1}, t_{k+1}) , \tag{5.190}$$

equation (5.189) (neglecting subscript L) can be converted into

$$i_{k+1} = i_k + h\frac{1}{L}v_{k+1} \tag{5.191}$$

[8] The network consists of conductance-type branches only (except for current and voltage sources).

where $i_{k+1}$, $i_k$ – inductance currents in steps $k+1$ and $k$,
$h$ – integration step,
$v_{k+1}$ – inductance voltage in step $k+1$.

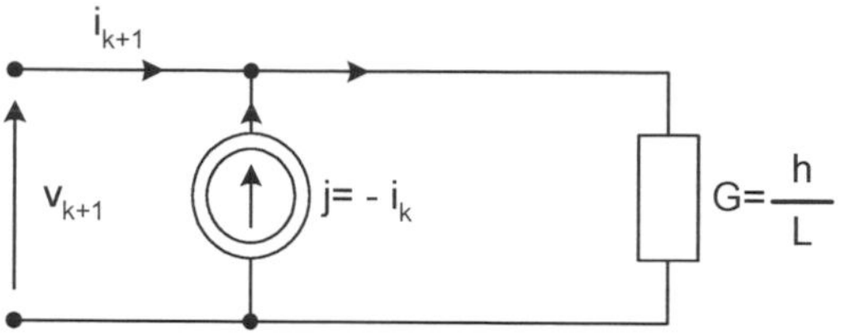


**Fig. 5.34.** Linear inductance model associated with Euler's interpolation algorithm

Comparing (5.191) with the model presented in Fig. 5.33 and (5.173) (when $e = 0$), we obtain the values of the current source $j$ and the conductance $G$, respectively, equal to $j = -i_k$ and $G = h/L$. Then the associated model of linear inductance combined with the Euler's interpolation algorithm takes the form as shown in Fig. 5.34.

For a linear capacitance C, the current as a function of voltage is described by

$$i_C(t) = C\frac{dv_C(t)}{dt} \tag{5.192}$$

or

$$\frac{dv_C(t)}{dt} = \frac{1}{C}i_C(t). \tag{5.193}$$

Utilizing Euler's interpolating algorithm (5.190), (5.193) (neglecting subscript C) can be converted into

$$v_{k+1} = v_k + h\frac{1}{C}i_{k+1}. \tag{5.194}$$

Rearranging (5.194), we obtain the final form of the associated model of capacitance:

$$i_{k+1} = -\frac{C}{h}v_k + \frac{C}{h}v_{k+1}. \tag{5.195}$$

Comparing (5.195) with the model presented in Fig. 5.33 and (5.173) (when $e = 0$) we obtain the values of the current source $j$ and the conductance $G$ according to, respectively:

$$j = \frac{C}{h}v_k, \qquad G = \frac{C}{h}. \tag{5.196}$$

Then the associated model of linear capacitance combined with Euler's interpolation algorithm takes the form shown in Fig. 5.35.

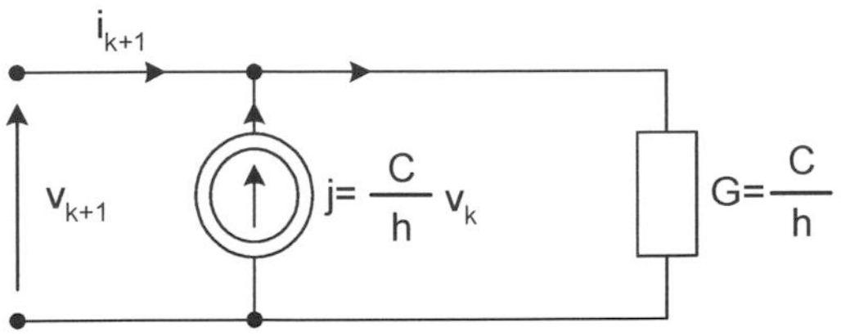


**Fig. 5.35.** Linear capacitance model associated with Euler's interpolation algorithm

It is also possible to combine RL (inductor with serial resistor) and RC (capacitor with shunt resistor) elements in one associated branch model. This operation enables us to reduce the number of associated branches and in the same way reduces the number of nodes and simultaneously reduces the dimension of the impedance matrix.

For a linear RL element, the voltage as a function of current is described by

$$v(t) = L\frac{\mathrm{d}i(t)}{\mathrm{d}t} + Ri(t) \tag{5.197}$$

or

$$\frac{\mathrm{d}i(t)}{\mathrm{d}t} = \frac{1}{L}v(t) - \frac{R}{L}i(t)\,. \tag{5.198}$$

Utilizing Euler's interpolating algorithm (5.190), (5.198) can be converted into

$$i_{k+1} = i_k + h(\frac{1}{L}v_{k+1} - \frac{R}{L}i_{k+1})\,. \tag{5.199}$$

Rearranging (5.199), we obtain the final form of the associated model:

$$i_{k+1} = \frac{L}{L+hR}i_k + \frac{h}{L+hR}v_{k+1}\,. \tag{5.200}$$

Comparing (5.195) with the model presented in Fig. 5.33 and (5.173) (when $e = 0$) we obtain the values of the current source $j$ and the conductance $G$ as, respectively:

$$j = -\frac{L}{L+hR}i_k\,, \qquad G = \frac{h}{L+hR}\,. \tag{5.201}$$

Therefore, the associated model of linear inductance with serial resistance combined with Euler's interpolation algorithm takes the form shown in Fig. 5.36.

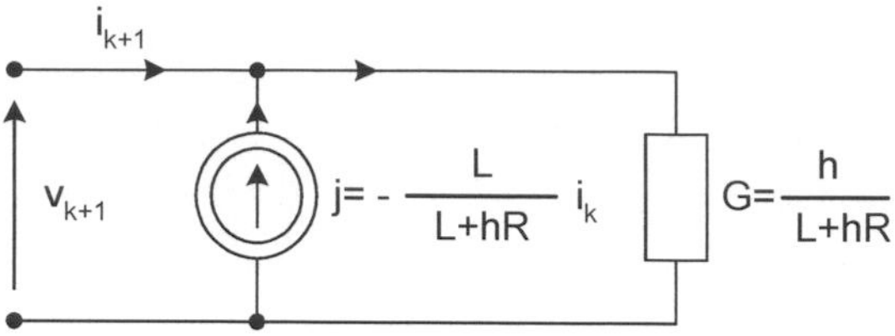


**Fig. 5.36.** Linear impedance RL model associated with Euler's interpolation algorithm

Like the above, for a linear capacitance with shunt resistance, the current as a function of voltage is described by

$$i(t) = C\frac{\mathrm{d}v(t)}{\mathrm{d}t} + \frac{1}{R}v(t) \tag{5.202}$$

or

$$\frac{\mathrm{d}v(t)}{\mathrm{d}t} = \frac{1}{C}i(t) - \frac{1}{RC}v(t)\,. \tag{5.203}$$

Utilizing Euler's interpolating algorithm (5.190), (5.203) can be converted into

$$v_{k+1} = v_k + h(\frac{1}{C}i_{k+1} - \frac{1}{RC}v_{k+1})\,. \tag{5.204}$$

Rearranging (5.204), we obtain the final form of the associated model of capacitance with shunt reactance:

$$i_{k+1} = -\frac{C}{h}v_k + \frac{h+RC}{hR}v_{k+1}\,. \tag{5.205}$$

Comparing (5.205) with the model presented in Fig. 5.33 and (5.173) (when $e = 0$) we obtain the following values of the current source $j$ and the conductance $G$, respectively:

$$j = \frac{C}{h}v_k\,, \qquad G = \frac{h+RC}{h}\,. \tag{5.206}$$

Therefore, the associated model of linear capacitance with shunt reactance combined with the Euler's interpolation algorithm takes the form shown in Fig. 5.37.

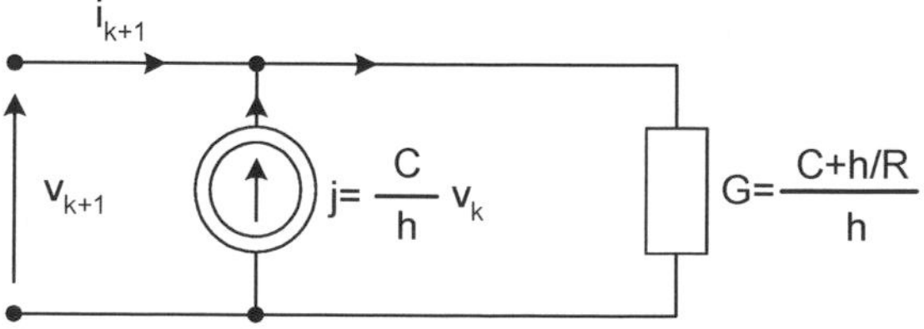


**Fig. 5.37.** Linear impedance RC model associated with Euler's interpolation algorithm

Generally, the differential equations (5.188) and (5.192) describing the energy storage elements can be solved utilizing any other polynomial-approximation algorithm of the type:

$$x_{k+1} = \sum_{i=0}^{I}[a_i x_{k-i} + hb_i f(x_{k-i}, t_{k-i})] + hb_{-1} f(x_{k+1}, t_{k+1}) . \tag{5.207}$$

For such an interpolation algorithm, the linear model of impedance is described by the conductance $G$ and the source current $j$, which are defined as follows:

$$G = \frac{hb_{-1}}{L} \tag{5.208}$$

$$j = -\sum_{i=0}^{I}(a_i i_{k-i} + \frac{h}{L} b_i v_{k-i}) , \tag{5.209}$$

while the linear model of capacitance describes the conductance $G$ and the current source $j$ defined as follows:

$$G = \frac{C}{hb_{-1}} \tag{5.210}$$

$$j = \frac{C}{hb_{-1}} \sum_{i=0}^{I}(a_i v_{k-i} + \frac{h}{C} b_i i_{k-i}) . \tag{5.211}$$

For example, for Gear's second-order interpolation algorithm:

$$x_{k+1} = \frac{4}{3} x_k - \frac{1}{3} x_{k-1} + \frac{2}{3} hf(x_{k+1}, t_{k+1}) \tag{5.212}$$

the coefficients $a_i$ and $b_i$ are equal to $a_0$ = 4/3, $a_1$ = -1/3, $b_{-1}$ = 2/3, $b_0$ = 0 and $b_1$ = 0. Therefore, the parameters of the inductance model associated with Gear's second-order interpolation algorithm are given by

$$G = \frac{2h}{3L} \tag{5.213}$$

$$j = -\frac{4}{3} i_k + \frac{1}{3} i_{k-1} \tag{5.214}$$

and the parameters of the capacitance model are given by

$$G = \frac{3C}{2h} \tag{5.215}$$

$$j = \frac{C}{2h}(4v_k - v_{k-1}) . \tag{5.216}$$

The organization of the time-domain process simulation based on a typical network (see Fig. 5.38) is presented below. The network consists of a voltage source $e(t) = E_{max}\sin\omega t$, a feeding system consisting of a line with parameters $R_1$ and $L_1$, a thyristor T and a load comprising elements $R_0$, $L_0$ and $C_0$. The thyristor is described by two conductances related to passing and blocking states $G_T = \{G_{on}, G_{off}\}$.

When modeling the system, in the first step, the AC network from Fig. 5.38 should be transformed into the DC network as a result of replacing inductances and capacitance by suitable discrete circuit models. By choosing models associated with Gear's interpolation algorithm, we achieve the DC equivalent of the considered AC network in the form presented in Fig. 5.39.

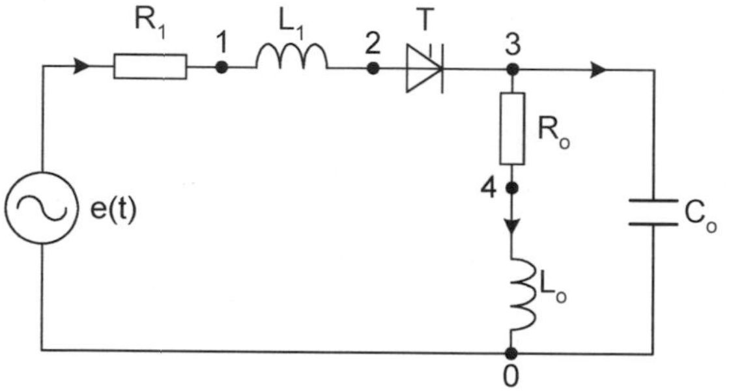


**Fig. 5.38.** Example of a simple system with rectifier ($E_{max}$ = 100 V, $\omega$ = 314.1 rad/s, $R_1$ = 0.1 Ω, $L_1$ = 1 mH, $G_T = \{10^5, 10^{-5}\}$ S, $R_0$ = 100 Ω, $L_0$ = 0.5 H, $C_0$ = 10 μF)

For such a model, the impedance matrix $\boldsymbol{Y}_n$ in $k$th step of simulation (the values of the matrix coefficient can change when the switching element (thyristor) changes state) is defined as follows:

$$\boldsymbol{Y}_{n(k)} = \begin{bmatrix} G_1 + \dfrac{h}{L_1} & -\dfrac{h}{L_1} & 0 & 0 \\ -\dfrac{h}{L_1} & G_1 + \dfrac{h}{L_1} & -G_{T(k)} & 0 \\ 0 & -G_{T(k)} & G_{T(k)} + \dfrac{h}{L_0} + \dfrac{C_0}{h} & -\dfrac{h}{L_0} \\ 0 & 0 & -\dfrac{h}{L_0} & G_0 + \dfrac{h}{L_0} \end{bmatrix}. \tag{5.217}$$

The vector of nodal currents for the model at the $k$th step is given by

$$\boldsymbol{J}^T_{n(k)} = \begin{bmatrix} G_1 e_{(kh)} - i_{L_1(k)} & i_{L_1(k)} & -i_{L_0(k)} + \dfrac{C_0}{h} v_{3(k)} & i_{L_0(k)} \end{bmatrix}. \tag{5.218}$$

The nodal voltages are computed by utilizing the nodal equation:

$$\boldsymbol{v}_{n(k+1)} = \boldsymbol{Y}^{-1}_{n(k)} \boldsymbol{j}_{n(k)}, \tag{5.219}$$

where

$$v_{n(k+1)}^{T} = \left[ v_{1(k+1)} \quad v_{2(k+1)} \quad v_{3(k+1)} \quad v_{4(k+1)} \right] \tag{5.220}$$

is the nodal voltage vector. Then the currents across the inductances and capacitance are computed by utilizing the following formulae:

$$i_{L_1(k+1)} = i_{L_1(k)} + \frac{h}{L_1}\left(v_{1(k+1)} - v_{2(k+1)}\right) \tag{5.221}$$

$$i_{L_0(k+1)} = i_{L_0(k)} + \frac{h}{L_0}\left(v_{3(k+1)} - v_{4(k+1)}\right) \tag{5.222}$$

$$i_{C_0(k+1)} = \frac{C_0}{h}\left(v_{3(k+1)} - v_{3(k)}\right) \tag{5.223}$$

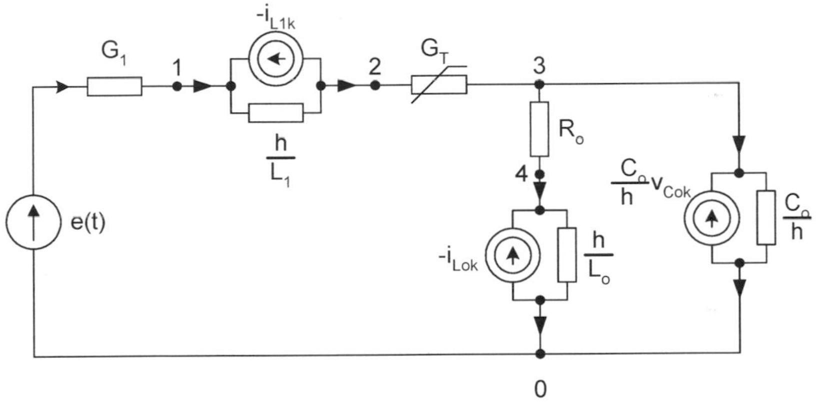


**Fig. 5.39.** Model of a typical rectifier system associated with Euler's interpolation algorithm

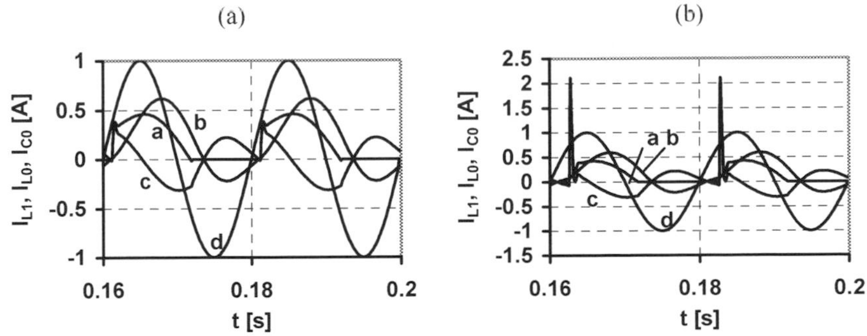


**Fig. 5.40.** Currents in the analyzed network. Thyristor firing angle: (a) $\alpha = 0$; (b) $\alpha = \pi/4$. Lines: a – $i_{L1}$; b – $i_{L0}$; c – $i_{C0}$; d – $0.01e(t)$.

As an example of computation, the currents in the considered network for two values of the thyristor firing angle are presented in Fig. 5.40. The computation has been made for the step length $h = 0.0002$ s, which gives 100 samples in a single period of the feeding voltage $e(t)$. The initial state of the system, i.e. the inductor current and capacitor voltage, have been set equal to zero.

The circuit-oriented model can be also described by a set of equations in the form of state equations [48, 49]:

$$\begin{aligned} \boldsymbol{E}\dot{\boldsymbol{x}} &= \boldsymbol{A}\boldsymbol{x} + \boldsymbol{B}\boldsymbol{u} \\ \boldsymbol{y} &= \boldsymbol{C}\boldsymbol{x} + \boldsymbol{D}\boldsymbol{u}\,. \end{aligned} \tag{5.224}$$

In such a case, it is useful to choose the inductor current and the capacitor voltage as the state variables. For the network in Fig. 5.38, the inductor currents $i_{\mathrm{L1}}$, $i_{\mathrm{L0}}$ and the capacitor voltage $v_{\mathrm{C0}}$ should be defined as the state variables. Then the state vector $\boldsymbol{x}$ and matrices $\boldsymbol{E}$, $\boldsymbol{A}$, $\boldsymbol{B}$ take the following form:

$$\begin{gathered} \boldsymbol{x}^T = \begin{bmatrix} i_{\mathrm{L1}} & i_{\mathrm{L0}} & v_{\mathrm{C0}} \end{bmatrix} \quad \boldsymbol{u} = \begin{bmatrix} e(t) \end{bmatrix} \\ \boldsymbol{E} = \begin{bmatrix} L_1 & L_0 & 0 \\ 0 & L_0 & 0 \\ 0 & 0 & C_0 \end{bmatrix} \quad \boldsymbol{A} = \begin{bmatrix} -R_1 - R_{\mathrm{T}} & -R_0 & 0 \\ 0 & -R_0 & 1 \\ 1 & -1 & 0 \end{bmatrix} \quad \boldsymbol{B} = \begin{bmatrix} 1 \\ 0 \\ 0 \end{bmatrix}. \end{gathered} \tag{5.225}$$

Equation (5.224) can be solved by using any known method suitable for the state equation defined models. Matrices $\boldsymbol{C}$ and $\boldsymbol{D}$ depend on the information that should be needed for switching operation (the thyristor state and the resistance $R_{\mathrm{T}}$ change) and should, for example, give voltages in the nodes of the network. For example, the nodal voltages can be computed as follows:

$$\boldsymbol{v} = \boldsymbol{F}\dot{\boldsymbol{x}} + \boldsymbol{G}\boldsymbol{x} + \boldsymbol{H}\boldsymbol{u}\,, \tag{5.226}$$

where the matrices $\boldsymbol{F}$, $\boldsymbol{G}$, $\boldsymbol{H}$ are given by

$$\boldsymbol{F} = \begin{bmatrix} 0 & 0 & 0 \\ -L_1 & 0 & 0 \\ 0 & 0 & 0 \\ 0 & 0 & 0 \end{bmatrix} \quad \boldsymbol{G} = \begin{bmatrix} -R_1 & 0 & 0 \\ -R_1 & 0 & 0 \\ 0 & 0 & 1 \\ 0 & R_0 & 0 \end{bmatrix} \quad \boldsymbol{H} = \begin{bmatrix} 1 \\ 1 \\ 0 \\ 0 \end{bmatrix}. \tag{5.227}$$

In some cases, e.g. for more complex networks, it can be useful to model the switches as a serial resistance $R_{\mathrm{T}} = \{R_{\mathrm{on}}, R_{\mathrm{off}}\}$ and reactance $L_{\mathrm{T}} = \{L_{\mathrm{on}}, L_{\mathrm{off}}\}$ [85]. This approach increases the state vector size, but simultaneously the switch currents can become the state variables with all of the consequences, e.g. direct computation of the variables, decrease (or elimination) of the numerical problems during switching operations, etc.

As a recapitulation of the considerations related to converter modeling, the circuit-oriented model of a typical converter with a voltage-source inverter (typical for modern WTGSs) is presented in Fig. 5.41.

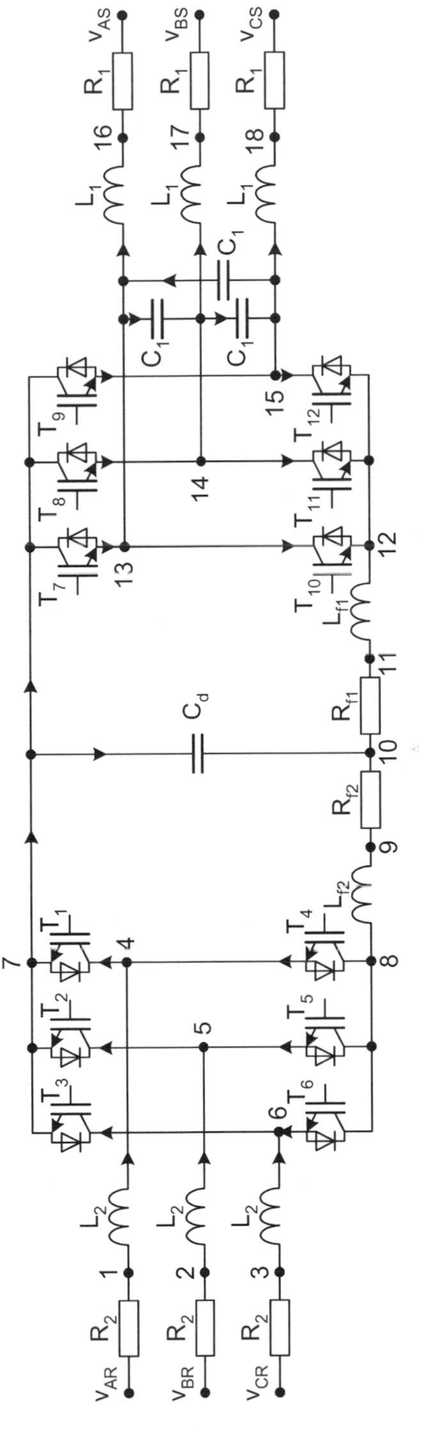


**Fig. 5.41.** Example circuit-oriented model of a WTGS converter

The left-hand side (in the figure) of the converter is connected to the asynchronous machine rotor terminal (Fig. 5.22) or to the synchronous machine stator terminal (Fig. 5.25), depending on the WTGS machine type. The right-hand side of the converter in the case of a WTGS equipped with asynchronous generator is connected to the converter transformer, while in the case of a WTGS equipped with synchronous generator it is connected to the step-up transformer.

## 5.6 Control System Modeling

In the case of classical sources of power (i.e. synchronous generators) driven by steam or water turbines, control system modeling is relatively simple because the objects are well known and the control system structures are standardized. The IEEE standard models are widely utilized here in the power system stability analyses. Even when the structure of the given controller (e.g. voltage controller) is different from that defined by the standard model – when it is possible – for many analytical purposes, the controller parameters are converted to fit the standard IEEE model structure.

Utilisation of the standard models is useful because, among other things, it enables exchange of the model data between transmission companies and researchers, and thus provides opportunities for comparing the considered systems and their control algorithms.

This is possible mainly because the power systems with classical sources of power "exist" for a long time. Additionally, the requirements related to the power system safety (e.g. stability) have forced the producers to disclose some data for their products. Therefore, the parameters of many objects (transmission lines, transformers, generators, turbines and their controllers) are relatively easily available.

In the case of the considered wind turbine generator systems, it is quite the opposite, i.e. the data necessary for WTGS modeling are not easily available – especially for researchers who do not cooperate with WTGS manufacturers. The main reason is that the considered objects (especially wind generators equipped with converters) are relatively new constructions and the competition among producers is high, the result being that the control system descriptions of the existing WTGS almost do not exist.

The researchers who study and model WTGS operation usually model the controllers in an extremely simple way, which can cause not only quantitative but also qualitative differences between the real objects and the models. The other group of researchers develop their "own" controllers and study the WTGSs equipped with such controllers.

The situation is not satisfactory for the power system, whose safe operation is one of the basic requirements. Therefore, the standardization of WTGS models and the availability of the model data is a must.

### 5.6.1 General Idea of Control

Wind turbine generator systems are designed to convert the energy of the wind into electrical energy. The more effective the conversion process is the better is the WTGS – of course, the cost aspect also influences the WTGS evaluation. It means that the main aim of a WTGS is to capture the maximum (allowable in certain weather circumstances) amount of energy, at the same time providing for the safety of the WTGS construction.

This can be achieved by proper design of the object (especially blades) and by the utilization of properly designed control systems. For WTGSs utilizing the stall-type of control system and asynchronous generators directly connected to the electric power system the above aim is achieved only by proper design. In the case of other types of WTGSs, the control systems actively participate in the process of wind energy conversion.

The WTGS control system, in general, is a hierarchic, two-level system (Fig. 5.42). A supervisory control system (also called a management system) is located at the main level. The lower level of control is the power and speed control system(s).

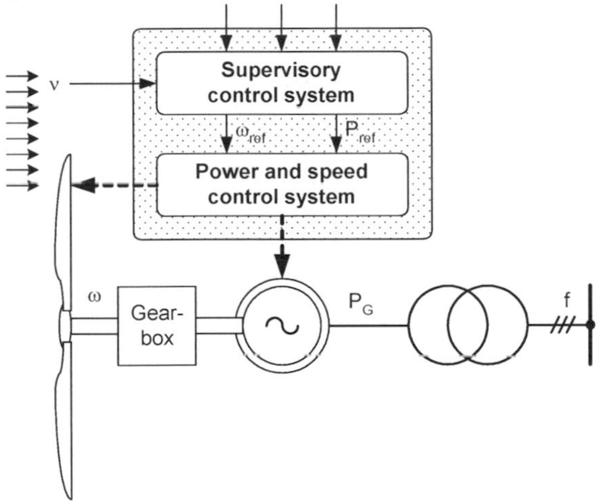


**Fig. 5.42.** WTGS as an object of control

The supervisory control system, on the basis of various measured data, generates:

- reference values for the power and speed control system,
- sequentially controls signals, making it possible for the WTGS to pass from one operating state to another,
- sequentially checks the WTGS's state, performing protective functions.

The power and speed control system can be considered as two systems: while the first acts on the wind wheel (turbine), the other acts on the generator (in practice, on the converter).

The first system controls the blade pitch angle (the speed and power are the regulated parameters) and its purpose is to:

- capture as much energy as possible in given weather conditions,
- protect the rotor, the generator and the power electronic equipment from overloading in high wind,
- protect the mechanical part of the WTGS after load shedding (the impetuous increase of the rotor speed takes place then).

The second system controls the torque (the power, currents, etc. are the regulated parameters) and its purpose is to smoothen the WTGS power output and (sometimes) to damp electromechanical oscillations.

Generally, two basic schemes of WTGS control systems can be considered:

- Constant tip-speed ratio scheme. This is based on the fact that the maximum energy is extracted from the wind when the optimum tip-speed ratio is achieved by the WTGS. The rotor power characteristic $c_P(\lambda, \vartheta)$ of the wind turbine is stored in the control system memory. Wind velocity and rotor speed are continuously measured. Then the tip-speed ratio is computed and compared with the optimal value (and reference) $\lambda_{opt}$. The error signal is fed to the control system, which changes the turbine speed to minimize the error. The main disadvantage of the scheme is inaccuracy of the wind velocity measurement while the measurement system is located on the nacelle (close behind the wind wheel in the up-wind WTGS). Additionally, the characteristic $c_P(\lambda, \vartheta)$ changes considerably as a result of the blade surface change. Nevertheless, in practice, such a scheme of control is applied to present WTGSs.
- Maximum point power tracking (MPPT) scheme. This is based on the fact that the power versus speed curve has a single well-defined peak $dP/d\omega = 0$. During WTGS operation, the rotor speed is increased or decreased in small increments, while the power is continuously measured. If $dP/d\omega > 0$, then the rotor speed is further increased. If $dP/d\omega < 0$, then the rotor speed is further decreased. And finally, if $dP/d\omega \approx 0$, then the rotor speed is not changed – the maximum power is extracted from the wind. This scheme is insensitive to the errors in the wind velocity measurement and to the change of the blade characteristic (e.g. blades covered by dirt or ice). The control scheme is also utilized in modern WTGSs.

Both control schemes are utilized by WTGS control systems at the turbine partial load only (Fig. 5.43). WTGS partial load is a state of operation with the wind velocity $v$ located between the cut-in speed $v_{cut\text{-}in}$ and the rated speed $v_n$ (usually 3–5 m/s < $v$ < 12–15 m/s). The rated wind speed means the wind velocity at which the wind turbine generates rated power. The second possible state of operation in a power system is a full-load state. The wind turbine operates at the full load when the wind velocity $v$ is located between the rated speed $v_n$ and the cut-off speed $v_{cut\text{-}off}$ (12–15 m/s < $v$ < 25 m/s).

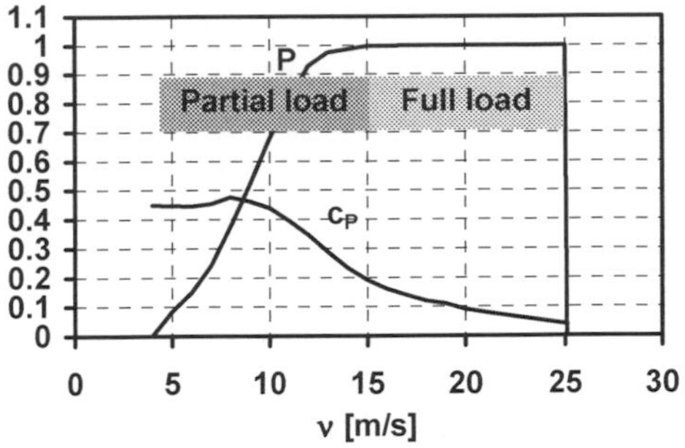


**Fig. 5.43.** Power versus wind velocity characteristic (the real power $P$ and rotor power coefficient $c_P$ data in the wind speed range 4 m/s < v < 16 m/s are for a Vestas V80-2.0 MW turbine)

The control system structure depends also on the WTGS type. For constant speed systems (but with pitch control), the control system consists of a turbine controller only, while for variable speed systems it consists of a turbine and generator controllers. The types of control systems utilized by various types of WTGS are presented in Table 5.3.

In some cases, separate speed controllers are utilized for start-up, shut-down and idle-operation purposes.

In the case of WTGSs utilizing the constant tip-speed ratio scheme, the control algorithm performed depends on the current values of the wind velocity and the turbine rotor speed. The generator control algorithm in graphical form is presented in Fig. 5.44, and for the given operating point is defined as [53]:

- At partial load.
  (a) When the wind velocity $v$ is lower than the value equal to $v_1$, but higher than the cut-in speed and defined as:

$$v < v_1 = \frac{\omega_{\min} R}{\lambda_{\mathrm{opt}}}, \tag{5.228}$$

where $\omega_{\min}$ – minimum value of rotor speed,
$\lambda_{\mathrm{opt}}$ – optimum value of the tip-speed ratio (controlled quantity),
$R$ – wind wheel radius,
the controller keeps the rotor speed equal to the minimum:

$$\omega_{\mathrm{ref}} = \omega_{\min} \,. \tag{5.229}$$

  (b) When the wind velocity fulfils the condition:

$$v_1 \le v < v_2 = \frac{\omega_{\max} R}{\lambda_{\mathrm{opt}}}, \tag{5.230}$$

where $\omega_{\max}$ is the maximum value of the rotor speed (usually equal to the rated value $\omega_n$), and the controller keeps the optimum value of the tip-speed ratio:

**Table 5.3.** Control systems of various types of WTGS

| Turbine control | Generator control | |
|---|---|---|
| | Constant speed systems<br>($\omega$ = const) | Variable speed systems<br>($\omega$ = var) |
| Stall control<br>(power limiting) | 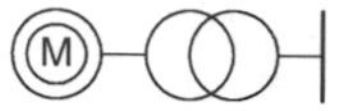<br>Without control | - |
| Pitch control<br><br>(Partial load)<br>Power control<br>$P_{\mathrm{ref}} = P(\frac{\mathrm{d}P}{\mathrm{d}\omega} = 0)$<br>or blade pitch<br>angle control<br>$\vartheta_{\mathrm{ref}} = \vartheta_{\mathrm{opt}}(\omega)$<br>or constant blade<br>pitch angle, e.g.<br>$\vartheta_{\mathrm{ref}} = 0$<br><br>(Full load)<br>Speed control<br>$\omega_{\mathrm{ref}} = \omega_{\mathrm{n}}$ | 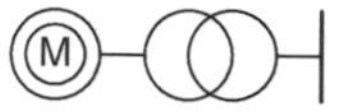<br>Without control | 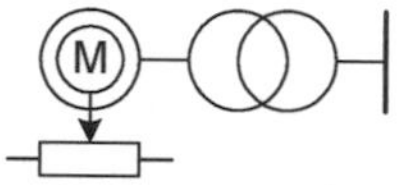<br>Power control<br>$P_{\mathrm{ref}} = \begin{cases} P(\omega) \text{ partial load} \\ P_{\mathrm{n}} \quad \text{full load} \end{cases}$<br>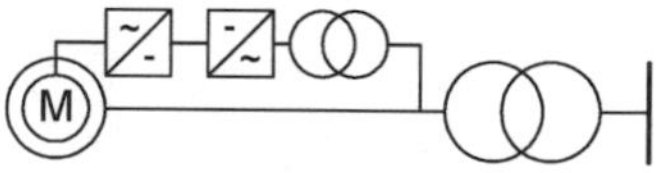<br>Speed control (partial load)<br>$\omega_{\mathrm{ref}} = \omega(P)$<br>Power control (full load)<br>$P_{\mathrm{ref}} = P_{\mathrm{n}}$<br>Reactive power control $Q_{\mathrm{ref}} = 0$<br>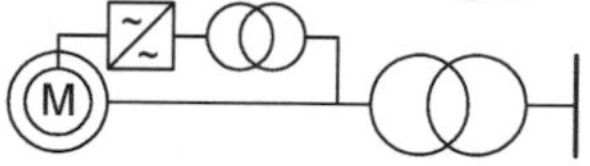<br>Speed control (partial load)<br>$\omega_{\mathrm{ref}} = \omega(P)$<br>Power control (full load)<br>$P_{\mathrm{ref}} = P_{\mathrm{n}}$<br>Reactive power control $Q_{\mathrm{ref}} = 0$ |
| | 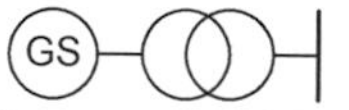<br>Voltage control $V_{\mathrm{ref}}$ | 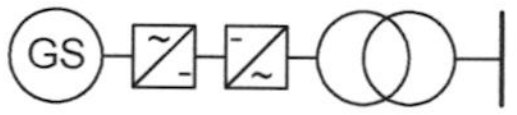<br>Generator voltage control $V_{\mathrm{ref}}=f(\omega)$<br>Converter power control<br>$P_{ref} = \begin{cases} P(\omega) \text{ partial load} \\ P_{\mathrm{n}} \quad \text{full load} \end{cases}$<br>Converter reactive power $Q_{\mathrm{ref}}$ or<br>power factor $\cos\varphi_{\mathrm{ref}}$ control |

$$\lambda = \lambda_{\text{opt}} , \qquad (5.231)$$

which means that the rotor speed reference is given by

$$\omega_{\text{ref}} = \frac{v\lambda_{\text{opt}}}{R} . \qquad (5.232)$$

(c) When the wind velocity is higher than $v_2$, then the generator controller keeps maximum (rated) value of the rotor speed[9]:

$$\omega_{\text{ref}} = \omega_{\max} . \qquad (5.233)$$

- At full load. The generator controller keeps the rotor speed constant, equal to the rated (see (5.233)) of the rotor speed.

At the same time, the turbine controller realizes its control algorithm, which can take the following form:

- At partial load, the algorithm depends on the WTGS type. The blade pitch angle can be kept constant (e.g. equal to zero or equal to a small negative value) or can be optimized during the wind turbine operation (according to the rule $dP/d\omega = 0$) or can be varied and dependent on the operating point (resulting from an off-line optimization); see Fig. 5.45b.
- At full load, the blade pitch angle is adjusted to keep the turbine speed and/or the power at set (usually rated) values.

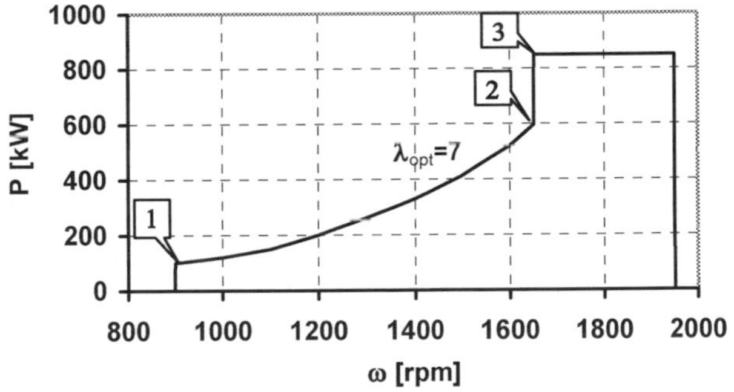


**Fig. 5.44.** Power versus rotor speed characteristics of Vestas V52m-850 wind turbine (points: **1** – $v = 5.5$ m/s, $\omega = 900$ rpm ($n_W = 14.5$ rpm), $P = 100$ kW; **2** – $v = 10$ m/s, $\omega = 1650$ rpm ($n_W = 26.72$ rpm), $P = 596$ kW; **3** – $v = 16 \div 25$ m/s, $\omega = 1650$ rpm ($n_W = 26.72$ rpm), $P = 850$ kW)

[9] In some types of control systems the WTGS characteristic between points 2 and 3 in Fig. 5.44 is defined by the function $P = K \cdot \omega$ with high slope (high value of coefficient $K$).

An example of the turbine controller characteristics is presented in Fig. 5.45.

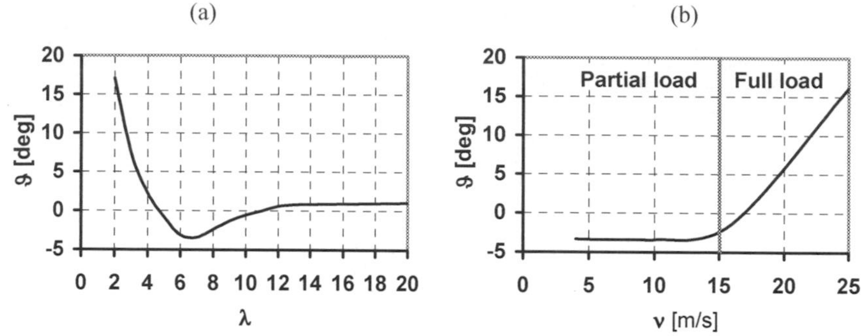


**Fig. 5.45.** Blade pitch angle as a function of the tip-speed ratio and the wind velocity in a WTGS with OptiTip control system [52]

The idea of WTGS control is presented in Figs. 5.46–5.48. Figure 5.46 presents for variable speed WTGSs equipped with an asynchronous generator (and converter) the response to a change of the wind velocity. The bold line represents the wind turbine power versus rotor speed characteristics. The two basic operating points (partial and full load) are considered.

Let us assume that at partial load the initial operating point is located at the point marked as 1. The power produced by the unit at the wind velocity $v_1$ is equal to $P_1$. The rotor shaft speed is equal to $\omega_1$ and the blade pitch angle is equal to $\vartheta_1$. Point 1 is located at the maximum of the power extracted from the wind curve $P_{WW}(v_1, \vartheta_1, \omega_1)$. Let us assume that the wind velocity step increase from value $v_1$ to value $v_2$ appears. Then the operating point shifts from point 1 to point 2 (step change of the rotor shaft speed is impossible because of the non-zero rotor inertia constant). Then the power generated increases up to $P_2$. This causes the turbine $P_{WW} = P_2$ and the generator $P_G = P_1$ power imbalance and further, according to the equations of motion, the rotor-shaft speed change (here increase). The new operating point 3, in a system in which at partial load the blade pitch angle is kept constant, is located on the intersection of the $P_G(\omega)$ (bold line) and $P_{WW}(v_2, \vartheta_1, \omega)$ curves. At this point, the WTGS produces power $P_3$ at the rotor-shaft speed equal to $\omega_3$.

This dynamic process of passing from point 1 to 3 consists of two steps. In the first one, as a result of the wind-velocity step change, the power extracted from the wind changes in a step way (Fig. 5.46b) from $P_1$ to $P_2$.[10] In the next step, the power and the rotor-shaft speed change relatively slowly (from time $t_1$ to $t_3$) and the power increase (from $P_2$ to $P_3$) is relatively small. During the considered process, the speed increases as shown in Fig. 5.46c, but in general, an oscillatory response is also possible.

At partial load, because the WTGS operates in the area where the power versus speed characteristics $P_G(\omega)$ passes by maxim of the $P_{WW}(v, \vartheta, \omega)$ curves, it is al-

[10] When a non-realistic step change of the wind velocity is assumed.

most impossible to smooth out the power variations resulting from the wind velocity variations. This state is also visible in Figs. 2.6, 2.8, 2.10, which present the data measured from a real system.

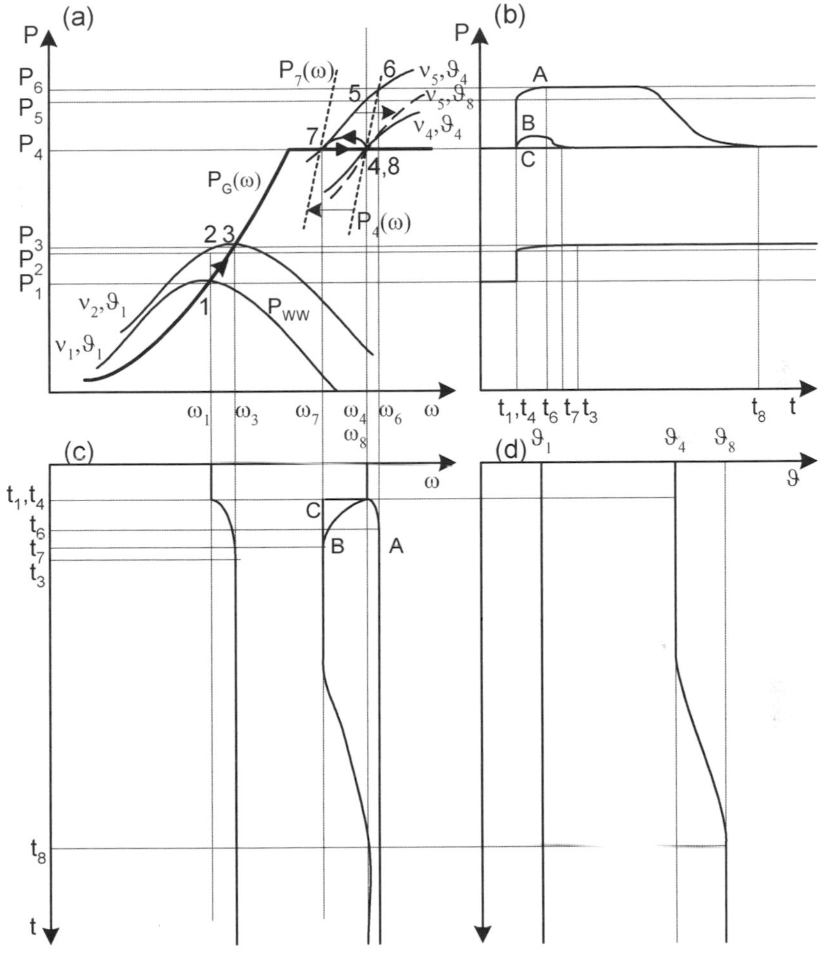


**Fig. 5.46.** Power and speed control for a variable speed WTGS

At full load the process looks different. Let us assume that the initial operating point is located at point 4. The power produced by the unit at wind velocity $v_4$ is equal to $P_4$. The rotor shaft speed is equal to $\omega_4$ and the blade pitch angle is equal to $\vartheta_4 > \vartheta_1$. Point 4 is not located at the maximum of the power extracted from the wind curve $P_{WW}(v_4, \vartheta_4, \omega)$. Let us assume that the wind velocity step increase appears from value $v_4$ to value $v_5$. Then the operating point shifts from point 4 to point 5. The power generated increases from $P_4$ to $P_5$. The power imbalance ($P_5 - P_4$ in time $t_{4+}$) causes the rotor shaft speed to increase.

The next part of the process considered depends on the WTGS type. In the case when the mechanical characteristics $P_4(\omega)$ cannot be changed, the system reaches point 6 with the rotor shaft speed equal to $\omega_6$ and the power generated equal to $P_6$ (curve A in Fig. 5.46b). Because the power generated is greater than the rated power, the turbine controller has to change the blade pitch angle and has to move the $P_{WW}(v, \vartheta, \omega)$ characteristics back to the point 4. This can be achieved by increasing the blade pitch angle, and here it is achieved for a blade pitch angle equal to $\vartheta_8 > \vartheta_4$. Because the pitch angle control is slow, wind velocity gusts (in general, all wind velocity variations) cause power variations.

To avoid this (power variation), it is necessary to use a system which enables us to change the mechanical characteristics of generator $P(\omega)$. In the considered case, a fast change of the characteristics from $P_1(\omega)$ to $P_7(\omega)$ can cause a fast transition from point 4 to point 7. This action would not increase the power generated (curve C in Fig. 5.46b). But this requires a step change of the rotor speed, which is of course impossible.

In practice, during an increase of the wind velocity (which is not of the step type) the transition to point 7 is achieved by the continuous change of the rotor speed and power characteristics. Depending on the inertia constant, the slope of the wind speed increase and the control system parameters, it is achieved with a smaller or higher power overload. In Fig. 5.46a this is shown by the curve with an arrow directed from point 4 to 7, and in Fig. 5.46b by curve B. In this state of operation (full load), when utilizing the fast control system of the generator, it is possible to reduce (smooth) the power variations significantly.

Simultaneously in the considered process, but with a delay, the turbine controller changes the blade pitch angle from $\vartheta_4$ to $\vartheta_8$, to finish the movement of the equilibrium point back to point 4.

For a WTGS with a power versus speed characteristic similar to the one presented in Fig. 5.44 (typical in systems with DASM) the above considerations related to the wind turbine operation at full load and the control system reaction to the wind velocity change are not valid. The reason is that in the steady state, the turbine operates with constant (and rated) speed $\omega_n$ (point 3 in Fig. 5.44 and point 4 in Fig. 5.47). Point 4 defines the rated rotor speed $\omega_4 = \omega_n$ and the rated power $P_4 = P_n$. The wind velocity step increase causes an increase of the power extracted from the wind from value $P_4$ to value $P_5$. When the generator control system cannot change the machine mechanical characteristics $P_4(\omega)$, then the operating point passes through point 5 to point 6 (curve A). Then after the turbine control system react by increasing the blade pitch angle from $\vartheta_4$ to $\vartheta_8$, the power generated is reduced to the rated value (point 8). In this case, the power variations as a result of the wind variation are not damped.

When, after the wind velocity change, the generator control system would change (move) the machine mechanical characteristics, e.g. to $P_5(\omega)$, then the new (but temporary) operating point becomes point 5. Then, after the turbine controller reaction, the WTGS state comes back to point 4 (curve B). In this case, the power variations as a result of the wind variation are not damped but are lower than in the previous case.

And finally, if the generator control system could move the mechanical characteristic to point 7 rapidly (characteristics $P_7(\omega)$), then the power generated would not change. On the contrary, the rotor speed would increase in a step way (curve C).

In this case, the power variations as a result of the wind variation can be completely damped. This unfortunately needs the rotor speed to achieve a step change, which in practice is not possible.

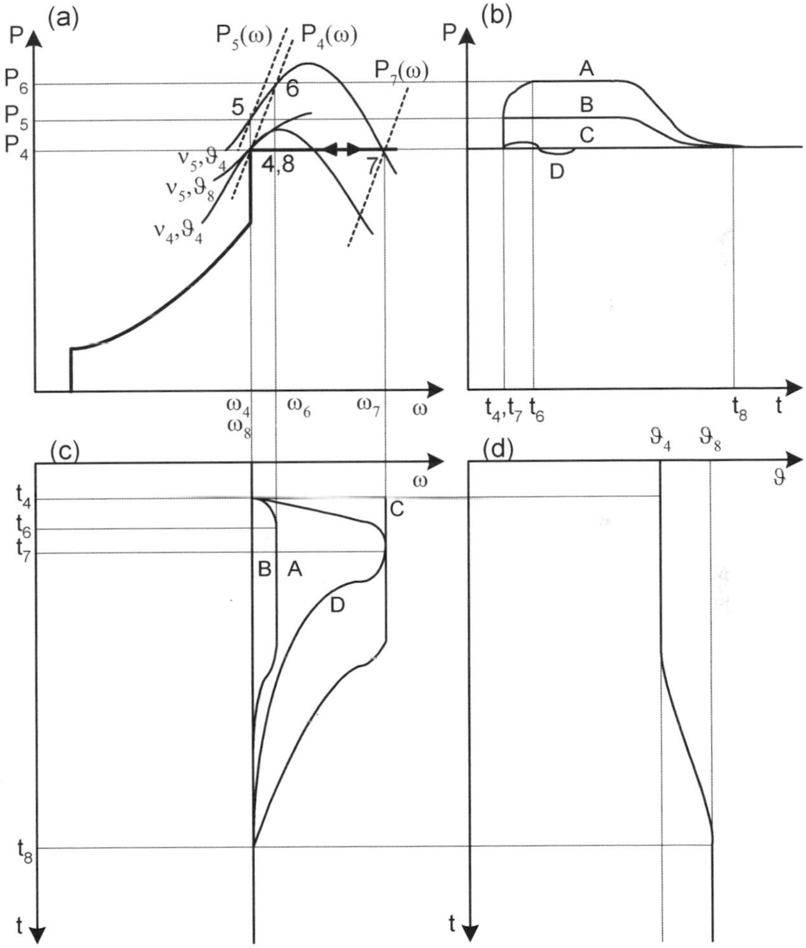


**Fig. 5.47.** Power and speed control for a variable speed WTGS with DASM

Despite this, real control systems utilize the latter method for power variation damping. In such a case, the rotor speed increases with limited slope, which causes the appearance of small power variations (curve D) only. This effect can be seen in Fig. 2.7, which presents the data measured form a real system.

For the sake of completeness, the reaction of the WTGS without a control systems (asynchronous generator without converter and with stall power control) to the wind velocity change is worth considering. The relevant characteristics are presented in Fig. 5.48. In this case, the WTGS (constant speed system) power versus rotor speed characteristic is equal to the mechanical characteristic of the asynchronous generator $P_e(\omega)$.

When the wind turbine operates at partial load, the reaction of the system to a wind-velocity step change is similar to the one considered above (points 1, 2, 3 in Fig. 5.48). The wind change causes the power change and there is no way to efficiently decrease the level of power variations (excluding additional systems located outside the WTGS, e.g. energy storages[11]).

When the wind turbine operates at full load, for example at point 4 ($P_4$ is equal to the "rated" power) the increase of the wind velocity from $v_4$ to $v_5$ causes the shift and decrease of the $P_{WW}(v_5,\omega)$ characteristics. It causes the step decrease of the power generated by the WTGS to a value equal to $P_5$. Next, the power generated decreases more, to achieve value $P_6$, where the curves $P_{WW}(v_5,\omega)$ and $P_e(\omega)$ intersect.

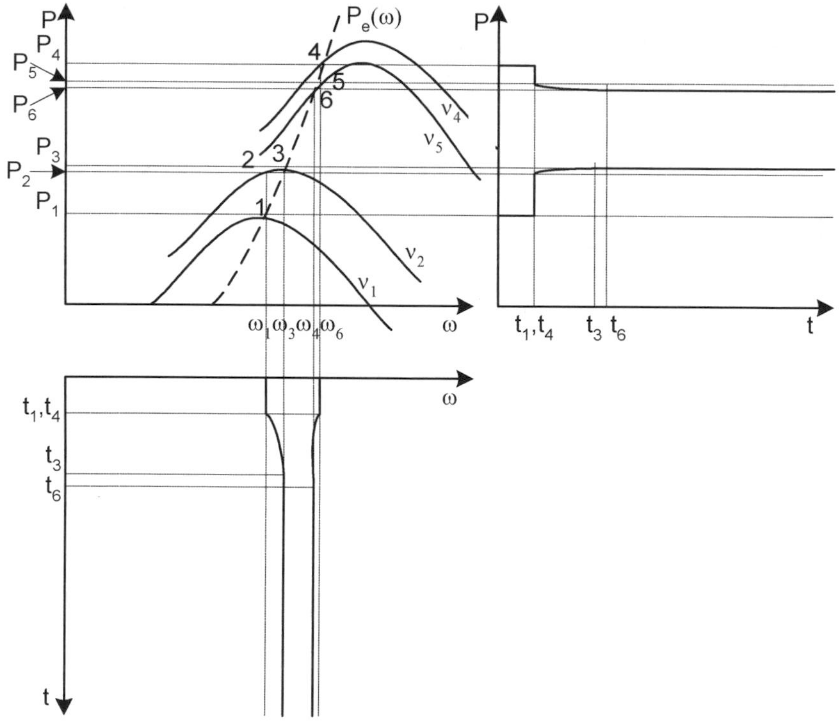


**Fig. 5.48.** Power control for a constant speed and stall control WTGS

[11] In fact, energy storage is not utilized for this purpose today.

Also here, at full load, it is not possible to damp power variations resulting from wind velocity variations. But because of the turbine "flat" characteristics, the variations are relatively small – smaller than at partial load. This state can be seen in Fig. 2.5.

## 5.6.2 Supervisory Control

The goal of a supervisory control system is to ensure automatic operation of the WTGS in various operating states without human control (realized directly and on line). Additionally, the system has to perform protection tasks.

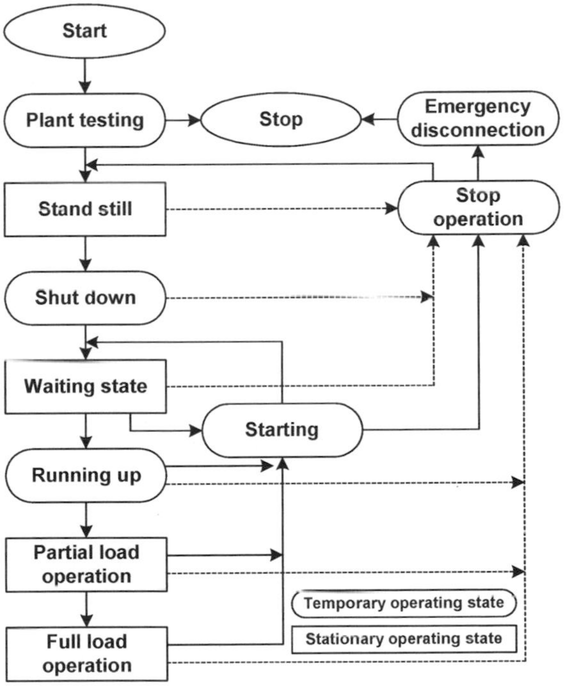


**Fig. 5.49.** Operating states of a WTGS [31]

WTGSs can operate in various states (Fig. 5.49): some of them are stationary and some of them temporary (with limited time of performance). Usually, the following operating states are identified:

- *plant testing* (temporary). In this state, the configuration, various signal states and the reaction of various components (e.g. actuators) are tested. The tests are recorded. Errors lead to an interruption of further operations.
- *stop (standstill) operation* (stationary). In this state, the rotor stands still. Breakers are active. Rotor blades are in their feathering position. The generator is disconnected from the grid. Depending on the reason for the stop, various tests

tests are carried out and the possibilities of starting the WTGS are tested.

- *starting* (temporary). In this state the WTGS rotor is driven by the wind but generator is not connected to the grid. The rotor speed is controlled by changing the pitch angle. When the minimum waiting speed (see next point) is achieved by the rotor (and the protection conditions allow it) the transition to the stationary waiting state is achieved.
- *waiting state* (stationary). In this state, the WTGS is ready to be connected to the grid. The rotor speed is controlled by adjusting the pitch angle. The possibilities of running up are tested. When fulfilled, the WTGS passes to the running-up state.
- *running up* (temporary). In this state, the rotor speed is increased to the speed necessary for synchronization. When achieved, the synchronization process is performed. After synchronization, the minimum reference power is set. Protection functions are performed all the time.
- *partial load operation* (stationary). In this state (for various speed systems), the blade pitch angle is adjusted to the optimal value (sometimes it is kept constant) and the optimal value of the generator power in relation to the rotor speed is computed. The power reference value is altered (according to the power speed characteristics) when the control reserve ($x$ in Fig. 5.51) is reached. The reference values are sent to the turbine and generator controllers. Simultaneously, protection functions are performed.
- *full load operation* (stationary). In this state, the rated value of power is set as the reference. The algorithm performs protection functions.
- *shutting down* (temporary). This is a state (process) of going from running up, partial load or full load to the waiting state (when no fault occurred). During this state, the appropriate control signals are sent to the turbine and generator controllers. After the controller action (the generator power and currents are reduced to zero) the WTGS is disconnected from the grid. When a fault occurs, the fault disconnection procedure is performed.
- *stop operation* (temporary). This is a state (process) allowing the WTGS to pass to the standstill state from any other operating state.
- *fault and emergency disconnection* (temporary). This is a process of stopping the WTGS operation because of a fault (without the possibility to restart automatically).

In each state, a separate algorithm is performed by the control supervisory system. Most of the algorithm is related to the protection of the system. Various parameters are measured and compared to maximal, minimal or critical ones. Any case of exceeding the measured values beyond the border values makes the WTGS pass to the shut-down operation, stop operation, or fault/emergency disconnection.

Figures 5.50–5.52 present three algorithms related to the three basic states[12] that can be considered while analyzing the WTGS operation in the electric power system, including the procedures of connecting to the grid and disconnecting.

[12] Various types of WTGS utilize supervisory algorithms that can differ from that presented here. Those algorithms should be considered as examples only.

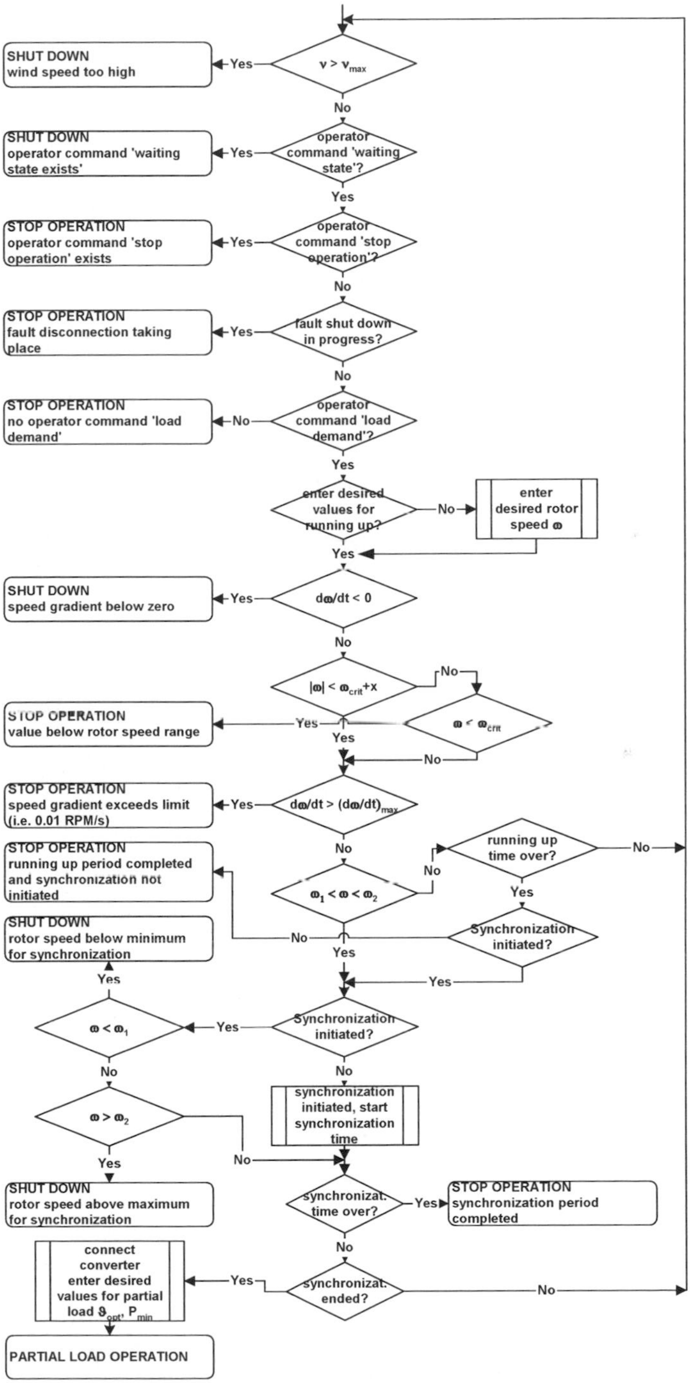


**Fig. 5.50.** Algorithm of WTGS running-up operation [31]

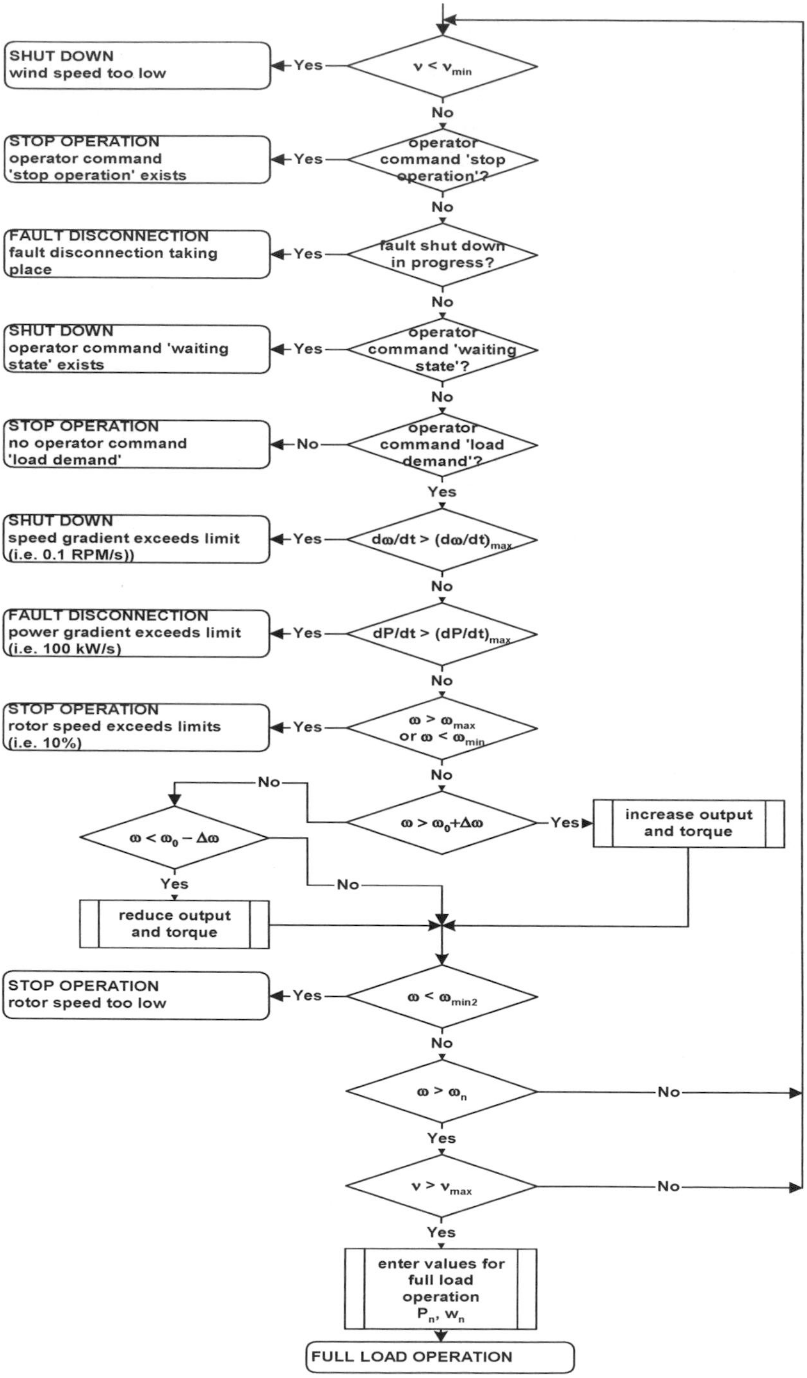


**Fig. 5.51.** Algorithm of WTGS partial load operation [31]

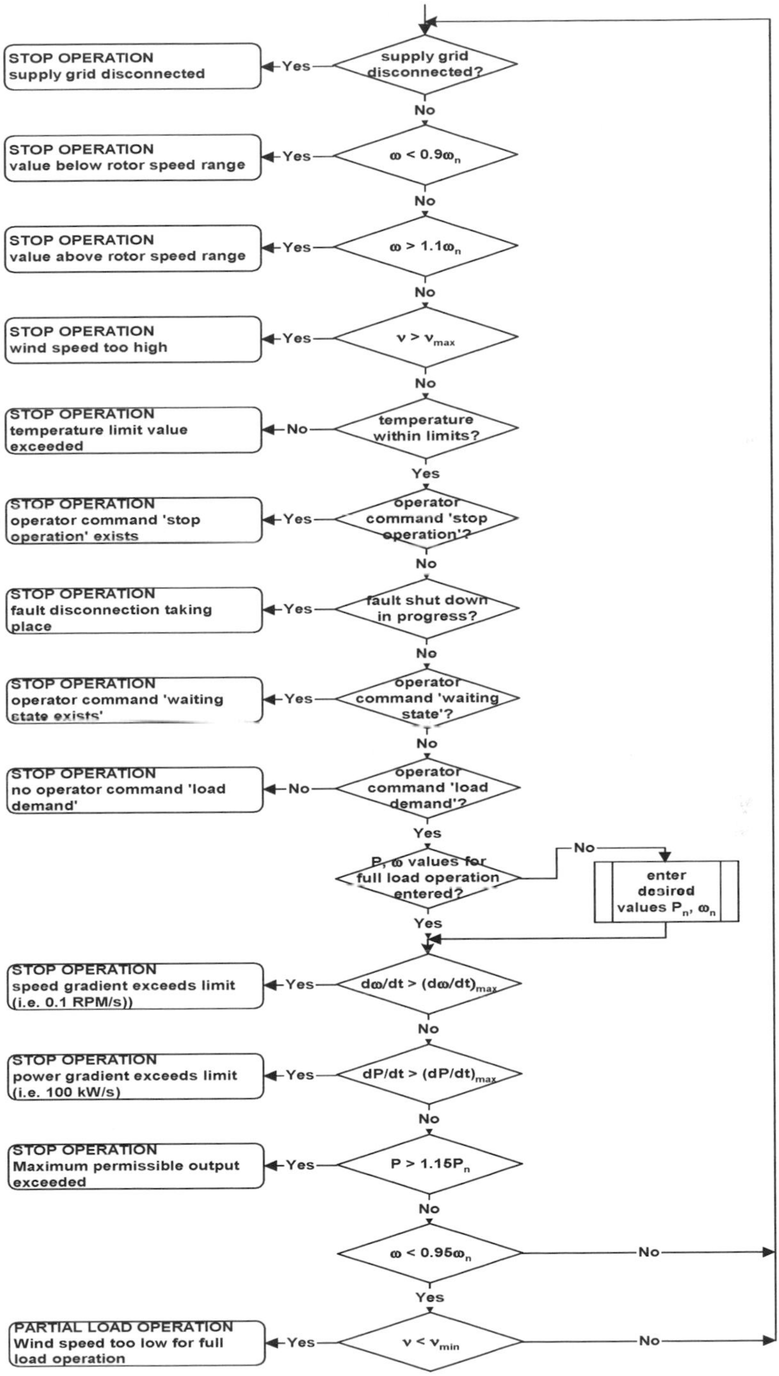


**Fig. 5.52.** Algorithm of WTGS full load operation [31]

The figures show that the main[13] tasks that are performed by the algorithms are protection tasks. The control task is performed practically only by the partial load operation algorithm. During the full load operation, the control algorithm keeps the rated value of active power.

That leads to the following conclusions related to the modeling of a WTGS operating in an electric power system:

- when considering the full load operation of the WTGS under standard wind variations, which does not lead to the WTGS being shut down, the supervisory system can be neglected,
- when considering the partial load operation of the WTGS (under standard wind variations) near the equilibrium point, the supervisory system can be neglected as well,
- when considering high wind velocity changes at partial load or wind changes near the rated wind velocity (passing from partial load to full load and in the opposite direction), the parts of the supervisory system algorithm related to the reference values changing should be included,
- when considering extremely strong wind gusts, the relevant parts of the supervisory control system, e.g. the parts relating to protection systems, should be included.

### 5.6.3 Turbine Control

The task of the wind wheel (turbine) control system is to keep the value of the rotor shaft speed and the active power at a set level. Then the system usually consists of speed, power and blade pitch angle regulators. The control system structure depends on the WTGS specific construction, year of production, manufacturer, etc. Generally, two basic variants of the system can be considered:

- turbine control in a system with an asynchronous generator,
- turbine control in a system with a synchronous generator.

#### *5.6.3.1 Turbine Control in a System with an Asynchronous Generator*

The turbine control system of WTGS with a directly grid-coupled asynchronous generator is presented in Fig. 5.53. The system consists of speed, power and blade pitch angle controllers. The pitch angle controller acts directly on the blades. The other controllers form the reference value of the pitch angle. Small WTGSs usually have a simplified pitch angle controller, which does not use the inner loops i.e. loop from the pitch angle derivative and/or from the pitch angle presented in Fig. 5.53.

The speed controller operates as the main controller in idle operation or during start-up, shutdown or during a power system malfunction (failure). Then the refer-

[13] Taking into account the number of conditions being checked.

ence speed $\omega_{ref}$ varies according to need. An algorithm of the control system is defined as follows. When the generator is not connected to the grid, the generator power $P_G$ is equal to zero and, in general, is less than the power reference $P_{ref}$. Then the power controller output signal $\vartheta_{ref}$ reaches the band that is equal to $\vartheta_{min}$ (computed by the supervisory control system) or that is equal to the output signal from the speed controller.[14] When the minimum pitch angle $\vartheta_{min}$ is set below the speed controller output signal, then the speed controller operates as the main controller. The same mechanism is used after load shedding (disconnection from the grid), when the WTGS speed increases.

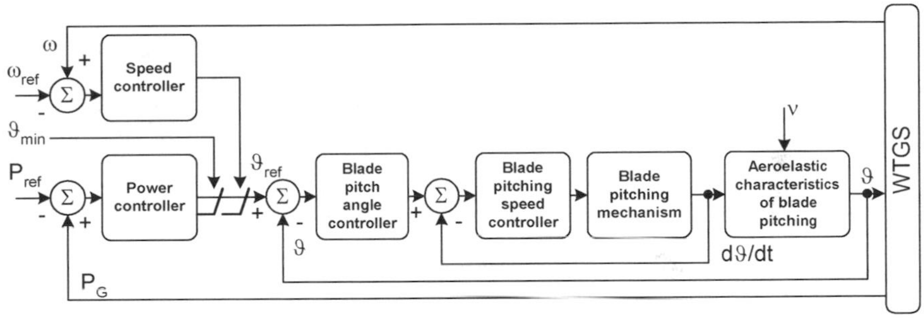


**Fig. 5.53.** Turbine control in a system with a directly grid-coupled asynchronous generator [31]

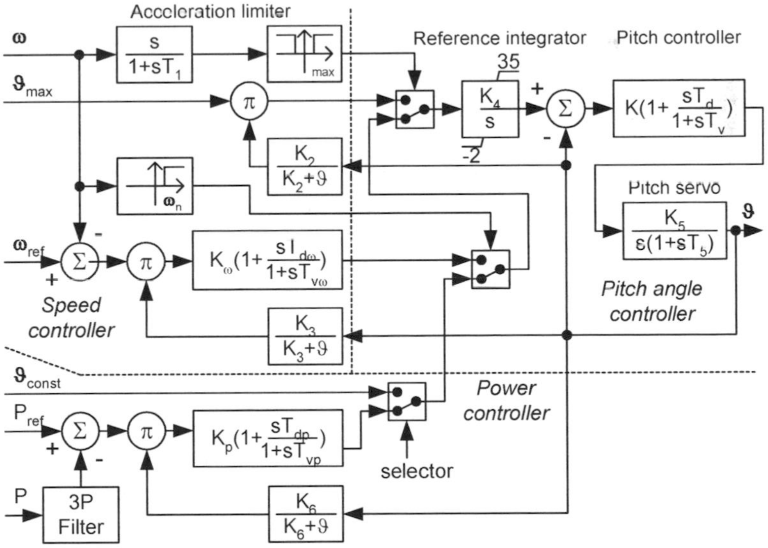


**Fig. 5.54.** Typical wind turbine control system model [27]

---

[14] The controller reaction to a given input signal depends, of course, on the controller structure. Here, a controller with integration block is considered.

When the WTGS operates on the grid, the reference speed $\omega_{ref}$ is set up a few percent above the speed resulting from the grid frequency or, for DASM, it is set equal to the rated speed. At partial load, it makes the output signal from the speed controller reach its lower band, e.g. zero, which in practice means that the controller is switched off. The power controller operates then as the main controller.

At full load, the speed controller can operate as the main controller again, but it depends on the generator control system structure and algorithm.

An example of the controller mathematical model is presented in Fig. 5.54. The characteristic feature of the controller is the dependence of the gains of the power and speed controllers on the blade pitch angle $\vartheta$.

### 5.6.3.2 Turbine Control in a System with a Synchronous Generator

The turbine control system of a WTGS with a directly grid-coupled synchronous generator is presented in Fig. 5.55. The system consists (like the system described above) of speed, power and blade pitch angle controllers. The pitch angle controller is located in the control system as is the one in the WTGS equipped with an asynchronous generator.

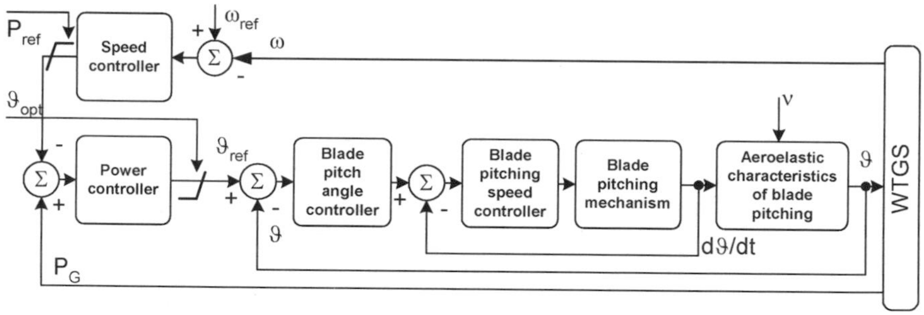


**Fig. 5.55.** Turbine control in a system with a directly grid-coupled synchronous generator [31]

Like the above-described one, the speed controller operates as the main controller in idle operation or during start-up, shutdown or during a power system malfunction (failure).

When the WTGS operates on the grid, the control algorithm depends on the WTGS operating point:

- At partial load – the reference speed $\omega_{ref}$ is set up a few percent above the speed resulting from the grid frequency.[15] That is why the speed controller output reaches its upper limit $P_{ref}$. When the $P_{ref} = P_n$, the power controller output reaches its lower border because $P_{ref} > P_G$. Therefore, the control system operates as the pitch angle controller, where $\vartheta_{ref} = \vartheta_{opt}$. The optimal pitch angle reference is computed by the supervisory control system (sometimes it is kept

[15] The control system can determine the turbine power–speed characteristics.

constant). When $P_{ref}$ is set adequately to the power extracted from the wind, the power controller operates as the main controller.

- At full load – as at partial load, the reference speed $\omega_{ref}$ is set up a few percent above the speed resulting from the grid frequency. The speed controller output reaches its upper border $P_{ref} = P_n$ and the control system operates as the power controller. In this state, the blade pitch angle $\vartheta_{opt}$ does not usually limit the controller output.

### 5.6.4 Generator Control

Like turbine control systems, the generator control systems of a modern WTGS vary and their structure and parameters depend on the wind turbine type (rated power, construction, turbine manufacturer, etc.). Therefore the description presented below should be treated as typical for the given type of wind turbine.

#### *5.6.4.1 Asynchronous Generator with Dynamic Slip Control*

The dynamic slip-control system influences the torque and power transferred from the rotor to the stator (and to the grid) by changing the rotor slip according to the following equations (assuming that losses are neglected):

$$\begin{aligned} P_R &= P_{airgap} s \\ P_S &= P_{airgap}(1-s) \approx P_G \\ P_{airgap} &\approx P_{WW}, \end{aligned} \tag{5.234}$$

where $P_R$ – machine rotor power,
$P_S$ – machine stator power,
$P_{WW}$ – wind wheel (turbine) power,
$s$ – rotor slip.

In squirrel-cage rotor asynchronous machines, the slip variations are small (usually below 2–3% – dotted line in Fig. 5.56) and, therefore, the mechanical power (and the torque) variations pass into the stator side of the machine almost directly (without smoothing). Using a wounded-rotor (slip-ring) asynchronous machine – by adding resistors to the rotor – it is possible to change the machine's mechanical characteristics by changing the rotor slip (e.g. dashed line in Fig. 5.56). Changing the resistance, it is possible to change the rotor slip and simultaneously (with the given torque) the power transferred to the grid. Then the variations of the power transferred to the grid in the system with variable resistors connected to the rotor winding can also be smoothed – which is the main goal of using these additional resistors. The second of the (5.234) equations shows how it works. When the power $P_{WW}$ extracted from the wind changes, the controller changes the rotor slip $s$ to compensate the variation and to keep the electrical power $P_G$ constant. The surplus of the energy extracted from the wind is converted into the energy of the rotating turbine rotor – causing the rotor speed to increase.

When the power extracted from the wind decreases, the rotor kinetic energy is converted back into electrical energy – causing the turbine rotor speed to decrease.

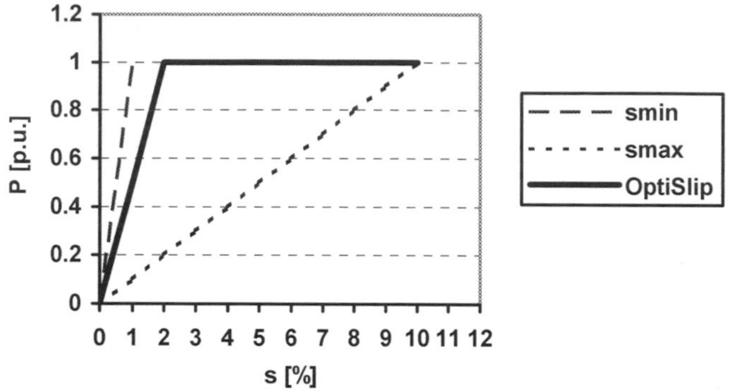


**Fig. 5.56.** Power reference for a variable slip generator

The Vestas OptiSlip system (Fig.5.57), whose characteristic is presented in Fig. 5.56, is an example of such a control. The resistors and the rotor current controller are located in the generator rotor. The control signal from the WTGS controller is sent by optical link. The major task of the controller located in the machine rotor is the rotor current control, assuming that the current is proportional to the power output. The required resistance is achieved by pulse-wave modulation of the resistors using IGBT (Insulated Gate Bipolar Transistors) as DC switches. The switch frequency is approximately 3 kHz.

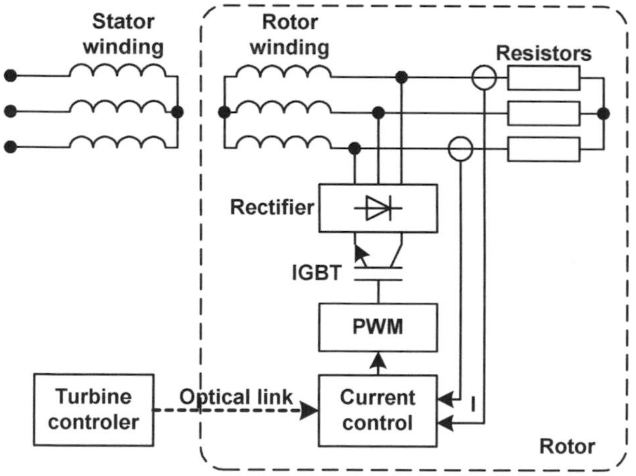


**Fig. 5.57.** Asynchronous generator with wounded rotor and dynamic slip control system (Vestas OptiSlip control system)

The WTGS turbine and generator controller is presented in Fig. 5.58. The controller consists of two speed controllers and a power controller (the rotor current controller is not shown here). The speed controller A is used when the WTGS operates idle (also before cut-in and during cut-out of the generator), while the speed controller B is used when the WTGS operates in the grid. The power controller in the steady-state produces the mechanical characteristic presented in Fig. 5.56 by the bold continuous line. The rotor slip can be changed from 1 to 10%.

The tasks and control abilities of the WTGS control system in the two main areas of operation vary and are as follows:

- Partial load – the speed controller maintains optimal speed of the wind wheel by adjusting the blade pitch (optimal means the speed that gives maximum power extracted from the wind). The rotor slip is kept below 2% and varies according to the increasing part of characteristic shown in Fig. 5.56 (continuous bold line). The generator power is controlled by changing the power reference according to the actual slip resulting from the above characteristics. The power controller's ability to damp the power variations are determined here by the distance between the mechanical characteristics: the dashed line $s_{min}$ (characterized by rated slip equal to 2%) and the dotted line $s_{max}$ (characterized by rated slip equal to 10%). The ability for power variation damping is not great but does exist. The damping ability increases when the power generated increases. For example, the 3P power variations with amplitude up to 10% are effectively damped.
- Full load – when the wind speed exceeds the nominal velocity, the generator speed increases above 1.02 [p.u.] of the synchronous one (the slip is above 2% – Fig. 5.56). The power reference becomes constant and equal to the rated one. By adjusting the blade pitch angle, the speed controller B (relatively slow) keeps the average slip equal to 4–5%. The power controller (fast), with the rotor-current controller, is then able to change the rotor slip within the range of about ±5% from the equilibrium point and, therefore, can effectively influence the power transferred to the grid, and at the same time it can effectively damp the power variations.

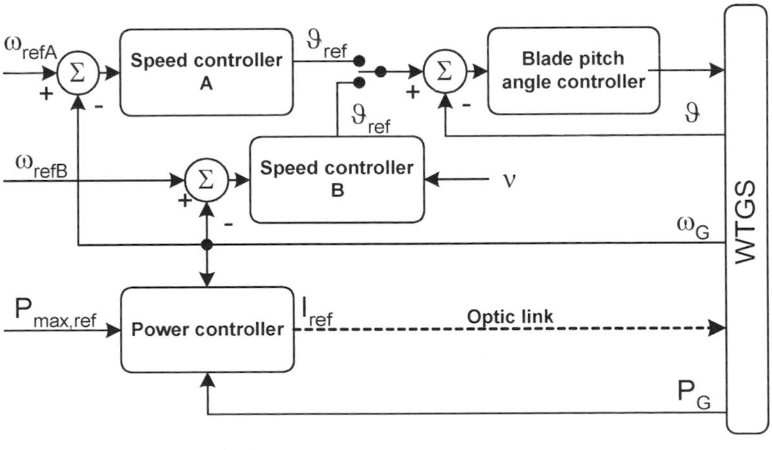


**Fig. 5.58.** Control system structure of a WTGS with dynamic slip control (A – idle operation, B – grid operation)

### 5.6.4.2 Asynchronous Generator with Over-Synchronous Cascade Control

Utilization of an asynchronous generator with wounded rotor and with additional resistors connected to the rotor winding enables it to influence the rotor speed (slip) but at the same time this action causes an increase of the power losses (losses in the additional resistors). To eliminate this, a converter allowing energy transfer from the rotor to the grid can be used.

An example of such a system – called an over-synchronous cascade – is presented in Fig. 5.59. The converter consists of a diode rectifier, reactor (as intermediate element) and current-source inverter. The converter enables transfer of the power in one direction and, therefore, the generator can operate only at a speed above that of the synchronous one. The advantage of this system is that it is a relatively simple controller, which consists of a speed regulator (SR) and an intermediate circuit current regulator (ICCR). The reference speed $\omega_{ref}$ is calculated by using the speed versus active power nonlinear function (optimized for the given WTGS). The speed regulator computes the intermediate circuit current reference $i_{dref}$. The ICCR generates signals controlling the inverter valves (thyristors or transistors).

The asynchronous generator with over-synchronous cascade does not permit reactive power control. Other schemes, e.g. switched capacitors, should be used for that purpose.

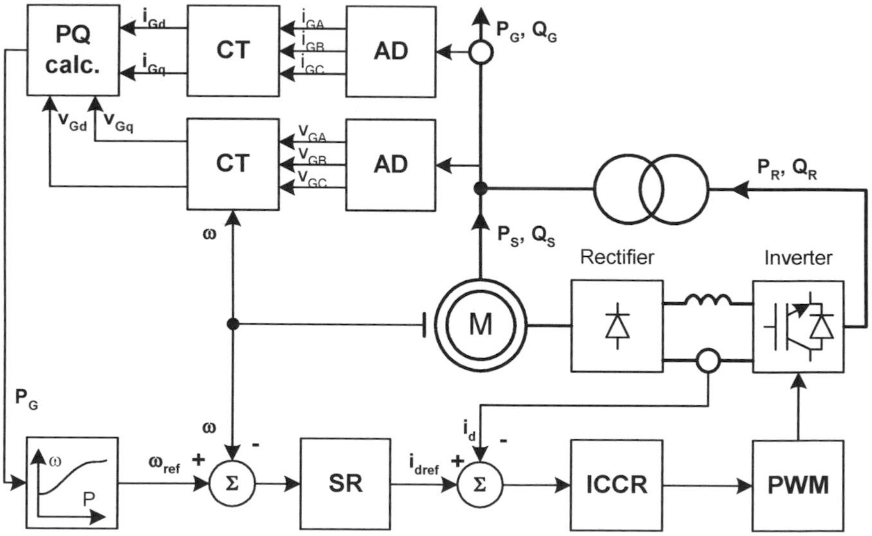


**Fig. 5.59.** Asynchronous generator with over-synchronous cascade control system (AD – analogue–digital converter, CT – coordinate translation, SR – speed regulator, ICCR – intermediate circuit current regulator, PWM – pulse-width modulation)

The tasks and control abilities of the WTGS control system in the two main areas of operation are similar to those of the previously considered system (the dynamic slip control system):

- Partial load – the generator speed controller maintains the optimal speed of the wind wheel (turbine rotor). At the same time, the turbine controller adjusts the blade pitch angle to the optimal (in some cases, the pitch angle in this operation area is kept constant). The generator power results from the speed versus power characteristics. The ability of the generator controller to damp the power variations is limited here as well. But the 1P and 3P power variations can be effectively damped.
- Full load – when the wind speed increases above the nominal value, the generator speed increases as well. By adjusting the blade pitch angle, the turbine controller (relatively slow) keeps the pre-set rotor speed while the generator controller keeps the power according to the speed versus power characteristics, which in this area should have a very low slope. Flat characteristics can also be utilized, but then another structure of the power controller should be utilized. At full load, the generator controller, which is very fast, is able to change the rotor speed in the appropriate range and to influence effectively the variations of the active power transferred to the grid. Then the power variations can be effectively damped.

### 5.6.4.3 Doubly-Fed Asynchronous Machine Control

The doubly-fed asynchronous machine – with a converter making it possible to transfer energy in both directions (4-quadrant converter) – enables the generator to operate above and below the synchronous speed [11, 45, 54, 73, 76, 78, 86, 87, 92]. When the machine operates at over-synchronous speed, the power flows from the rotor to the grid (Fig. 5.60a). When the machine operates at sub-synchronous speed, then the power flows from the stator (or grid) to the rotor (Fig. 5.60b). By controlling the rotor current (magnitude and phase) it is possible to influence highly (over a wide range) the machine slip and the active and reactive power transferred to the grid.

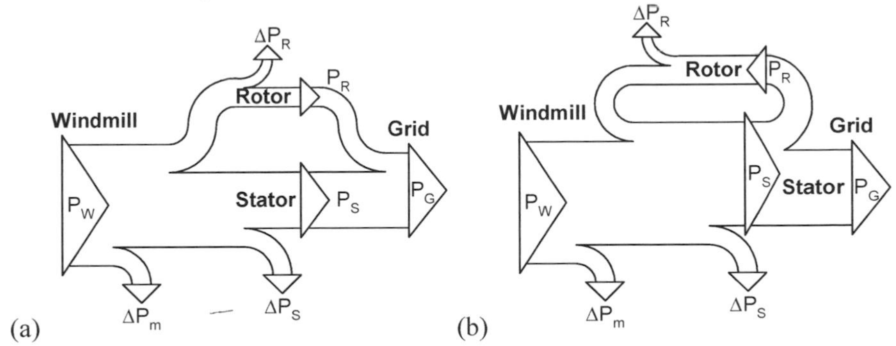


**Fig. 5.60.** Power flow in a doubly-fed asynchronous generator: (a) over-synchronous operation, (b) sub-synchronous operation ($\Delta P_m$ – mechanical losses; $\Delta P_S$, $\Delta P_R$ – losses in stator and rotor, respectively)

The ability to change the generator rotor speed over a wide range enables us to increase the power production by:

- Optimization of the power output – the inherent energy in the wind is extracted more effectively when the rotor speed is allowed to vary.
- Exploitation of the rotational energy in the generator rotor – storing the energy excess as rotational (kinetic) energy and using it when the wind velocity is lowered (smoothing the power transferred to the grid).
- Rapid adjustment to different wind speeds – allows the rotor speed to vary rapidly and, therefore, to exploit energy in transient gusts. The energy would be lost when a relatively slow pitch angle control mechanism is used. On the other hand, the converter allows the pitch angle control system to operate slowly, decreasing the stresses and overloading in the blades and the rotor shaft.

The additional advantages of the system include the following:

- possibilities of the reactive power control with no need to use capacitors,
- lower noise emission,
- reduction of wear and tear on gearbox, blades and tower,
- improvement of power quality,
- active damping of oscillations and harmonics.

The control system of the doubly-fed asynchronous generator is more complex than the control system of the over-synchronous cascade. The system consists of a grid frequency converter control system and a machine converter control system (Fig. 5.61).

The grid frequency converter controller regulates the intermediate-circuit parameters: voltage (pre-set constant value with given hysteresis band) or DC current (when the reactor is used as an intermediate element) and the current of the grid side of the converter – as in Fig. 5.61. The reactive power on the grid side of the converter is usually kept close to zero. Then the machine side of the converter enables control of the reactive power consumed/produced by the WTGS.

The machine converter control system consists of a set of controllers that allow control of the real power/speed and the reactive power. The doubly-fed asynchronous machine controllers usually utilize the concept of disconnection of the real and reactive power controls by transformation of the machine parameters (currents, voltages, fluxes, etc.) into the *dq*-reference frame and by separate forming of the rotor voltages $v_{dR}$, $v_{qR}$. Then, the real power (and speed) can be controlled by influencing the *d*-axis[16] component of the rotor current $i_{dR}$ while the reactive power can be controlled by influencing the *q*-axis component of the rotor current $i_{qR}$.

[16] Whether the real power is proportional to the *d*-axis rotor current and the reactive power is proportional to the *q*-axis rotor current, or whether the opposite state takes place depends on the *dq*-reference frame definition.

In the steady-state, the values of the real and reactive power at the machine stator terminal are proportional to the above currents and are given by

$$P_S = -\frac{3}{2}\frac{L_m}{L_S} v_S i_{dR} \tag{5.235}$$

$$Q_S = -\frac{3}{2}\left(\frac{v_S^2}{\omega L_S} + \frac{L_m}{L_S} v_S i_{qR}\right), \tag{5.236}$$

where (5.235) and (5.236) are derived from (5.81), (5.86), (5.89) and (5.90) with the assumption $v_{dS} = v_S$. The rotor terminal real and reactive power is proportional to the slip and stator power product. Then the WTGS power, which is the sum of the stator and rotor power, can be controlled effectively by the rotor current components. The above is valid in the steady state only and does not comprise the cross-axis interdependencies, which occur during transient processes. Fortunately, despite this, the above approach enables an effective control scheme to be developed and today it is widely used for doubly-fed asynchronous machine control.

The controller presented in Fig. 5.61 utilises the speed versus power characteristics that enable us to compute the reference speed as a function of the WTGS real power.[17]

An example of such characteristics – for the Vestas OptiSpeed[18] generator control system – is presented in Fig. 5.62. The asynchronous generator operates here at the speeds from 900 to 1950 rpm. The $\omega$ = 1680 rpm speed is the rated one. During steady-state operation, this speed is maintained when the power generation is higher than 50% of the rated power. Higher speed (up to 1950 rpm) is reached by the rotor during transient processes.

In general, there are many various types of doubly-fed asynchronous generator control systems, e.g. with and without positioning sensors (with observers). They are still the subject of intensive investigations. But independently of the type of regulator the considered control system makes possible effective smoothing of the real power variation, mainly at full load operation of the WTGS (at rated power). During partial load operation the active power variations are smoothed to a relatively small degree.

The main principle of control (and cooperation of the turbine and generator controllers) for the oversynchronous cascade is similar to that for the doubly-fed asynchronous generator (described in the previous section).

[17] In such a case, the maximum point power tracking (MPPT) scheme, based on the $dP/d\omega = 0$ rule, is not utilized.

[18] The generator stator winding can be (and is) switched between delta and star connection. The area of operation of the generator with the given connection is marked in Fig. 5.61. The generator operation with the star connection reduces losses when the wind speed is lower. At higher winds, when the rated power is achieved, the generator operates with delta-connected stator windings.

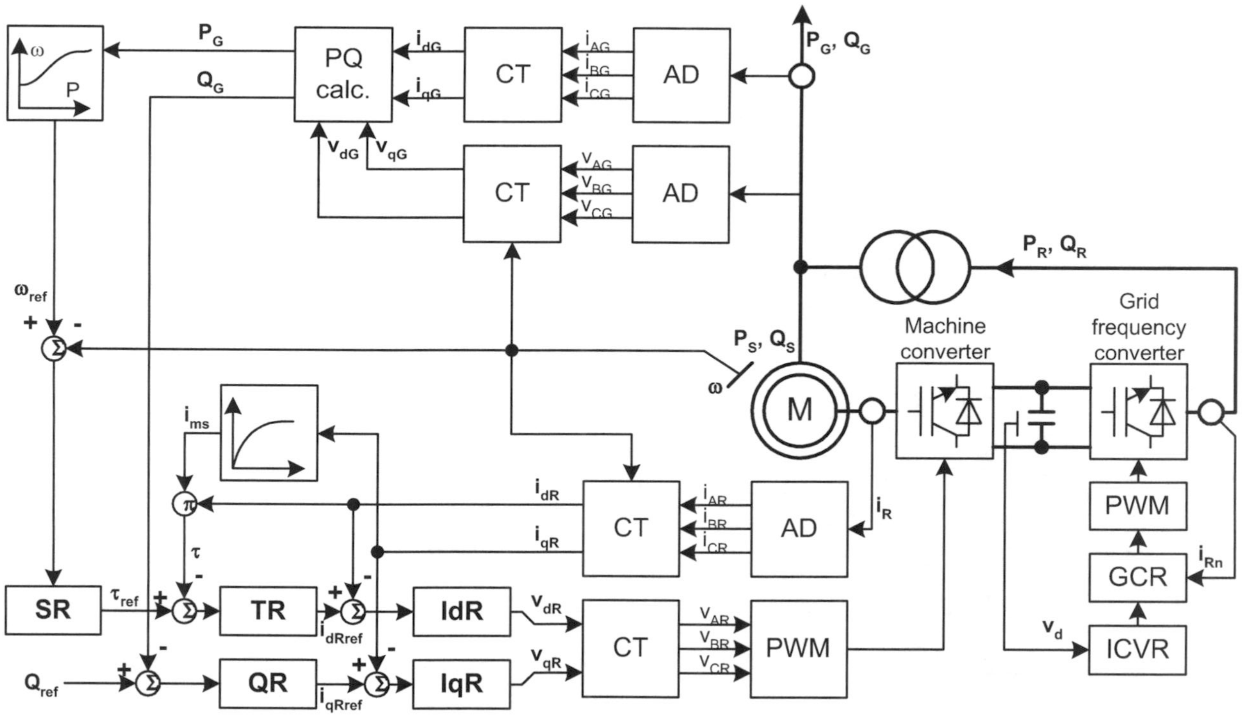


**Fig. 5.61.** Doubly-fed asynchronous generator control system (AD – analogue–digital converter, CT – coordinate translation, SR – speed regulator, TR – torque regulator, QR – reactive power regulator, IdR – rotor direct axis current regulator, IqR – rotor quadrature axis current regulator, GCR – grid current regulator, ICVR – intermediate circuit voltage regulator, PWM – pulse width modulation)

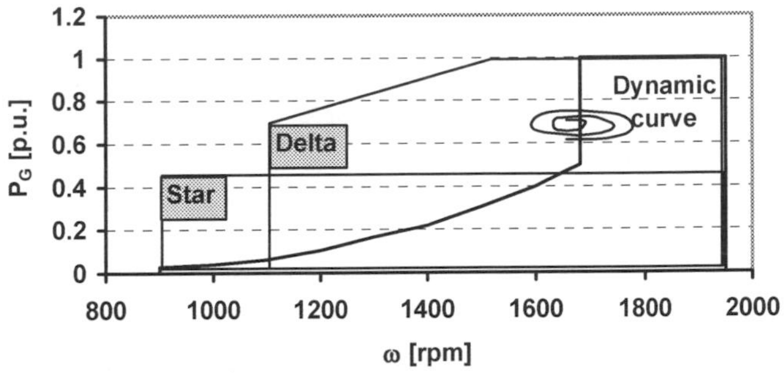


**Fig. 5.62.** Typical generator power versus rotor speed static characteristics (Vestas OptiSpeed control system [88])

### 5.6.4.4 Synchronous Generator Control

The structure of the synchronous generator control system depends on the WTGS type – mainly on one connection to the power system [30, 31, 46, 64, 79].

WTGSs equipped with synchronous generators directly connected to the grid are equipped with excitation control systems (excitation system and voltage controller). The static excitation systems are usually used here. The static excitation systems consist of an exciter transformer and a controlled rectifier. The excitation power is supplied through an excitation transformer from the generator terminals, the result being that the voltage, which feeds the thyristor rectifier, is directly proportional to the generator terminal voltage. The voltage controller can be additionally equipped with a power system stabilizer (PSS). The aim of the control system is to control the terminal voltage and to damp the electromechanical (real power, rotor speed, etc.) oscillations.

For modeling purposes, the standard IEEE models of the excitation system (e.g. ST1) and the power system stabilizer can be used. Block diagrams of both control elements are presented in Figs. 5.67 and 5.69. It is worth mentioning that the presented structure of the PSS is more convenient for damping of the electromechanical oscillations whose source is located in the power system (e.g. resulting from the generator power change) than for damping oscillations caused by the turbine (resulting from the turbine power change). If the PSS model is used, it is necessary to choose its parameters carefully.

The excitation system model ST1 (Fig. 5.67), and the thyristor rectifier, is highly simplified. The rectifier is modeled only as a variable limit of output signal, which depends on the terminal voltage $V_g$ and the field current $I_f$, the result being that when the field voltage is between these bands, the field voltage (controller output) does not depend on the terminal voltage. It is not true in the static excitation systems, in which the terminal voltage variations directly influence the controller output.

For more detailed modeling purposes, the excitation rectifier can be described by the following equation:

$$V_{\mathrm{f}} = \begin{cases} \dfrac{3\sqrt{2}}{\pi} V_{\gamma} \cos \alpha_{\mathrm{SE}} - \dfrac{3}{\pi} I_{\mathrm{f}} X_{\gamma} & \text{when} \quad \dfrac{I_{\mathrm{f}} X_{\gamma}}{V_{\gamma}} \le \dfrac{\sqrt{6}}{3} \\ 0 & \text{when} \quad \dfrac{I_{\mathrm{f}} X_{\gamma}}{V_{\gamma}} > \dfrac{\sqrt{6}}{3} \end{cases}, \tag{5.237}$$

where $V_{\mathrm{f}}$ – excitation voltage,
$V_{\gamma}$ – line-to-line voltage of secondary winding of excitation transformer,
$I_{\mathrm{f}}$ – field current,
$X_{\gamma}$ – commutating reactance (usually modeled by the reactance of the excitation transformer),
$\alpha_{\mathrm{SE}}$ – firing angle.

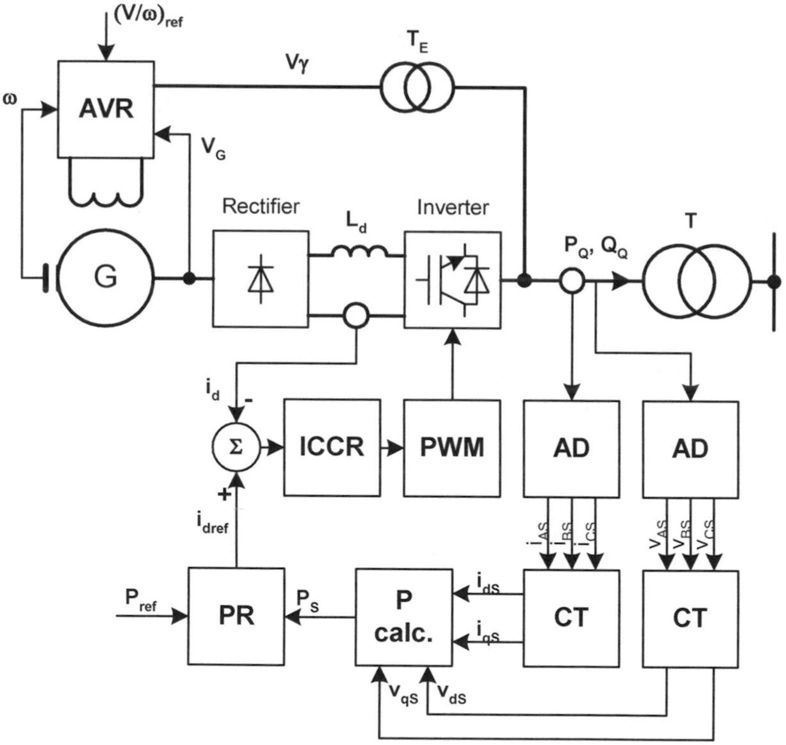


**Fig. 5.63.** Synchronous generator with current-source inverter control system (AVR – voltage regulator, AD – analogue–digital converter, CT – coordinate translation, PR – active power regulator, ICCR – intermediate circuit current regulator, PWM – pulse-width modulation, $T_E$ – excitation transformer, T – step-up transformer)

The field voltage (rectifier output) defined by (5.237) is limited. The boundaries of the rectifier working area are defined by following equations

$$A = \frac{3\sqrt{2}}{\pi} V_\gamma - \frac{3}{\pi} I_f X_\gamma$$
$$C = \frac{3\sqrt{6}}{\pi} V_\gamma - \frac{9}{\pi} I_f X_\gamma, \tag{5.238}$$

where the curves A($V_\gamma$, $I_f$), -A($V_\gamma$, $I_f$), C($V_\gamma$, $I_f$) and -C($V_\gamma$, $I_f$) create the full set of the boundaries.

For the WTGS equipped with a synchronous generator, whose field winding is fed by an AC machine excitation system, the standard IEEE model AC1 presented in Fig. 5.68 can be used.

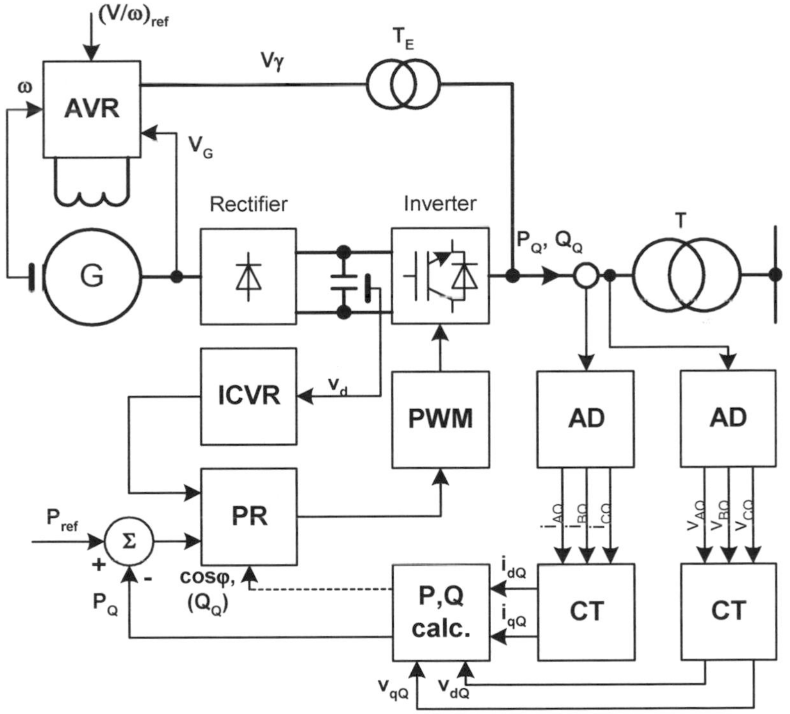


**Fig. 5.64.** Synchronous generator with voltage-source converter control system (AD – analogue–digital converter, CT – coordinate translation, PR – power regulator, ICVR – intermediate circuit voltage regulator, AVR – voltage regulator, PWM – pulse width modulation, $T_E$ – excitation transformer, T – step-up transformer)

WTGSs equipped with synchronous generators connected to the grid indirectly (through the power electronic converter) are equipped with an excitation control system and a converter control system. The excitation control system in these cases is similar to the one described above. Static exciters are usually utilized. The difference lies in the other controlled parameter. The synchronous generator operates here at various speeds. The generator rotor speed range is about 50% of the rated speed (e.g. ENERCON wind turbine generator system). This high variation of the generator speed causes high variations of the generator terminal voltage, which is proportional to the speed (for constant field voltage). Keeping the terminal voltage constant can lead to over-excitation. Therefore, the voltage reference should be a function of the rotor speed, or the voltage controller should control the $V_g/\omega$ parameter – terminal voltage and rotor speed quotient. This control quantity protects the generator from over-excitation. This type of control is in fact the flux control.

The structure of the converter control system depends on the converter type used.[19] For the current-source inverter, as presented in Fig. 5.63, the control possibilities are limited control of the real power. Additionally, a function of electromechanical oscillation damping can be implemented.

For the voltage-source inverter, as presented in Fig. 5.64, the control possibilities are wider. The converter enables control of the real and reactive power or the real power and the power factor. The power system stabilizer function is usually utilized. In general, the structure of the controller is similar to that of the doubly-fed asynchronous machine converter controller. The *dq*-reference frame concept is used to control the real and reactive power separately.

## 5.7 Power System Modeling

### 5.7.1 Power System as an Object of Control

The electric power system, Fig. 5.65, is a set of interconnected devices, designed to generate, transport and distribute electric energy [32]. Its aim is to supply electric energy with specified parameters (power quality) and in a reliable way to consumers. The frequency and voltage are the basic quality parameters. Therefore, the basic control processes in an electric power system can be defined as:

- Frequency control process in which the frequency is the controlled parameter, and the disturbing parameters are real power load changes, and the aim of control is to balance the real power through the real power generation change.
- Voltage control processes, in which the controlled parameters are the voltages at the power system nodes, the disturbing parameters are reactive power load changes, and the aim of control is to balance the reactive power. In practice, this aim is realized by maintaining the node voltages in the desired ranges.

[19] Because the power flows here in one direction only, converters with non-controlled rectifiers can be (and usually are) utilized.

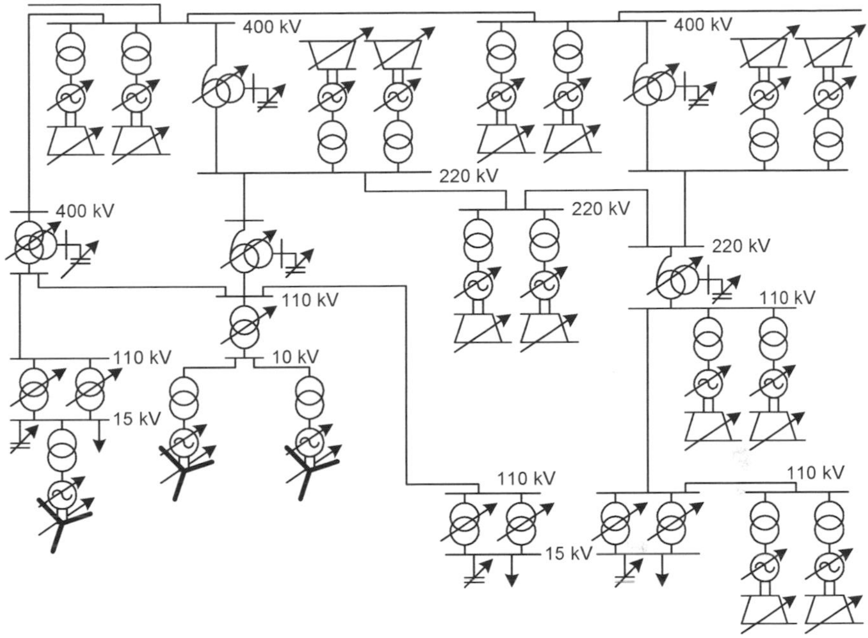


**Fig. 5.65.** Structure of the electric power system (the arrows show the controlled elements of the electric power system)[20]

To make these considerations clear, this model disregards the coincidences between real and reactive power in the power system. But in fact, the real and the reactive power generation and flow, the node voltages, and the frequency variation are closely related. Therefore, both control processes in the electric power system should be (and usually are) closely related too. The process of power system control is realized taking into account not only the technical but the economic requirements as well.

The electric power system, from a geographic point of view, is a widely spread system. At the same time, from the point of view of the control system structure, the system is hierarchical and multilevel. The system is characterized by a large number of controllable elements of various degrees of importance to the system operation.

The frequency control process is usually coordinated by the highest level dispatching center, and all power stations (including WTGSs and wind farms) are involved, directly or indirectly, in the control process. The executive elements of the process are the turbine controllers (governors) are the executive elements of the process.

The voltage-level control processes are coordinated at all levels of the power control system, from the power stations to the individual consumers of electric energy. The following control elements are utilized for voltages-control purposes:

[20] The rated voltage of the network depends on the power system.

- synchronous generators,
- transformers and autotransformers in EHV, HV, MV and LV networks,
- shunt reactors (in high-level voltage networks),
- capacitor banks (in medium-level voltage networks, and individually on load devices),
- flexible AC transmission systems (FACTS). Including: static condenser (STATCON), phase-angle regulators (PAR), unified power flow controllers (UPFC), energy storage devices (SMES, BES), static var compensators (SVC) with: thyristor-controlled reactors (TCR), thyristor-switched capacitors (TCS), controlled parallel reactances (CPR), or controlled series capacitors (CSC, TCSC)),
- high voltage DC systems (HVDC),
- doubly-fed asynchronous machines in the WTGS (when the voltage control algorithm is utilized).[21]

At the control stage, the reliability of electric energy distribution is achieved by the power grid (network) configuration and by the load flow control in the system steady-states. It is achieved through the protection systems and proper control of generating units, rectifiers and inverters of DC systems, and FACTS systems during transient states.

### 5.7.2 Power System Model

The electric power system is a nonlinear system. The nonlinearity mainly results from:

- existence of limitations, saturation, dead zones, etc. in continuous-control systems,
- existence of step-type characteristics in non-continuous control systems (e.g. tap controllers, capacitor bank controllers),
- employment of electronic switching devices (i.e. thyristors, diodes),
- existence of saturation effect of magnetic circuits (i.e. synchronous generators, motors, transformers, reactors, etc.),
- existence of nonlinear dependability between the power (flowing into and from the network and flowing through transmission elements) and the nodal voltages,
- nature of the flow phenomena in turbines.

Because of these nonlinearities, the model of the power system can be described by the nonlinear function:

$$\boldsymbol{F}(\boldsymbol{x}, \dot{\boldsymbol{x}}, \ddot{\boldsymbol{x}},..., t) = 0\,, \tag{5.239}$$

where $\boldsymbol{x}$ – time-dependent state vector,

[21] Usually, in such a type of WTGS, the power factor $\cos\varphi$ is controlled, which means that present wind turbines are not utilized for voltage control.

$t$ – time,
$\boldsymbol{F}$ – nonlinear function.

In practical implementations, the power system model is simplified and takes the form of a set of nonlinear first-order differential equations and a set of algebraic equations. Then the model, in general, can be described by

$$\begin{aligned}\dot{\boldsymbol{x}} &= \boldsymbol{f}(\boldsymbol{x},\boldsymbol{u},t)\\ \boldsymbol{g}(\boldsymbol{x},\boldsymbol{y},\boldsymbol{u},t) &= \boldsymbol{0}\end{aligned} \tag{5.240}$$

where $\boldsymbol{y}$ – vector of outputs,
$\boldsymbol{u}$ – vector of inputs,
$\boldsymbol{f} = [f_1 f_2 \dots f_n]^T$ – vector of nonlinear functions,
$\boldsymbol{g} = [g_1 g_2 \dots g_m]^T$ – vector of algebraic functions.

The state vector $\boldsymbol{x}$, output vector $\boldsymbol{y}$ and input vector $\boldsymbol{u}$ consist of subvectors related to the modeled elements of the power system, which can be defined as

$$\begin{aligned}
\boldsymbol{x}=[&\boldsymbol{x}_{\mathrm{G1}}\dots\boldsymbol{x}_{\mathrm{G}N_{\mathrm{g}}}\ \boldsymbol{x}_{\mathrm{GC1}}\dots\boldsymbol{x}_{\mathrm{GC}N_{\mathrm{g}}}\ \boldsymbol{x}_{\mathrm{T1}}\dots\boldsymbol{x}_{\mathrm{T}N_{\mathrm{T}}}\ \boldsymbol{x}_{\mathrm{TC1}}\dots\boldsymbol{x}_{\mathrm{TC}N_{\mathrm{TC}}},\\
&\boldsymbol{x}_{\mathrm{Tr1}}\dots\boldsymbol{x}_{\mathrm{Tr}N_{\mathrm{Tr}}}\ \boldsymbol{x}_{\mathrm{TrC1}}\dots\boldsymbol{x}_{\mathrm{TrC}N_{\mathrm{TrC}}}\ \boldsymbol{x}_{\mathrm{L1}}\dots\boldsymbol{x}_{\mathrm{L}N_{\mathrm{L}}},\\
&\boldsymbol{x}_{\mathrm{LO1}}\dots\boldsymbol{x}_{\mathrm{LO}N_{\mathrm{LO}}}\ \boldsymbol{x}_{\mathrm{M1}}\dots\boldsymbol{x}_{\mathrm{M}N_{\mathrm{M}}}\ \boldsymbol{x}_{\mathrm{AE}}]^T\\
\boldsymbol{y}=[&\boldsymbol{y}_{\mathrm{G1}}\dots\boldsymbol{y}_{\mathrm{G}N_{\mathrm{g}}}\ \boldsymbol{y}_{\mathrm{GC1}}\dots\boldsymbol{y}_{\mathrm{GC}N_{\mathrm{g}}}\ \boldsymbol{y}_{\mathrm{T1}}\dots\boldsymbol{y}_{\mathrm{T}N_{\mathrm{T}}}\ \boldsymbol{y}_{\mathrm{TC1}}\dots\boldsymbol{y}_{\mathrm{TC}N_{\mathrm{TC}}},\\
&\boldsymbol{y}_{\mathrm{Tr1}}\dots\boldsymbol{y}_{\mathrm{Tr}N_{\mathrm{Tr}}}\ \boldsymbol{y}_{\mathrm{TrC1}}\dots\boldsymbol{y}_{\mathrm{TrC}N_{\mathrm{TrC}}}\ \boldsymbol{y}_{\mathrm{L1}}\dots\boldsymbol{y}_{\mathrm{L}N_{\mathrm{L}}},\\
&\boldsymbol{y}_{\mathrm{LO1}}\dots\boldsymbol{y}_{\mathrm{LO}N_{\mathrm{LO}}}\ \boldsymbol{y}_{\mathrm{M1}}\dots\boldsymbol{y}_{\mathrm{M}N_{\mathrm{M}}}\ \boldsymbol{y}_{\mathrm{AE}}]^T\\
\boldsymbol{u}=[&\boldsymbol{u}_{\mathrm{G1}}\dots\boldsymbol{u}_{\mathrm{G}N_{\mathrm{g}}}\ \boldsymbol{u}_{\mathrm{GC1}}\dots\boldsymbol{u}_{\mathrm{GC}N_{\mathrm{g}}}\ \boldsymbol{u}_{\mathrm{T1}}\dots\boldsymbol{u}_{\mathrm{T}N_{\mathrm{T}}}\ \boldsymbol{u}_{\mathrm{TC1}}\dots\boldsymbol{u}_{\mathrm{TC}N_{\mathrm{TC}}},\\
&\boldsymbol{u}_{\mathrm{Tr1}}\dots\boldsymbol{u}_{\mathrm{Tr}N_{\mathrm{Tr}}}\ \boldsymbol{u}_{\mathrm{TrC1}}\dots\boldsymbol{u}_{\mathrm{TrC}N_{\mathrm{TrC}}}\ \boldsymbol{u}_{\mathrm{L1}}\dots\boldsymbol{u}_{\mathrm{L}N_{\mathrm{L}}},\\
&\boldsymbol{u}_{\mathrm{LO1}}\dots\boldsymbol{u}_{\mathrm{LO}N_{\mathrm{LO}}}\ \boldsymbol{u}_{\mathrm{M1}}\dots\boldsymbol{u}_{\mathrm{M}N_{\mathrm{M}}}\ \boldsymbol{u}_{\mathrm{AE}}]^T
\end{aligned} \tag{5.241}$$

where $\boldsymbol{x}_{\mathrm{G}i}, \boldsymbol{y}_{\mathrm{G}i}, \boldsymbol{u}_{\mathrm{G}i}$ – state, output and input vector of $i$th synchronous generator,
$\boldsymbol{x}_{\mathrm{GC}i}, \boldsymbol{y}_{\mathrm{GC}i}, \boldsymbol{u}_{\mathrm{GC}i}$ – state, output and input vector of $i$th generator controller,
$\boldsymbol{x}_{\mathrm{T}i}, \boldsymbol{y}_{\mathrm{T}i}, \boldsymbol{u}_{\mathrm{T}i}$ – state, output and input vector of $i$th turbine,
$\boldsymbol{x}_{\mathrm{TC}i}, \boldsymbol{y}_{\mathrm{TC}i}, \boldsymbol{u}_{\mathrm{TC}i}$ – state, output and input vector of $i$th turbine controller,
$\boldsymbol{x}_{\mathrm{Tr}i}, \boldsymbol{y}_{\mathrm{Tr}i}, \boldsymbol{u}_{\mathrm{Tr}i}$ – state, output and input vector of $i$th transformer,
$\boldsymbol{x}_{\mathrm{TrC}i}, \boldsymbol{y}_{\mathrm{TrC}i}, \boldsymbol{u}_{\mathrm{TrC}i}$ – state, output and input vector of $i$th transformer controller,
$\boldsymbol{x}_{\mathrm{L}i}, \boldsymbol{y}_{\mathrm{L}i}, \boldsymbol{u}_{\mathrm{L}i}$ – state, output and input vector of $i$th transmission line,
$\boldsymbol{x}_{\mathrm{LO}i}, \boldsymbol{y}_{\mathrm{LO}i}, \boldsymbol{u}_{\mathrm{LO}i}$ – state, output and input vector of $i$th load,
$\boldsymbol{x}_{\mathrm{M}i}, \boldsymbol{y}_{\mathrm{M}i}, \boldsymbol{u}_{\mathrm{M}i}$ – state, output and input vector of $i$th motor,
$\boldsymbol{x}_{\mathrm{AE}i}, \boldsymbol{y}_{\mathrm{AE}i}, \boldsymbol{u}_{\mathrm{AE}i}$ – state, output and input vector of $i$th other component,
$N_j$ – number of $j$th type components.

The nonlinear mathematical model of an electric power system, defined by (5.240), is widely used for time domain analysis.[22] Unfortunately, some of the known mathematical methods of system analysis – mainly related to the system steady-state stability evaluation, frequency analysis, controller synthesis (LQR, LQG, $H_2$, $H_\infty$, μ-synthesis, etc.) – do not accept nonlinear models. Then, the linear model of the power system must be utilized, usually in the form of *state equations*:

$$\begin{aligned} \Delta\dot{x} &= \boldsymbol{A}\Delta\boldsymbol{x} + \boldsymbol{B}\Delta\boldsymbol{u} \\ \Delta y &= \boldsymbol{C}\Delta\boldsymbol{x} + \boldsymbol{D}\Delta\boldsymbol{u} \ , \end{aligned} \tag{5.242}$$

where

$$\boldsymbol{A} = \begin{bmatrix} \frac{\partial f_1}{\partial x_1} & \cdots & \frac{\partial f_1}{\partial x_n} \\ \vdots & \ddots & \vdots \\ \frac{\partial f_n}{\partial x_1} & \cdots & \frac{\partial f_n}{\partial x_n} \end{bmatrix} \qquad \boldsymbol{B} = \begin{bmatrix} \frac{\partial f_1}{\partial u_1} & \cdots & \frac{\partial f_1}{\partial u_r} \\ \vdots & \ddots & \vdots \\ \frac{\partial f_n}{\partial u_1} & \cdots & \frac{\partial f_n}{\partial u_r} \end{bmatrix}$$

$$\boldsymbol{C} = \begin{bmatrix} \frac{\partial g_1}{\partial x_1} & \cdots & \frac{\partial g_1}{\partial x_n} \\ \vdots & \ddots & \vdots \\ \frac{\partial g_m}{\partial x_1} & \cdots & \frac{\partial g_m}{\partial x_n} \end{bmatrix} \qquad \boldsymbol{D} = \begin{bmatrix} \frac{\partial g_1}{\partial u_1} & \cdots & \frac{\partial g_1}{\partial u_r} \\ \vdots & \ddots & \vdots \\ \frac{\partial g_m}{\partial u_1} & \cdots & \frac{\partial g_m}{\partial u_r} \end{bmatrix}$$

where $\Delta\boldsymbol{x} = [\Delta x_1\ \Delta x_2 \ldots \Delta x_n]^T$ – state vector,
$\Delta\boldsymbol{y} = [\Delta y_1\ \Delta y_2 \ldots \Delta y_m]^T$ – output vector,
$\Delta\boldsymbol{u} = [\Delta u_1\ \Delta u_2 \ldots \Delta u_r]^T$ – input vector,
$\boldsymbol{A}, \boldsymbol{B}, \boldsymbol{C}, \boldsymbol{D}$ – state matrix of size $n \times n$, input matrix of size $n \times r$, output matrix of size $m \times n$, output matrix of size $m \times r$, respectively, which defines the proportion of the input which appears directly in the output,
$n, m, r$ – number of state variables, number of output variables, number of input variables, respectively.

Various types of the power system models are utilized for power system analysis. The single machine and the multi-machine model [2, 32, 50–51, 58–61, 66, 91, 94, 95] are the two main types. The models differ in the number of simplifications introduced and the domain of applicability. Therefore, the use of a given model depends on the problem analyzed.

For example, because of its simplicity and some advantages (e.g. mathematical model of the synchronous generator with fewer simplifications than in the multi-machine system), the single-machine power system model is often used for:

[22] The power network is modeled by a set of algebraic equations, while the generating units (and sometimes loads), FACTS and AC/DC systems are modeled by a set of differential and algebraic equations.

- the structure and parameters of synchronous generator and/or turbine controllers optimization,
- preliminary tests of newly established control systems,
- analysis of dynamic properties of turbine generator unit control systems,
- steady-state and transient stability analysis.

The multi-machine models of a power system are usually used for:

- verification of control systems algorithms in "near real" conditions (where "near real" means a multi-modal system[23]),
- steady state and transient stability analysis.

### 5.7.3 Power System Components Models

A power system model, as mentioned before, can be divided into two parts: (1) the generating units, FACTS, AC/DC systems, and sometimes motors and their controllers (modeled as dynamic objects) and (2) the power network model (usually modeled as a set of algebraic equations).

The dynamic models (excluding the asynchronous and synchronous generator models presented in Sects. 5.3 and 5.4) are the subject of the current chapter. The power network models are considered in the next two sections.

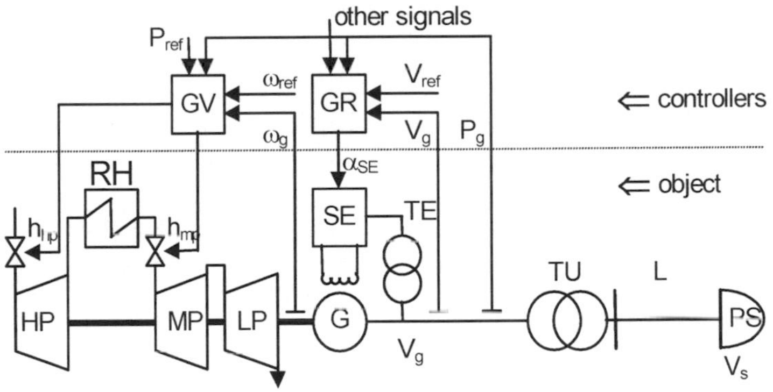


**Fig. 5.66.** Structure of a generating unit connected to the power system (G – synchronous generator; SE – static exciter; TE – excitation transformer; TU – step-up transformer; L – transmission line; PS – power system; GR – synchronous generator regulator (AVR + PSS); HP, MP, LP – high, medium and low pressure section of turbine; RH – reheater, GV – governor; $V_g$, $V_{ref}$ – synchronous generator voltage and reference voltage; $\omega_g$, $\omega_{ref}$ – shaft speed and reference speed; $P_g$, $P_{ref}$ – real power and reference power; $\alpha_{SE}$ – firing angle of thyristor rectifier; $h_{hp}$, $h_{mp}$ – control and intercept valve position)

[23] Its multi-modality is the positive feature of the multi-machine power system, which is impossible to obtain in a single-machine system. This model can produce many more complex and difficult operating conditions (characteristic of the real system) for turbine and generator controllers than those obtainable in the single-machine system.

A generating unit is the basic component of a power system. Its model – for the type of analysis considered here – is usually limited to a turbine with a governor and a generator excitation system (including controller). The generator is connected to the power network bus through a step-up transformer and transmission line. Such a mathematical model (shown here as a single-machine power system) is presented in Fig. 5.66. Models of the components of the generating unit are presented below.

### 5.7.3.1 Excitation System Model

The standard IEEE models of the power system components are utilized in many analyses [37–44]. These types of model are presented below. Additionally, the more detailed models of the static and AC excitation systems are included. The reason is as follows. The value of the field voltage in the IEEE static excitation system model (e.g. ST1) is limited and the limits depend on the terminal voltage. Unfortunately, only the ceiling voltage is limited in this model. The terminal voltage does not influence the exciter (thyristor rectifier) when the field voltage is between ceilings. It can lead to differences between the dynamic properties of the real generating unit and the one modeled by the IEEE type-ST1 model. The use of the IEEE excitation system model makes the modeled system more stable than the real system. The excitation system, presented in this section after the IEEE excitation system, regards the feature of the real system, i.e. dependence of static rectifier output on terminal voltage.

*IEEE Excitation system model*

Synchronous generators in an electric power system, in general, are equipped with two types of excitation systems: static excitation systems and AC excitation systems.

The static excitation system (for example, the IEEE type-ST1 excitation system) represents a voltage controller and a potential-source controlled rectifier excitation system, whose AC power source is a simple power transformer fed from the generator terminals. The exciter ceiling voltage is directly proportional to the generator terminal voltage. The voltage drop on the excitation transformer and rectifier are proportional to the parameter $K_c$. A block diagram of the model is presented in Fig. 5.67.

The inputs of the model are the compounded generator terminal voltage $V_g$ and the generator field current $I_f$. The output is an excitation electromotive force $E_f$ proportional to the excitation voltage $V_f$.

The AC excitation system (for example, the IEEE type-AC1 excitation system) represents a field-controlled alternator excitation system with non-controlled rectifiers, and is applicable to a brushless excitation system. The diode rectifier characteristic imposes a lower limit of zero on the exciter output voltage. The exciter field winding is supplied by a pilot exciter. Then the voltage regulator power supply is not affected by external transients. This model is suitable for an excitation system using an alternator, non-controlled stationary rectifiers and slip rings.

A block diagram of the model is presented in Fig. 5.68. The inputs of the model are the compounded generator terminal voltage $V_g$ and the generator field current $I_f$. The output is an excitation electromotive force $E_f$ proportional to the excitation voltage $V_f$.

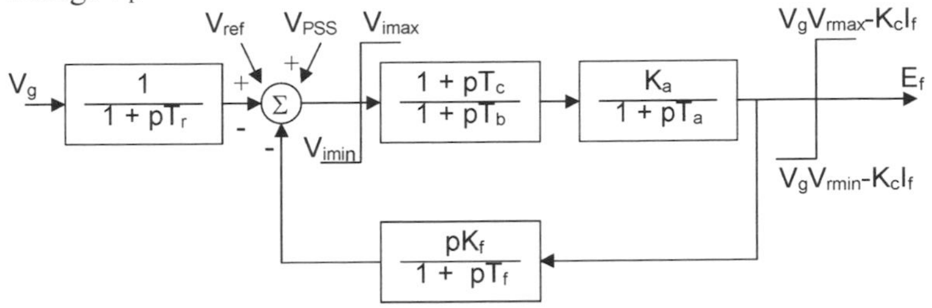


**Fig. 5.67.** Block diagram of the voltage controller with static exciter model [37] ($V_g$ – terminal voltage; $V_{ref}$ – reference voltage; $I_f$ – field current; $E_f$ – excitation electromotive force proportional to excitation voltage $V_f$; $T_r$ – filter time constant; $V_{imax}$, $V_{imin}$ – maximum and minimum error; $T_c$, $T_b$ – lead and lag time constant $K_a$ – gain; $T_a$ – time constant; $V_{rmax}$, $V_{rmin}$ – maximum and minimum controller output; $K_c$ – excitation system regulation factor; $K_f$ – feedback gain; $T_f$ – rate feedback time constant; $V_{PSS}$ – signal from PSS)

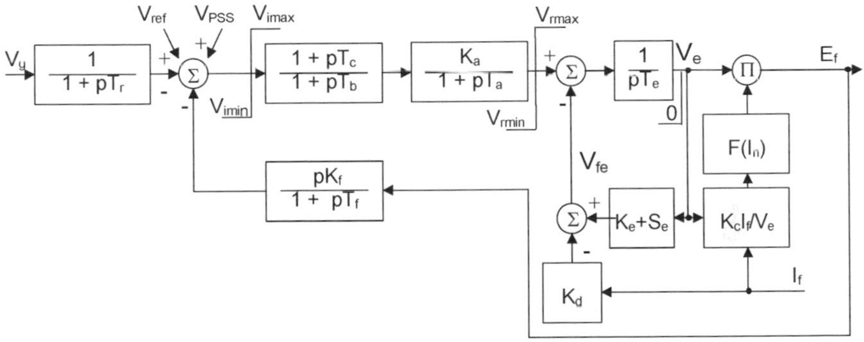


**Fig. 5.68.** Block diagram of the voltage controller with AC machine exciter model [37], ($E_f$ – excitation electromotive force proportional to excitation voltage $V_f$; $T_r$ – filter time constant; $T_b$, $T_c$ - lag and lead time constant; $K_a$ – voltage regulator gain; $T_a$ – time constant; $V_{imax}$, $V_{imin}$ – maximum and minimum error; $V_{rmax}$, $V_{rmin}$ – maximum and minimum control element output; $T_e$ – exciter time constant; $K_f$ – rate feedback gain; $T_f$ – rate feedback time constant; $K_c$ – excitation system regulation factor; $K_d$ – exciter internal reactance; $K_e$ – exciter field resistance constant; $S_e$ – saturation factor; $V_{PSS}$ – signal from PSS)

Synchronous generators operating in the power system are usually equipped with power system stabilizers (PSS), whose aim is to damp the electromechanical oscillations. A standard structure of the stabilizer is presented in Fig. 5.69. The following can be used as the input of the PSS model: generator shaft speed, frequency, generator real power or accelerating power, or armature current. Usually, the real power is introduced as the input to the power system stabilizers.

Two input power system stabilizers are utilized today in power systems as well. In this case, the turbine shaft speed and terminal real power are utilized as input signals.

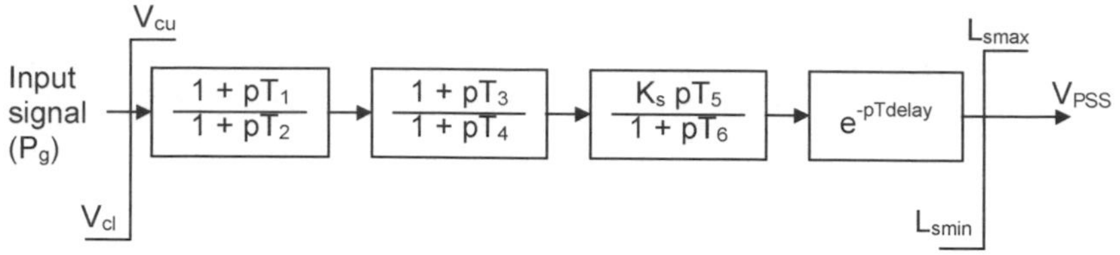


**Fig. 5.69.** Block diagram of the power system stabilizer model ($T_1$–$T_4$ – lead/lag time constant; $T_5$, $T_6$ – washout numerator and denominator time constant; $K_s$ – stabilizer gain; $L_{smax}$, $L_{smin}$ – maximum and minimum stabilizer output; $V_{cu}$, $V_{cl}$ – stabilizer input upper and lower cut-off threshold; $T_{delay}$ – time delay; $V_{PSS}$ – stabilizer output)

### *Static excitation system model*

The static excitation system consists of an excitation transformer and a controlled rectifier. The excitation power is supplied through a transformer from the generator terminals. Therefore, the exciter voltage, which feeds the thyristor rectifier, depends proportionally on the generator terminal voltage.

A mathematical model of the static excitation system [32, 59, 71, 95] consists of an equation for the controlled rectifier only (5.243).[24]

$$V_f = \begin{cases} A(V_\gamma, I_f, \alpha_{SE}) & \text{when } 0 \le \dfrac{I_f X_\gamma}{V_\gamma} < \dfrac{\sqrt{6}}{4} \\ \min\{A(V_\gamma, I_f, \alpha_{SE}), C(V_\gamma, I_f, \alpha_{SE})\} & \text{when } \dfrac{\sqrt{6}}{4} \le \dfrac{I_f X_\gamma}{V_\gamma} < \dfrac{\sqrt{6}}{3}, \\ 0 & \text{when } \dfrac{I_f X_\gamma}{V_\gamma} \ge \dfrac{\sqrt{6}}{3} \end{cases} \quad (5.243)$$

where

$$\begin{aligned} A(V_\gamma, I_f, \alpha_{SE}) &= \frac{3\sqrt{2}}{\pi} V_\gamma \cos\alpha_{SE} - \frac{3}{\pi} I_f X_\gamma \\ B(V_\gamma, I_f) &= \frac{3\sqrt{6}}{2\pi}\sqrt{V_\gamma^2 - 2(I_f X_\gamma)^2} \\ C(V_\gamma, I_f, \alpha_{SE}) &= \frac{3\sqrt{6}}{\pi} V_\gamma \cos(\alpha_{SE} - \frac{\pi}{6}) - \frac{9}{\pi} I_f X_\gamma, \end{aligned} \quad (5.244)$$

where $V_f$ – excitation (field) voltage,
$V_\gamma$ – voltage-feeding rectifier equal to line-to-line voltage of the low-side voltage of excitation transformer,
$I_f$ – field current,

[24] The presented controlled rectifier model described by (5.237) is a simplified model of the one defined by (5.243).

$\alpha_{SE}$ – firing angle,
$X_\gamma$ – commutating reactance (reactance of excitation transformer).

The excitation transformer is considered as a non-inertia element, and its serial reactance is included in the rectifier equation. The rectifier operating area is limited by curves $A(V_\gamma, I_f, 0)$, $-A(V_\gamma, I_f, 0)$, $B(V_\gamma, I_f)$, $-B(V_\gamma, I_f)$, $C(V_\gamma, I_f, \pi/6)$ and $-C(V_\gamma, I_f, \pi/6)$.

The commutating reactance $X_\gamma$ can be computed by using (5.245):

$$X_\gamma = u_{kn} \frac{V_{nTE}^2}{S_{nTE}} \tag{5.245}$$

where $V_{nTE}$ – rated secondary winding (low) voltage of excitation transformer,
$S_{nTE}$ – volt-ampere rating of excitation transformer,
$u_{kn}$ – short-circuit voltage of excitation transformer.

The voltage-current characteristics of the controlled rectifier is presented in Fig. 5.70.

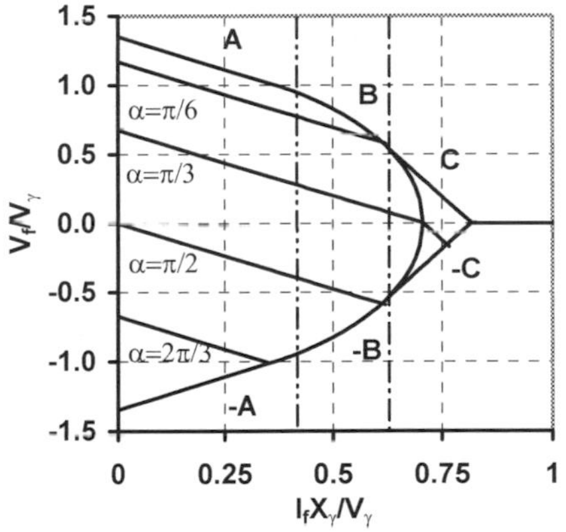


**Fig. 5.70.** Characteristics of controlled rectifier (when $L_f/R_f \rightarrow \infty$)

### *AC excitation system model*

The AC excitation system [32, 51, 60] consists of three basic elements (the voltage controller is not presented here):

- controlled rectifier feeding the field winding of the AC exciter,
- AC exciter (i.e. synchronous machine),
- non-controlled (diode) rectifier.

The structure of the system is presented in Fig. 5.71 and the mathematical models of its elements are presented below.

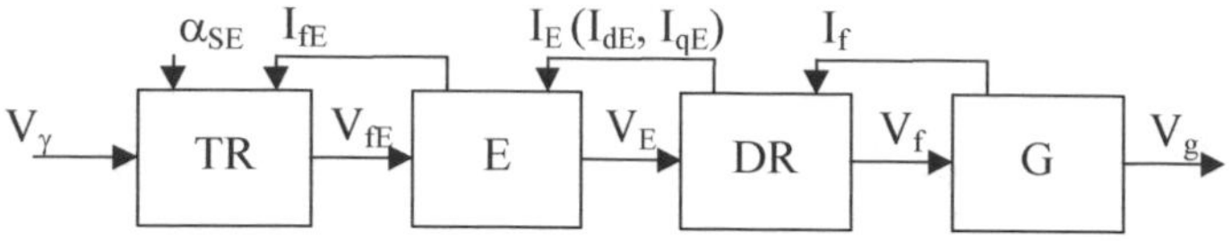


**Fig. 5.71.** AC excitation system (TR – controlled rectifier; E – synchronous exciter; DR – diode rectifier; G – synchronous generator; $V_g$ – synchronous generator voltage; $V_f$ – synchronous generator field voltage; $I_f$ – synchronous generator field current; $V_E$ – terminal voltage of synchronous exciter; $I_E$ – terminal current of synchronous exciter; $I_{fE}$ – field current of synchronous exciter; $V_{fE}$ – field voltage of synchronous exciter; $\alpha_{SE}$ – firing angle; $V_\gamma$ – voltage of a source feeding controlled rectifier)

*Mathematical model of synchronous exciter*
To model the synchronous exciter the synchronous generator model presented in Sect. 5.4 can be used. But because of the model configuration (see Fig. 5.71), where the input value is the exciter terminal current $I_E$ while the output is the terminal voltage $V_E$, in the synchronous generator model the transformer emfs p $\Psi_d$, p $\Psi_q$ have to be neglected [51, 61]. Additionally, because of the high range of possible synchronous exciter terminal voltages, especially during transients, the saturation effect cannot be neglected. Here, when modeling the round-rotor generator, the saturation effect in both the *d*-axis and the *q*-axis should be considered. But for the salient-rotor machine the saturation effect in the *q*-axis can be (and usually is) neglected.

Taking into consideration the above effects the mathematical model of the salient-rotor synchronous exciter can be described by the following equations (all values, except the rated speed $\omega_{nE}$, are expressed per unit related to the rated values of the machine):
Flux-current equation:

$$\boldsymbol{I}_E = \boldsymbol{L}_E^{-1}(\boldsymbol{\Psi}_E - \Delta\boldsymbol{\Psi}_E) \tag{5.246}$$

Current-voltage equation:

$$\mathrm{p}\boldsymbol{\Psi}_E = \omega_{nE}(\boldsymbol{V}_E - \boldsymbol{R}_E\boldsymbol{I}_E) \tag{5.247}$$

Other equations:

$$\begin{gathered}
\Psi_{adE} = L_{adE}(I_{dE} + I_{fE}/X_{adE} + I_{DE} - I_{satE}) \\
I_{satE} = F(\Psi_{adE}) \\
V_{qE} = \omega(\Psi_{adE} + L_{lE}I_{dE}) - R_{lE}I_{qE} \\
V_{dE} = -\omega(\Psi_{aqE} + L_{lE}I_{qE}) - R_{lE}I_{dE} \\
V_E = \sqrt{V_{dE}^2 + V_{qE}^2} \\
I_E = K_E I_f \, ,
\end{gathered} \tag{5.248}$$

where $\Delta\boldsymbol{\Psi}_E = [L_{adE}(I_{dE} - I_{satE}) \;\; L_{adE}(I_{dE} - I_{satE}) \;\; L_{aqE}I_{qE}]$ – vector of fluxes,
$\boldsymbol{\Psi}_E = [\,\Psi_{fE} \;\; \Psi_{DE} \;\; \Psi_{QE}]^T$ – vector of synchronous exciter fluxes,

$\boldsymbol{I}_{\mathrm{E}} = [I_{\mathrm{fE}}/X_{\mathrm{adE}}\ \ I_{\mathrm{DE}}\ \ I_{\mathrm{QE}}]^T$ – vector of synchronous exciter currents,
$\boldsymbol{V}_{\mathrm{E}} = [V_{\mathrm{fE}}R_{\mathrm{fE}}/X_{\mathrm{adE}}\ \ 0\ \ 0]^T$ – voltages vector,
$\boldsymbol{R}_{\mathrm{E}} = diag[R_{\mathrm{fE}}\ \ R_{\mathrm{DE}}\ \ R_{\mathrm{QE}}]$ – matrix of resistances,

$$\boldsymbol{L}_{\mathrm{E}} = \begin{bmatrix} L_{\mathrm{fE}} & L_{\mathrm{adE}} & 0 \\ L_{\mathrm{adE}} & L_{\mathrm{D}} & 0 \\ 0 & 0 & L_{\mathrm{Q}} \end{bmatrix}$$ – matrix of inductances,

$I_{\mathrm{f}}$ – field current of synchronous generator,
$I_{\mathrm{dE}}$, $I_{\mathrm{qE}}$ – currents of synchronous exciter in *d*-axis and *q*-axis,
$I_{\mathrm{fE}}$, $I_{\mathrm{DE}}$, $I_{\mathrm{QE}}$ – field and amortisseur circuit currents,
$I_{\mathrm{satE}}$ – fictitious current representing saturation effect,
$V_{\mathrm{fE}}$ – field voltage of synchronous exciter,
$V_{\mathrm{E}}$ – terminal voltage of synchronous exciter,
$L_{\mathrm{adE}}$, $L_{\mathrm{aqE}}$ – mutual inductances of exciter (unsaturated),
$L_{\mathrm{fE}}$, $L_{\mathrm{DE}}$, $L_{\mathrm{QE}}$ – field and amortisseur circuit inductances of exciter,
$L_{\mathrm{lE}}$ – stator leakage inductance of exciter,
$\omega$, $\omega_{\mathrm{nE}}$ – rotor speed and rated speed,
$\Psi_{\mathrm{adE}}$, $\Psi_{\mathrm{aqE}}$ – air-gap flux linkage in *d*-axis and *q*-axis,
$\Psi_{\mathrm{fE}}$, $\Psi_{\mathrm{DE}}$, $\Psi_{\mathrm{QE}}$ – exciter field and amortisseur (damper) circuit fluxes,
$K_{\mathrm{E}}$ – proportionality coefficient of a first harmonic of the synchronous exciter terminal current to the synchronous generator field current (Sect. 5.5.2).

Taking into consideration the saturation effect forces us to compute the nonlinear characteristic $I_{\mathrm{satE}} = F(\Psi_{\mathrm{adE}})$. The method of computing the function $F = G^{-1}$ from the air- gap (curve *A*) and saturation characteristic (curve *S*) is presented in Fig. 5.72.

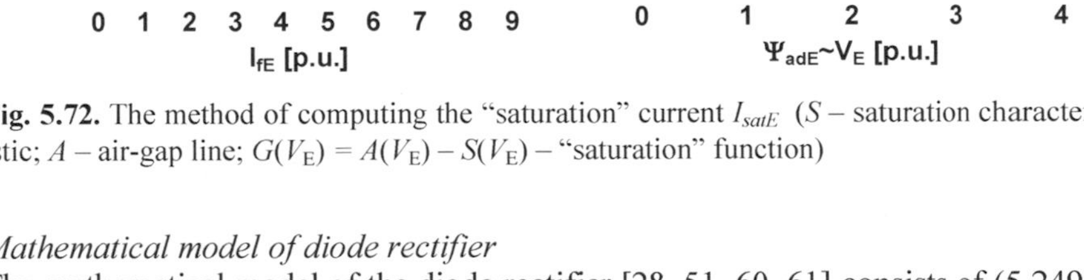

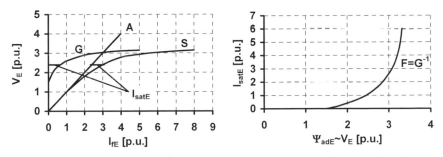

**Fig. 5.72.** The method of computing the "saturation" current $I_{satE}$ (*S* – saturation characteristic; *A* – air-gap line; $G(V_{\mathrm{E}}) = A(V_{\mathrm{E}}) - S(V_{\mathrm{E}})$ – "saturation" function)

*Mathematical model of diode rectifier*

The mathematical model of the diode rectifier [28, 51, 60, 61] consists of (5.249), which divides the operating area into four parts:

$$V_{\mathrm{f}} = \begin{cases} \dfrac{3\sqrt{2}}{\pi} V_{\gamma} - \dfrac{3}{\pi} I_{\mathrm{f}} X_{\gamma} & \text{when } 0 \le \dfrac{I_{\mathrm{f}} X_{\gamma}}{V_{\gamma}} \le \dfrac{\sqrt{2}}{4} \\ \dfrac{3\sqrt{6}}{2\pi} \sqrt{V_{\gamma}^{2} - 2(I_{\mathrm{f}} X_{\gamma})^{2}} & \text{when } \dfrac{\sqrt{2}}{4} < \dfrac{I_{\mathrm{f}} X_{\gamma}}{V_{\gamma}} \le \dfrac{\sqrt{6}}{4} \\ \dfrac{3\sqrt{6}}{\pi} V_{\gamma} - \dfrac{9}{\pi} I_{\mathrm{f}} X_{\gamma} & \text{when } \dfrac{\sqrt{6}}{4} < \dfrac{I_{\mathrm{f}} X_{\gamma}}{V_{\gamma}} \le \dfrac{\sqrt{6}}{3} \\ 0 & \text{when } \dfrac{I_{\mathrm{f}} X_{\gamma}}{V_{\gamma}} \ge \dfrac{\sqrt{6}}{3} \end{cases}, \tag{5.249}$$

where $X_{\gamma} = \frac{1}{2}(X''_{\mathrm{d}} + X''_{\mathrm{q}})$ – commutation reactance depending on the synchronous exciter subtransient reactances,

$V_{\mathrm{g}} = V_{\mathrm{E}}$ – rectifier feeding voltage (synchronous exciter terminal voltage).

The characteristics of the diode rectifier are presented in Fig. 5.73.

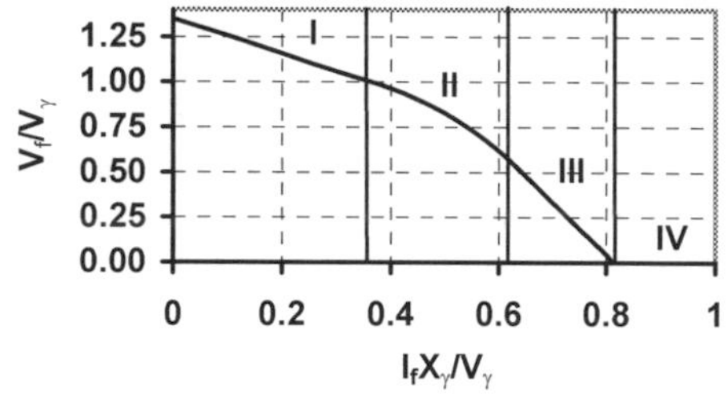


**Fig. 5.73.** Characteristics of diode rectifier (when $L_{\mathrm{f}}/R_{\mathrm{f}} \rightarrow \infty$)

### 5.7.3.2 IEEE Turbine and Governor Models

In the power system model, the turbine models are represented by the hydro turbine and the steam turbine with an adequate governor model. The model whose block diagram is presented in Fig. 5.74 can be considered as the basic IEEE turbine and governor model. The shaft speed $\omega_{\mathrm{g}}$ is the input of the model and the turbine mechanical power $P_{\mathrm{m}}$ is the output. The model, in fact enables the implementation of up to two turbines (e.g. driving two generators). The two turbines can be high- and low-pressure machines, respectively, of a cross-compound steam turbine set, or gas and steam turbine machines of a combined cycle plant. The gains $K_1$–$K_8$ and time constants $T_5$–$T_7$ describe the division of power output among the turbine stages and the transfer of energy in the boiler or combustion prime mover. The values of $K_1$, $K_3$, $K_5$, $K_7$ describe the proportionate development of power on the first turbine shaft and $K_2$, $K_4$, $K_6$, $K_8$ describe the second turbine shaft. Normally, $K_1+K_3+K_5+K_7=1$, $K_2+K_4+K_6+K_8=1$.

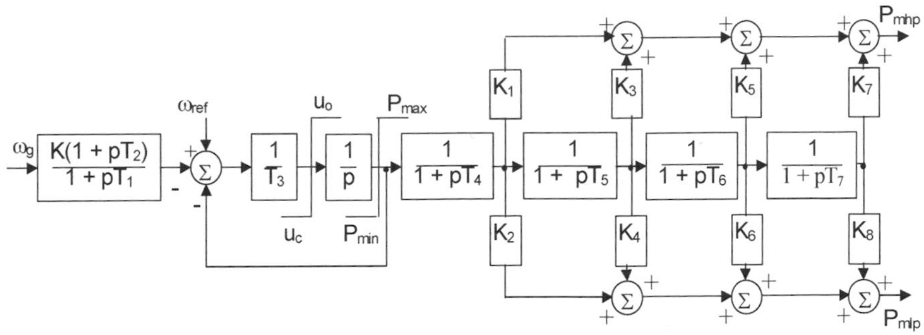


**Fig. 5.74.** Block diagram of the steam turbine model [36] ($P_{mhp}$ – mechanical power high pressure turbine; $P_{mlp}$ – mechanical power low pressure turbine; $K$ – governor gain (reciprocal of droop); $T_1$, $T_2$ – lead and lag governor time constant; $T_3$ – valve positioner time constant; $u_o$ – maximum valve opening velocity; $u_c$ – maximum valve closing velocity; $P_{max}$, $P_{min}$ – maximum and minimum valve opening; $T_4$ – inlet piping/steam bowl time constant; $K_1$, $K_2$ – fraction of high and low pressure turbine power developed after first boiler pass; $T_5$ – time constant of second boiler pass (i.e. reheater); $K_3$, $K_4$ – fraction of high and low pressure turbine power developed after second boiler pass; $T_6$ – time constant of third boiler pass; $K_5$, $K_6$ – fraction of high and low pressure turbine power developed after third boiler pass; $T_7$ – time constant of fourth boiler pass; $K_7$, $K_8$ – fraction of high and low pressure turbine power developed after fourth boiler pass)

The basic IEEE hydro turbine and governor model represents plants with straightforward penstock configurations and hydraulic-dashpot governors or electro-hydraulic governors that mimic dashpot governors. A block diagram of the model is presented in Fig. 5.75. The shaft speed $\omega_g$ is the input of the model and the turbine power $P_m$ is the output.

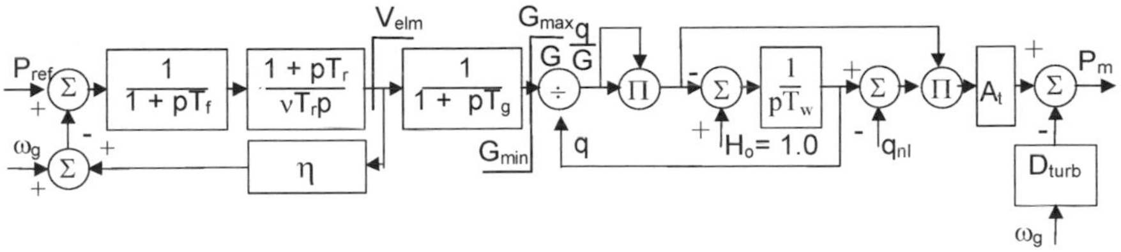


**Fig. 5.75.** Block diagram of the water turbine model [36] ($q$ – water flow; $H$ – hydraulic head at gate; $H_o$ – initial steady-state hydraulic head at gate; $G$ – ideal gate opening; $\eta$ – permanent droop; $\nu$ – temporary droop; $T_r$ – washout time constant; $T_f$ – filter time constant; $T_g$ – gate servo time constant; $V_{elm}$ – maximum gate velocity; $G_{max}$ – maximum gate opening; $G_{min}$ – minimum gate opening; $T_w$ – water inertia time constant; $A_t$ – turbine gain; $D_{turb}$ – turbine damping factor; $q_{nl}$ – no-load turbine flow at nominal head)

### 5.7.4 Power Network Model

In the mathematical model of the power system (described by linear or nonlinear differential equations), the network is always described as a set of algebraic equations. The network equation can be written in the following form of a *nodal equation*:

$$\underline{\boldsymbol{I}} = \underline{\boldsymbol{Y}}\,\underline{\boldsymbol{V}}, \tag{5.250}$$

where $\underline{\boldsymbol{I}} = [\underline{I}_1\ \underline{I}_2 \ldots \underline{I}_p]^T$ – vector of currents flowing into the network (currents injections) at nodes 1...*p*,

$\underline{\boldsymbol{V}} = [\underline{V}_1\ \underline{V}_2 \ldots \underline{V}_p]^T$ – vector of phase-to-neutral voltages at nodes 1...*p*,

$\underline{\boldsymbol{Y}}$ – nodal admittance matrix,

$p$ – number of nodes.

The nodal admittance matrix is a square (usually sparse) matrix with dimension $p \times p$:

$$\underline{\boldsymbol{Y}} = \begin{bmatrix} \underline{Y}_{11} & \underline{Y}_{12} & \cdots & \underline{Y}_{1p} \\ \underline{Y}_{21} & \underline{Y}_{22} & \cdots & \underline{Y}_{2p} \\ \vdots & \vdots & \ddots & \vdots \\ \underline{Y}_{p1} & \underline{Y}_{p2} & \cdots & \underline{Y}_{pp} \end{bmatrix}. \tag{5.251}$$

The matrix elements located on its diagonal $\underline{Y}_{ii}$ are called the self-admittance at node *i* and are equal to the sum of all admittances terminating at node *i* (including any shunt admittance $\underline{y}_{i0}$):

$$\underline{Y}_{ii} = \sum_{j=0}^{p} \underline{y}_{ij}, \tag{5.252}$$

where $\underline{y}_{ij}$ is the admittance of the element connecting nodes *i* and *j*.

The rest of the nodal admittance matrix elements $\underline{Y}_{ij}$ are called the mutual admittances between nodes *i* and *j* and are equal to the negative value of the branch series admittance between nodes *i* and *j*:

$$\underline{Y}_{ij} = -\underline{y}_{ij}. \tag{5.253}$$

The current injection at node *i* ($i = 1, 2, \ldots, p$) can be computed as[25]:

$$\underline{I}_i = \frac{\underline{S}_i^*}{3\underline{V}_i^*} = \frac{P_i - jQ_i}{3\underline{V}_i^*}, \tag{5.254}$$

where $P_i$ – real power injected at node *i*,

$Q_i$ – reactive power injected at node *i*.

[25] This assumed that $\underline{S} = 3\underline{V}\,\underline{I}^*$, which causes the inductive power to be positive ($Q > 0$). When equation $\underline{S} = 3\underline{V}^*\underline{I}$ is used, the inductive power becomes negative ($Q < 0$). Coefficient 3 results from utilizing the phase-to-neutral rms voltage $V$.

The value of the power injected at the $i$th node depends on the nodal voltages $\underline{V}$ and the network structure (parameters of nodal matrix $\underline{Y}$). The following are one of the possible forms of equations that enable us to compute the real and reactive power (assuming $\underline{Y} = G + \mathrm{j}B$ and $\underline{V} = V\mathrm{e}^{\mathrm{j}\delta}$):

$$P_i = 3V_i \sum_{j=1}^{p} (G_{ij} V_j \cos \delta_{ij} + B_{ij} V_j \sin \delta_{ij}) \tag{5.255}$$

$$Q_i = 3V_i \sum_{j=1}^{p} (G_{ij} V_j \sin \delta_{ij} - B_{ij} V_j \cos \delta_{ij}) \,. \tag{5.256}$$

The current injection at the given node (equation (5.254)) depends on the power system element connected to the node. For the sources of energy (generators), the current injection is a function of the power generated. For the loads connected to the given node, the current injection depends on the type of load (and the method of load modeling). When the load is modeled as a dynamic object, the current injection is computed in the same way as that for the sources of energy. When the load is modeled as a static one, three approaches are possible:

- When the load is modeled as a constant admittance (power consumed is proportional to the square of the node voltage), the admittance is added to the nodal admittance matrix as a shunt element. Then the current injection at the considered node is equal to zero.
- When the load is modeled as a constant current (power consumed is proportional to the node voltage), the current injection at the considered node is constant (the load current has the opposite sign to the source current).
- When the load is modeled as a constant power or is nonlinear, the current injection at the considered node is defined by a function of the node voltage. For current injection computing, equation (5.254) in the first case and equation $\underline{I}_i = f(\underline{V}_i)$ in the second case are utilized.

When no load or source of energy is connected to the given node, the current injection is equal to zero.

### 5.7.5 Radial Power Network Model

Wind turbine generator systems are often connected to medium voltage networks[26] or – especially wind farms – to the high-voltage distribution network. These medium-voltage networks and some of the high-voltage networks (parts of the distribution networks) usually operate as radial networks.[27] This permits a relatively simple modeling of the WTGS (or the wind farm) operation in the radial power network, especially when the following assumptions are made:

[26] That is 10kV, 15kV, 20kV or 60kV, depending on the country grid type (voltage level and the WTGS location).

[27] It is usually possible to make these networks closed.

- The bus feeding the radial network (bus *F* in Fig. 5.76) is considered as an infinite bus. It is worth emphasizing that such an approach eliminates the dynamics of the power system, which means that some oscillatory modes do not appear during electromechanical oscillations. To avoid this, when the WTGS rated power is high in comparison to the rated power of the power plants located in the considered system or the WTGS rated power is comparable with the rated power of power plant located near the feeding node, the modeled power system should include these power plants.
- The loads are modeled as a constant admittance and/or a constant current.
- The WTGS or wind farm is connected to the network at one point (e.g. point *G* in Fig. 5.76).

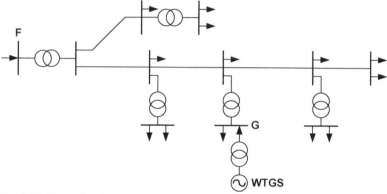

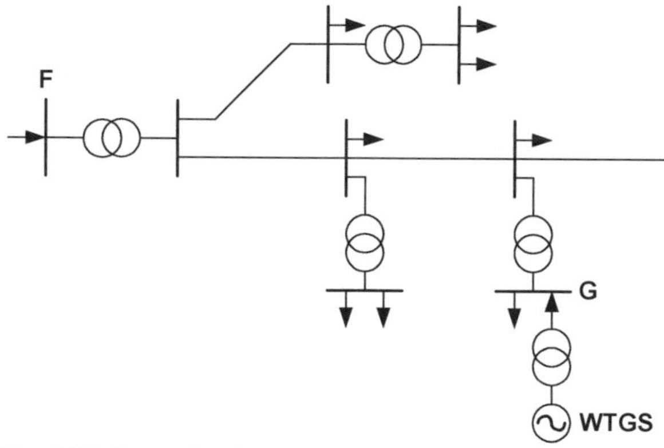


**Fig. 5.76.** Example of radial network

For a network defined in such a way, the nodal equation (equation (5.250)) can be written in the following form:

$$\begin{bmatrix} \underline{\boldsymbol{I}}_\alpha \\ \underline{\boldsymbol{I}}_\beta \end{bmatrix} = \begin{bmatrix} \underline{\boldsymbol{Y}}_a & \underline{\boldsymbol{Y}}_b \\ \underline{\boldsymbol{Y}}_c & \underline{\boldsymbol{Y}}_d \end{bmatrix} \begin{bmatrix} \underline{\boldsymbol{V}}_\alpha \\ \underline{\boldsymbol{V}}_\beta \end{bmatrix} \tag{5.257}$$

where $\underline{\boldsymbol{I}}_\alpha = [\underline{I}_G\ \underline{I}_F]^T$ – vector of currents injected in buses *G* and *F*,
$\underline{\boldsymbol{I}}_\beta = [\underline{I}_3\ \underline{I}_4 \ldots \underline{I}_p]^T$ – vector of currents injected at other nodes, e.g. by loads defined as constant current,
$\underline{\boldsymbol{V}}_\alpha = [\underline{V}_G\ \underline{V}_F]^T$ – vector of voltages at nodes *G* and *F*,
$\underline{\boldsymbol{V}}_\beta = [\underline{V}_3\ \underline{V}_4 \ldots \underline{V}_p]^T$ – vector of voltages at other nodes.

The currents injected by loads $\underline{\boldsymbol{I}}_\beta$ are known here, e.g. when the loads are modeled as constant admittances only, the currents injected are equal to zero. The voltages $\underline{\boldsymbol{V}}_\beta$ are not known.

The nodal equation is utilized here for two purposes: for computing the initial operating point and for computing the appropriate voltages and currents during transient process simulations.

For the first purpose, it is useful to assume that the WTGS voltage $\underline{V}_G$ and current $\underline{I}_G$ are known. Then the feeding bus voltage $\underline{V}_F$ and current $\underline{I}_F$, which are unknown, can be computed in the following way.[28]

Equation (5.257) can be written as a set of equations:

$$\underline{\boldsymbol{I}}_\alpha = \underline{\boldsymbol{Y}}_a \underline{\boldsymbol{V}}_\alpha + \underline{\boldsymbol{Y}}_b \underline{\boldsymbol{V}}_\beta \tag{5.258}$$

$$\underline{\boldsymbol{I}}_\beta = \underline{\boldsymbol{Y}}_c \underline{\boldsymbol{V}}_\alpha + \underline{\boldsymbol{Y}}_d \underline{\boldsymbol{V}}_\beta \,. \tag{5.259}$$

After eliminating the unknown vector of voltages $\underline{\boldsymbol{V}}_\beta$, the vector of source currents $\underline{\boldsymbol{I}}_\alpha$ can be computed as

$$\underline{\boldsymbol{I}}_\alpha = (\underline{\boldsymbol{Y}}_a - \underline{\boldsymbol{Y}}_b \underline{\boldsymbol{Y}}_d^{-1} \underline{\boldsymbol{Y}}_c) \underline{\boldsymbol{V}}_\alpha + \underline{\boldsymbol{Y}}_b \underline{\boldsymbol{Y}}_d^{-1} \underline{\boldsymbol{I}}_\beta \,. \tag{5.260}$$

Equation (5.260) can be written in the following form:

$$\begin{bmatrix} \underline{I}_G \\ \underline{I}_F \end{bmatrix} = \begin{bmatrix} \underline{Y}_{p11} & \underline{Y}_{p12} \\ \underline{Y}_{p21} & \underline{Y}_{p22} \end{bmatrix} \begin{bmatrix} \underline{V}_G \\ \underline{V}_F \end{bmatrix} + \begin{bmatrix} \underline{I}_{p1} \\ \underline{I}_{p2} \end{bmatrix}, \tag{5.261}$$

which enables us to calculate the unknown voltage in the feeding bus $\underline{V}_F$ and feeding current $\underline{I}_F$:

$$\underline{V}_F = \underline{Y}_{p12}^{-1} (\underline{I}_G - \underline{Y}_{p11} \underline{V}_G - \underline{I}_{p1}) \tag{5.262}$$

$$\underline{I}_F = (\underline{Y}_{21} - \underline{Y}_{22} \underline{Y}_{12}^{-1} \underline{Y}_{11}) \underline{V}_G + \underline{I}_{p2} - \underline{Y}_{22} \underline{Y}_{p12}^{-1} (\underline{I}_G - \underline{I}_{p1}) \,. \tag{5.263}$$

During the simulation, the WTGS current $\underline{I}_G$ and the feeding bus $\underline{V}_F$ voltage are known.[29] The other voltages are computed in each step of the simulation. The WTGS voltage (the only one necessary for simulation) can be computed using the following equation:

$$\underline{V}_G = \underline{Y}_{p11}^{-1} (\underline{I}_G - \underline{Y}_{p12} \underline{V}_F - \underline{I}_{p1}) \,. \tag{5.264}$$

The feeding current $\underline{I}_F$ can be computed from (5.263). The other voltages $\underline{\boldsymbol{V}}_\beta$ can be computed as follows:

$$\underline{\boldsymbol{V}}_\beta = \underline{\boldsymbol{Y}}_b^{-1} (\underline{\boldsymbol{I}}_\alpha - \underline{\boldsymbol{Y}}_a \underline{\boldsymbol{V}}_\alpha) \,. \tag{5.265}$$

Generally, a few WTGSs or wind farms can operate in such a type of network. In that case, the power network can be described by (5.257)–(5.260) with $\underline{\boldsymbol{I}}_\alpha = [\underline{I}_{G1} \ldots \underline{I}_{Gk}\ \underline{I}_F]^T$ as the vector of currents injected by the WTGSs and the feeding bus

[28] When the feeding bus voltage value is imposed, the WTGS voltage and current can be computed iteratively by using the WTGS $f(P, Q, V)$ characteristics.

[29] Terminal voltage (and mechanical torque) is the generator model input and the current is the output. Inverse models of generator utilize current as the input and voltage as the output. But such models require other organization of power network computation than the one presented above.

$F$, and with $\underline{V}_\alpha = [\underline{V}_{G1} \ldots \underline{V}_{Gk}\ \underline{V}_F]^T$ as the vector of voltages at generating nodes $G_i$ and feeding node $F$ ($k$ is the number of nodes with WTGSs connected). After eliminating voltages $\underline{V}_\beta$, (5.261) for this case can be written in the following form:

$$\begin{bmatrix} \underline{I}_{G1} \\ \vdots \\ \underline{I}_{Gk} \\ \underline{I}_F \end{bmatrix} = \begin{bmatrix} \underline{Y}_{11} & \cdots & \underline{Y}_{1k} & \underline{Y}_{1,k+1} \\ \vdots & \ddots & \vdots & \vdots \\ \underline{Y}_{k1} & \cdots & \underline{Y}_{kk} & \underline{Y}_{k,k+1} \\ \underline{Y}_{k+1,1} & \cdots & \underline{Y}_{k+1,1} & \underline{Y}_{k+1,k+1} \end{bmatrix} \begin{bmatrix} \underline{V}_{G1} \\ \vdots \\ \underline{V}_{Gk} \\ \underline{V}_F \end{bmatrix} + \begin{bmatrix} \underline{I}_{p1} \\ \vdots \\ \underline{I}_{pk} \\ \underline{I}_{pk+1} \end{bmatrix}. \tag{5.266}$$

This form of equation does not allow us to compute the feeding voltage $\underline{V}_F$ directly, when the WTGS voltages and currents are defined a priori. For computing the initial operating point (equilibrium point), it is necessary to use any of the methods that are used for the load flow calculation. Then the network equation is solved iteratively (see the next section) using techniques such as the Gauss–Seidel, Newton–Raphson (N-R), Fast Decoupled Load Flow (FDLF), or other methods.

During time-domain simulation, the WTGS voltages $\underline{V}_{G1}$–$\underline{V}_{Gk}$ can be computed directly by solving the set of the following equations (derived from (5.266)):

$$\begin{bmatrix} \underline{I}_{G1} \\ \vdots \\ \underline{I}_{Gk} \end{bmatrix} = \begin{bmatrix} \underline{Y}_{11} & \cdots & \underline{Y}_{1k} & \underline{Y}_{1,k+1} \\ \vdots & \ddots & \vdots & \vdots \\ \underline{Y}_{k1} & \cdots & \underline{Y}_{kk} & \underline{Y}_{k,k+1} \end{bmatrix} \begin{bmatrix} \underline{V}_{G1} \\ \vdots \\ \underline{V}_{Gk} \\ \underline{V}_F \end{bmatrix} + \begin{bmatrix} \underline{I}_{p1} \\ \vdots \\ \underline{I}_{pk} \end{bmatrix}, \tag{5.267}$$

where the generators currents $\underline{I}_{G1}$–$\underline{I}_{Gk}$, and the feeding bus voltage $\underline{V}_F$ are known at each step of the simulation.

### 5.7.6 Meshed Power Network Model

Modeling a meshed power network for time-domain simulation, it is necessary to compute:

- initial load flow,
- initial values of the state variables of dynamic models,

and to organize the simulation process with equations allowing computation of the power network response to the power system state change (change of the currents injected into the network[30] or the power network or the change of the dynamic model parameters). This is in fact the load flow computation during each simulation step.

[30] A power system model in which the dynamic elements (e.g. generators) are modeled with current as output and voltage as input is considered.

### 5.7.6.1 Load Flow Problem

An electric power network is a nonlinear object which can be described by a set of equations defining the nodal power (power injected into the network at *i*th node)[31]:

$$P_i = V_i \sum_{j=1}^{p} V_j Y_{ij} \cos(\delta_i - \delta_j - \phi_{ij})$$
$$Q_i = V_i \sum_{j=1}^{p} V_j Y_{ij} \sin(\delta_i - \delta_j - \phi_{ij}), \tag{5.268}$$

where $\underline{V}_i = V_i e^{j\delta_i}$ is the nodal voltage and $\underline{Y}_{ij} = Y_{ij} e^{j\phi_{ij}}$ is the nodal admittance defined above (equations (5.251)–(5.253)). The explicit solution of this set of equations does not exist and therefore the iterative method must be used. One of these methods, which can be treated as a basic method, is the Newton-Raphson method[32]. The method is based on linearization of the nonlinear equations by using the Taylor series and by neglecting the nonlinear elements of the series. After this operation, (5.268) in the form of $\Delta \boldsymbol{g}(\boldsymbol{x}) = \boldsymbol{J} \Delta \boldsymbol{x}$, where $\boldsymbol{J}$ is the matrix of the first partial derivatives (Jacobian matrix), can be solved iteratively.

The process of solving for the load flow can be organized as follows [50, 60, 61]. Each node of the network is described by four quantities: real and reactive power, voltage magnitude and the voltage angle.[33] Because for the *p*-node network, the number of equations is equal to *2p* (*p* equations for real power and *p* equations for reactive power computing) and the total number of quantities describing the node state is equal to *4p,* then for each node two quantities must be set as the reference. Three types of nodes are defined (Table 5.4). For the given node type, the quantities defined as reference do not change during the computation process, while the quantities defined as initial[34] change during the computation process.

**Table 5.4.** Types of nodes in the load flow problem

| Node type | Real power | Reactive power | Voltage magnitude | Voltage angle |
|---|---|---|---|---|
| PV (source) | $P_{ref}$ | $Q_{init}$ | $V_{ref}$ | $\delta_{init}$ |
| PQ (load) | $P_{ref}$[35] | $Q_{ref}$ | $V_{init}$ | $\delta_{init}$ |
| slack | $P_{init}$ | $Q_{init}$ | $V_{ref}$ | $\delta_{ref} = 0$ |

---

[31] All quantities are per unit quantities.

[32] For load flow computation purposes (especially for computing bulk power systems), the methods using sparse matrix techniques are widely utilized.

[33] The problem can be solved also in rectangular coordinates *a*, *b* ($\underline{V} = V_a + jV_b$)

[34] There are set as initial in the computing procedure.

[35] When the loads are modeled as constant admittances then the power $P_{ref}$, $Q_{ref}$ is equal to zero.

Equation (5.268), after linearization, for the network with the number of nodes equal to $p$, can be written in a matrix form:

$$\begin{bmatrix} \Delta \boldsymbol{P} \\ \Delta \boldsymbol{Q} \end{bmatrix} = \begin{bmatrix} \boldsymbol{J}_{\mathrm{a}} & \boldsymbol{J}_{\mathrm{b}} \\ \boldsymbol{J}_{\mathrm{c}} & \boldsymbol{J}_{\mathrm{d}} \end{bmatrix} \begin{bmatrix} \Delta \boldsymbol{\delta} \\ \Delta \boldsymbol{V} \end{bmatrix}, \tag{5.269}$$

where the elements of the Jacobian matrix (submatrices $\boldsymbol{J}_{\mathrm{a}}$, $\boldsymbol{J}_{\mathrm{b}}$, $\boldsymbol{J}_{\mathrm{c}}$, $\boldsymbol{J}_{\mathrm{d}}$) are defined as follows:

- for $j \neq i$:

$$\begin{aligned} J_{\mathrm{a}ij} &= \frac{\partial P_i}{\partial \delta_j} = V_i V_j Y_{ij} \sin(\delta_i - \delta_j - \phi_{ij}) \\ J_{\mathrm{b}ij} &= \frac{\partial P_i}{\partial V_j} = V_i Y_{ij} \cos(\delta_i - \delta_j - \phi_{ij}) \\ J_{\mathrm{c}ij} &= \frac{\partial Q_i}{\partial \delta_j} = -V_i V_j Y_{ij} \cos(\delta_i - \delta_j - \phi_{ij}) \\ J_{\mathrm{d}ij} &= \frac{\partial Q_i}{\partial V_j} = V_i Y_{ij} \sin(\delta_i - \delta_j - \phi_{ij}) \end{aligned} \tag{5.270}$$

- for $j = i$:

$$\begin{aligned} J_{\mathrm{a}ii} &= \frac{\partial P_i}{\partial \delta_i} = -\sum_{\substack{j=1 \\ j\neq i}}^{p} V_i V_j Y_{ij} \sin(\delta_i - \delta_j - \phi_{ij}) \\ J_{\mathrm{b}ii} &= \frac{\partial P_i}{\partial V_i} = 2 V_i Y_{ii} \cos(-\phi_{ii}) + \sum_{\substack{j=1 \\ j\neq i}}^{p} V_j Y_{ij} \cos(\delta_i - \delta_j - \phi_{ij}) \\ J_{\mathrm{c}ii} &= \frac{\partial Q_i}{\partial \delta_i} = \sum_{\substack{j=1 \\ j\neq i}}^{p} V_i V_j Y_{ij} \cos(\delta_i - \delta_j - \phi_{ij}) \\ J_{\mathrm{d}ii} &= \frac{\partial Q_i}{\partial V_i} = 2 V_i Y_{ii} \sin(-\phi_{ij}) + \sum_{\substack{j=1 \\ j\neq i}}^{p} V_j Y_{ij} \sin(\delta_i - \delta_j - \phi_{ij}) \end{aligned} \tag{5.271}$$

For computation purposes, the elements related to the slack node must be eliminated from (5.269). Two rows and two columns are removed from the Jacobian matrix and two elements from the left-hand and right-hand side vectors. For example, if the slack node is numbered as 1, the rows and columns numbered as 1 and $p+1$ should be removed from the Jacobian matrix. At the same time, the first and the $(p+1)$th element should be removed from the relevant vectors.

The new Jacobian matrix $\boldsymbol{J'}_{2(p-1)\times 2(p-1)}$ and the new real/reactive power vector $[\Delta \boldsymbol{P'}\, \Delta \boldsymbol{Q'}]^T_{2(p-1)}$ and voltage angle/voltage magnitude vector $[\Delta \boldsymbol{\delta'}\, \Delta \boldsymbol{V'}]^T_{2(p-1)}$ are created. Then a new equation (converted from (5.268)) is created and can be used to compute the voltage magnitude and the voltage angle correction, when the power error (real or reactive) appears in the system:

$$\begin{bmatrix} \Delta\boldsymbol{\delta}' \\ \Delta\boldsymbol{V}' \end{bmatrix} = \begin{bmatrix} \boldsymbol{J}'_{\mathrm{a}} & \boldsymbol{J}'_{\mathrm{b}} \\ \boldsymbol{J}'_{\mathrm{c}} & \boldsymbol{J}'_{\mathrm{d}} \end{bmatrix}^{-1} \begin{bmatrix} \Delta\boldsymbol{P}' \\ \Delta\boldsymbol{Q}' \end{bmatrix}. \tag{5.272}$$

The nodal power errors (computed in step $k$) are given by

$$\begin{aligned} \Delta P_i^k &= P_i^{k-1} - P_i^k \\ \Delta Q_i^k &= Q_i^{k-1} - Q_i^k \end{aligned} \tag{5.273}$$

where for the source node the substitution $P_i^{k-1} = P_{\mathrm{ref}}$ should be used and for the load node the substitutions $P_i^{k-1} = P_{\mathrm{ref}}$, $Q_i^{k-1} = Q_{\mathrm{ref}}$ should be used.

The new values of the node voltage magnitudes and/or the voltage angles (in step $k$+1) are computed as:

- for the source type node:

$$\delta_i^{k+1} = \delta_i^k + \xi\Delta\delta_i' \tag{5.274}$$

- for the load type node:

$$\begin{aligned} \delta_i^{k+1} &= \delta_i^k + \xi\Delta\delta_i' \\ V_i^{k+1} &= V_i^k + \xi\Delta V_i', \end{aligned} \tag{5.275}$$

where coefficient $\xi$ is introduced to ensure that the computing process converges to the equilibrium point.

The new voltages (magnitudes) and voltage angles are substituted in (5.268), which enables us to compute the new power injected into the nodes. Next, the new power errors are computed (equation (5.273)). When the values of the errors are less than the defined maximum error, the computing process is stopped. When they are not, the next iteration is made.

Finally, during the load flow computation process, various quantities can be computed using the following equations:

- nodal voltages (as complex numbers):

$$\underline{V}_i = V_i \cos\delta_i + \mathrm{j}V_i \sin\delta_i = V_i\mathrm{e}^{\mathrm{j}\delta_i}, \tag{5.276}$$

- power flowing through a transmission element (line, transformer, etc.) from node $i$ to node $j$ (at node $i$):

$$\underline{S}_{ij} = \underline{V}_i((\underline{V}_i - \underline{V}_j)\underline{y}_{ij})^* + \underline{V}_i(\underline{V}_i\underline{y}_{i0})^*, \tag{5.277}$$

  where $\underline{y}_{i0}$ is the shunt admittance of the network element (line, transformer, etc.),
- power consumed by the $i$th load (when the load is defined by admittance $\underline{y}_{\mathrm{L}i}$):

$$\underline{S}_{\mathrm{L}i} = \underline{V}_i (\underline{V}_i \underline{y}_{\mathrm{L}i})^* . \tag{5.278}$$

For the sake of the completeness of the considerations related to steady-state analysis (load flow problem,) the dynamic models should be initialized before the time-domain simulation starts. For this purpose, the assumed and computed voltages and the powers generated by the sources can be utilized to initialize the dynamic models.

The initial state variables of the synchronous generators (and further the turbine and appropriate control systems) can be computed directly, e.g. without iteration, which is presented in Sect. 5.4, while the asynchronous generators (and turbines and appropriate controllers) need iterative computing of the state variables (Sect. 5.3).

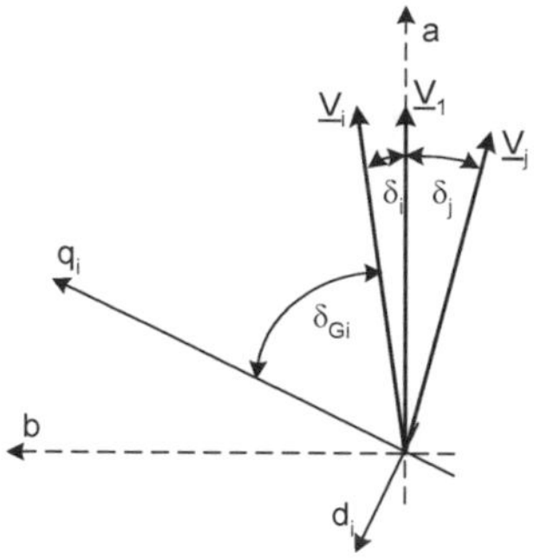


**Fig. 5.77.** Typical location of the network *ab*-reference frame and machine *dq*-reference frame ($\underline{V}_1$ – slack node voltage)

The parameters of the dynamic models of generators are usually expressed in relation to the *dq*-reference frame of each machine, while the nodal voltages of the network are expressed in relation (usually) to the frame related to the slack node voltage. A typical location of these frames is presented in Fig. 5.77. This state forces recalculating currents and voltages from the machine reference frame to the network reference frame and in opposite direction.

### 5.7.6.2 Time Domain Simulation

Time-domain simulation consists of parallel solving of the power network equations and the equations of the dynamic models (differential and algebraic). A typical sequence of simulation process steps can be listed as follows:

- For the given voltages $\underline{V}_i = V_{\mathrm{a}i} + \mathrm{j}V_{\mathrm{b}i}$ at the source nodes, computing voltages in the *dq*-reference frame related to each machine:

$$\begin{aligned} V_{\mathrm{d}i} &= -V_{\mathrm{a}i}\sin(\delta_{\mathrm{G}i} + \delta_i) + V_{\mathrm{b}i}\cos(\delta_{\mathrm{G}i} + \delta_i) \\ V_{\mathrm{q}i} &= V_{\mathrm{a}i}\cos(\delta_{\mathrm{G}i} + \delta_i) + V_{\mathrm{b}i}\sin(\delta_{\mathrm{G}i} + \delta_i) , \end{aligned} \tag{5.279}$$

  where $\delta_{\mathrm{G}i}$ is the power angle of the *i*th generator.

- Solve the differential equations of the dynamic models. The results (among other things) are the generators currents $I_{di}$, $I_{qi}$.
- Convert the machine currents in the $dq$-reference frame into the $ab$-reference frame related to the power network:

$$\begin{aligned} I_{ai} &= \zeta(-I_{di}\sin(\delta_{Gi}+\delta_i)+I_{qi}\cos(\delta_{Gi}+\delta_i)) \\ I_{bi} &= \zeta(I_{di}\cos(\delta_{Gi}+\delta_i)+I_{qi}\sin(\delta_{Gi}+\delta_i)), \end{aligned} \tag{5.280}$$

  where coefficient $\zeta$ is used when the models are per unit and the base power of the various models of synchronous generators (usually equal to the rated power of a machine) and the power network base power are various.
- After creating the new currents (vector $\underline{\boldsymbol{I}}$), compute the new voltages at the network nodes:

$$\underline{\boldsymbol{V}} = \underline{\boldsymbol{Y}}^{-1}\underline{\boldsymbol{I}}. \tag{5.281}$$

- Return to the beginning of the loop.

When the loads in the considered system are modeled as constant admittances, it is possible to simplify the process of computing (5.281) by decreasing the admittance matrix dimension to the number of nodes with sources of energy only. To do this, it is useful to express the nodal equation in the following form:

$$\begin{bmatrix} \underline{\boldsymbol{I}}_{\mathrm{G}} \\ \underline{\boldsymbol{I}}_{\mathrm{L}} \end{bmatrix} = \begin{bmatrix} \underline{\boldsymbol{Y}}_{\mathrm{a}} & \underline{\boldsymbol{Y}}_{\mathrm{b}} \\ \underline{\boldsymbol{Y}}_{\mathrm{c}} & \underline{\boldsymbol{Y}}_{\mathrm{d}} \end{bmatrix} \begin{bmatrix} \underline{\boldsymbol{V}}_{\mathrm{G}} \\ \underline{\boldsymbol{V}}_{\mathrm{L}} \end{bmatrix}, \tag{5.282}$$

where $\underline{\boldsymbol{I}}_{\mathrm{G}}$ is the vector of the source currents and $\underline{\boldsymbol{I}}_{\mathrm{L}}$ is the vector of the currents injected by loads. In this case the vector of the currents injected by the loads is equal to zero, $\underline{\boldsymbol{I}}_{\mathrm{L}} = \mathbf{0}$. By eliminating the vector of voltages $\underline{\boldsymbol{V}}_{\mathrm{L}}$, (5.282) can be converted to the following form:

$$\underline{\boldsymbol{V}}_{\mathrm{G}} = \underline{\boldsymbol{Z}}\,\underline{\boldsymbol{I}}_{\mathrm{G}}, \tag{5.283}$$

where the matrix of impedances $\underline{\boldsymbol{Z}}$ is given by

$$\underline{\boldsymbol{Z}} = (\underline{\boldsymbol{Y}}_{\mathrm{a}} - \underline{\boldsymbol{Y}}_{\mathrm{b}}\underline{\boldsymbol{Y}}_{\mathrm{d}}^{-1}\underline{\boldsymbol{Y}}_{\mathrm{c}})^{-1}. \tag{5.284}$$

Such an approach is also possible when the loads are modeled as a constant current (or partially as a constant admittance and a constant current). Then the following equation can be used:

$$\underline{\boldsymbol{V}}_{\mathrm{G}} = \underline{\boldsymbol{Z}}(\underline{\boldsymbol{I}}_{\mathrm{G}} - \underline{\boldsymbol{K}}\,\underline{\boldsymbol{I}}_{\mathrm{L}}), \tag{5.285}$$

where the impedance matrix $\underline{\boldsymbol{Z}}$ is defined by (5.284) and the matrix $\underline{\boldsymbol{K}}$ is given by

$$\underline{\boldsymbol{K}} = \underline{\boldsymbol{Y}}_{\mathrm{b}}\underline{\boldsymbol{Y}}_{\mathrm{d}}^{-1}. \tag{5.286}$$

The operations presented in (5.283) and (5.285) are based on complex values. When, for any reason, only operations on the real numbers are possible (for example in older versions of Simulink – without calling m-functions), the impedance matrix $\boldsymbol{Z}$ and the voltage and current vectors $\boldsymbol{V}$, $\boldsymbol{I}$ can be expressed in a non-complex form[36]:

$$\boldsymbol{Z} = \begin{bmatrix} R_{11} & -X_{11} & R_{12} & -X_{12} & \cdots & R_{1p} & -X_{1p} \\ X_{11} & R_{11} & X_{12} & R_{12} & \cdots & X_{1p} & R_{1p} \\ R_{21} & -X_{21} & R_{22} & -X_{22} & \cdots & R_{2p} & -X_{2p} \\ X_{21} & R_{21} & X_{22} & R_{22} & \cdots & X_{2p} & R_{2p} \\ \vdots & \vdots & \vdots & \vdots & \ddots & \vdots & \vdots \\ R_{p1} & -X_{p1} & R_{p2} & -X_{p2} & \cdots & R_{pp} & -X_{pp} \\ X_{p1} & R_{p1} & X_{p2} & R_{p2} & \cdots & X_{pp} & R_{pp} \end{bmatrix}$$

$$\boldsymbol{V} = [V_{1a} \quad V_{1b} \quad V_{2a} \quad V_{2b} \quad \cdots \quad V_{pa} \quad V_{pb}]^T \tag{5.287}$$

$$\boldsymbol{I} = [I_{1a} \quad I_{1b} \quad I_{2a} \quad I_{2b} \quad \cdots \quad I_{pa} \quad I_{pb}]^T .$$

Matrix $\boldsymbol{K}$, derived from (5.286), can be defined in the same way. Then the nodal equation can be expressed in the non-complex form.

[36] Such an approach can be applied to the full or reduced impedance matrix $\boldsymbol{Z}$.

# 6 Models of a WTGS Operating in a Power System

## 6.1 Power Network Model

The chapter presents models of various types of WTGS, whose components were considered in the previous chapters. In this chapter the following types of wind turbine models are considered:

- WTGS equipped with squirrel-cage rotor asynchronous generator,
- WTGS equipped with dynamic slip control system,
- WTGS equipped with doubly-fed asynchronous machine,
- WTGS equipped with synchronous generator (two types of converter are considered).

All the modeled WTGSs operate in the model of a radial network presented in Fig. 6.1. In each case, the WTGS is connected to node 5 through the step-up transformer, which is modeled separately, i.e. as an element of the grid. The dynamic model of the WTGS is always connected to node 1. The feeding node is 110 kV node 2. In the case when a compensating element is necessary (which exists in the real wind turbine), it is modeled by a capacitor connected to node 1.

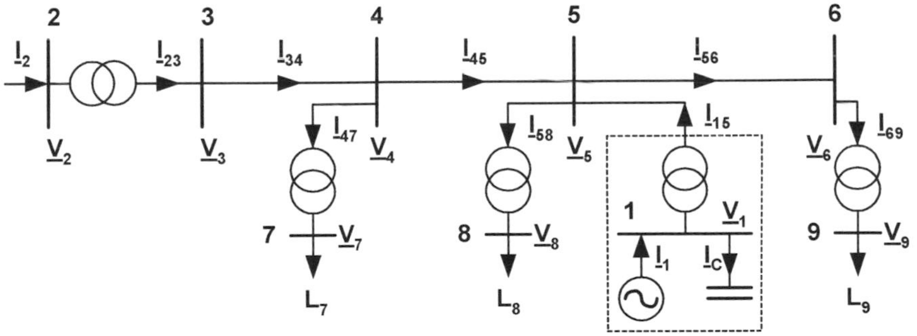


**Fig. 6.1.** Structure of the model radial network (the WTGS is bordered by the rectangle)

The radial network is modeled because of its simplicity, while the main attention is focused on the WTGS as a dynamic system. An additional reason for using the radial network is that all MV networks and many HV (e.g. 110 kV) networks operate as radial networks.

In the example considered here, the MV network with rated voltage equal to $V_n$ = 15 kV and overhead transmission lines is considered. The parameters of the lines and transformers are presented in Tables 6.1–6.4. This relatively weak network has been chosen to achieve a state where the WTGS rated power is near or above the limit defined by the often used condition, $S_n \leq S''_k / 20$ . In the considered network, the short-circuit apparent power at node 5 is equal to $S''_k$ = 13.6 MVA, which means that a WTGS with rated power below $S_n$ = 0.68 MVA can be connected to node 5.

In the following Sections, WTGSs with asynchronous generators with rated power equal to 0.9 MW, which is above the limit, and WTGSs with synchronous generators with rated power 0.5 MW, which is below the limit, are considered.

The LV loads (0.4 kV) are connected to nodes 7, 8, 9 (Table 6.5). Relatively small loads have been chosen because in such a type of network the voltage changes as a result of the WTGS generation changes are higher than in the case of the fully loaded grid. It is especially visible when the power generated by the wind turbine is similar to the total load.

The initial state of the network without the WTGS and after connecting the WTGS (for partial and full load) is presented in Tables 6.6 and 6.7.

**Table 6.1.** Transmission line model data (all lines)

| Parameter | Symbol | Value | Unit |
|---|---|---|---|
| Resistance | $R'$ | 0.24 | Ω/km |
| Reactance | $X'$ | 0.36 | Ω/km |
| Susceptance | $B'$ | 2.8 | μS/km |
| Length | $l$ | 20 | km |

**Table 6.2.** Load transformer data ($T_{47}$, $T_{58}$, $T_{69}$)

| Parameter | Symbol | Value | Unit |
|---|---|---|---|
| Rated apparent power | $S_n$ | 0.63 | MVA |
| Rated voltage of MV side | $V_{n,MV}$ | 15 | kV |
| Rated voltage of LV side | $V_{n,LV}$ | 0.4 | kV |
| Nominal short-circuit voltage | $U_{kn}$ | 6 | % |
| Copper losses at rated power | $\Delta P_{Cu}$ | 6 | kW |

**Table 6.3.** Step-up transformer data ($T_{15}$)

| Parameter | Symbol | Value | Unit |
|---|---|---|---|
| Rated apparent power | $S_n$ | 1.0 | MVA |
| Rated voltage of MV side | $V_{n,MV}$ | 15 | kV |
| Rated voltage of LV side | $V_{n,LV}$ | 0.69 | kV |
| Nominal short-circuit voltage | $u_{kn}$ | 6 | % |
| Copper losses at rated power | $\Delta P_{Cu}$ | 13.58 | kW |

**Table 6.4.** Feeding transformer data ($T_{23}$)

| Parameter | Symbol | Value | Unit |
|---|---|---|---|
| Rated apparent power | $S_n$ | 25 | MVA |
| Rated voltage of MV side | $V_{n,MV}$ | 110 | kV |
| Rated voltage of LV side | $V_{n,LV}$ | 15 | kV |
| Nominal short-circuit voltage | $u_{kn}$ | 11 | % |
| Copper losses at rated power | $\Delta P_{Cu}$ | 110 | kW |

**Table 6.5.** Load parameters

| Load | Symbol | Value | Unit |
|---|---|---|---|
| $L_1$ | $\underline{S}_{L1}$ | 0.150 + j0.147 | MVA |
| $L_2$ | $\underline{S}_{L2}$ | 0.150 + j0.147 | MVA |
| $L_3$ | $\underline{S}_{L3}$ | 0.150 + j0.147 | MVA |

**Table 6.6.** Voltages in the considered network

| Node | Rated voltage $V_n$ [kV] | Before WTGS connection to grid $V$ [p.u.] | After WTGS connection to grid (partial load, $v$ = 9 m/s) $V$ [p.u.] | After WTGS connection to grid (full load, $v$ = 15 m/s) $V$ [p.u.] |
|---|---|---|---|---|
| 1 | 0.69 | 0.990 | 1.000 | 1.000 |
| 2 | 110 | 1.030 | 1.030 | 1.047 |
| 3 | 15 | 1.028 | 1.028 | 1.043 |
| 4 | 15 | 1.005 | 1.010 | 1.026 |
| 5 | 15 | 0.989 | 1.000 | 1.018 |
| 6 | 15 | 0.982 | 0.994 | 1.011 |
| 7 | 0.4 | 0.989 | 0.994 | 1.010 |
| 8 | 0.4 | 0.974 | 0.985 | 1.002 |
| 9 | 0.4 | 0.966 | 0.978 | 0.994 |

**Table 6.7.** Currents in the considered network (0.4 kV side)

| Current | Before WTGS connection to grid $I$ [A] | After WTGS connection to grid (partial load, $v$ = 9 m/s) $I$ [A] | After WTGS connection to grid (full load, $v$ = 15 m/s) $I$ [A] |
|---|---|---|---|
| $\underline{I}_1$ | 0 | 666 – j347 | 1299 – j539 |
| $\underline{I}_2$ | 558 – j691 | – 82 – j775 | – 757 – j1020 |
| $\underline{I}_{23}$ | 558 – j691 | – 82 – j775 | – 757 – j1020 |
| $\underline{I}_{34}$ | 558 – j691 | – 82 – j775 | – 757 – j1020 |
| $\underline{I}_{45}$ | 369 – j458 | – 277 – j545 | – 757 – j1020 |
| $\underline{I}_{56}$ | 184 – j228 | 194 – j224 | 181 – j241 |
| $\underline{I}_{47}$ | 189 – j233 | 195 – j231 | 177 – j250 |
| $\underline{I}_{58}$ | 185 – j230 | 196 – j226 | 183 – j242 |
| $\underline{I}_{69}$ | 184 – j228 | 194 – j224 | 181 – j241 |
| $\underline{I}_{15}$ | 0 | 666 + j95 | 1299 + j287 |
| $\underline{I}_C$ | 0 | j252 | j252 |

Today the connection of the wind farm to the electric power system is considered more often than the connection of a single wind turbine. Then, while modeling the wind farm, we should model each WTGS in the farm separately or we can group the WTGSs and replace them by a single dynamic model. The method of modeling should depend on the type of analysis.

The MV network presented in Fig. 6.1 and defined by the data presented in Tables 6.1–6.5 can also be considered as a HV grid, for example, to which the wind farm is connected. The "conversion" of the network can be realized as follows. Let us assume that in the MV network (Fig. 6.1), the voltages, currents, impedances and powers are, respectively, equal to $\underline{V}_{MV}$, $\underline{I}_{MV}$, $\underline{Z}_{MV}$, $\underline{S}_{MV}$, and the base values of voltage and power are equal to $V_{baseMV}$, $S_{baseMV}$. Then the values of the network quantities per unit are equal to $\underline{V}$, $\underline{I}$, $\underline{Z}$, $\underline{S}$. The network expressed in per unit quantities can be considered as any network, taking into account the base values of voltage and power. Then, in such a case, choosing new base values $V_{baseHV}$, $S_{baseHV}$, we can "convert" the considered network to the "new" one by applying the following equations:

$$\underline{V}_{HV} = V_{baseHV}\underline{V} = \underline{V}_{MV}\frac{V_{baseHV}}{V_{baseMV}} \tag{6.1}$$

$$\underline{S}_{HV} = S_{baseHV}\underline{S} = \underline{S}_{MV}\frac{S_{baseHV}}{S_{baseMV}} \tag{6.2}$$

$$\underline{Z}_{HV} = \underline{Z}\frac{V^2_{baseHV}}{S_{baseHV}} = \underline{Z}_{MV}\frac{S_{baseMV}}{V^2_{baseMV}}\frac{V^2_{baseHV}}{S_{baseHV}} = \underline{z}'_{MV}l_{MV}\frac{S_{baseMV}}{V^2_{baseMV}}\frac{V^2_{baseHV}}{S_{baseHV}} = \underline{z}'_{HV}l_{HV}, \tag{6.3}$$

where from (6.3), by assuming the equality of $R/X$ (for transmission lines) in both networks, the length of the HV line can be computed as

$$l_{HV} = l_{MV}\frac{\underline{z}'_{MV}}{\underline{z}'_{HV}}\frac{S_{baseMV}}{V^2_{baseMV}}\frac{V^2_{baseHV}}{S_{baseHV}}. \tag{6.4}$$

For example, let the base values in the model MV network be equal to $V_{baseMV} = 15$ kV, $S_{baseMV} = 0.9$ MVA (e.g. equal to the rated power of the WTGS), and let the base values in the HV network be equal to $V_{baseHV} = 110$ kV, $S_{baseHV} = 50$ MVA (equal to rated power of the wind farm). Then the values of voltages and powers in the "new" network can be computed from (6.1) and (6.2). The "new" lengths of transmission lines can be computed from (6.4). For example, the line connecting nodes 3 and 4 with length $l_{MV} = 20$ km can be the equivalent of the 110 kV line (assuming $R_{MV}/X_{MV} = R_{HV}/X_{HV}$)[1] with length:

[1] This is, of course, a simplification because in practice the $R/X$ factor for HV transmission lines is lower than for MV lines. Additionally, the HV transmission line capacitance is higher than the MV line capacitance.

$$l_{\mathrm{HV}} = 20\frac{0.9}{15^2}\frac{110^2}{50} = 0.968 \cdot 20 = 19.36 \text{ km}. \tag{6.5}$$

Therefore, the MV network presented in Fig. 6.1, with WTGS with rated power $S_n = 0.9$ MVA, can also be treated as a 110 kV system (where the transmission lines lengths are retained) in which the wind farm with rated power $S_n = 50$MVA is connected to node 5. But it is worth remembering that this is a simplification only and for many types of analysis it is necessary to model the system exactly.

## 6.2 WTGS with Squirrel-Cage Rotor Asynchronous Generator

The model of the wind turbine generator system equipped with a squirrel-cage rotor asynchronous generator operating in the MV system described above is presented in Fig. 6.2 and consists of:

- wind variation model, which allows the introduction of a constant input value of wind velocity, wind gusts (with defined amplitude and period) and wind velocity harmonics (with defined amplitude and period) and allows making step changes (with defined slope) of the wind velocity,
- wind-wheel model (described in Sects. 5.2.5, 5.2.8),
- mechanical eigenswing model, which allows the introduction of 1P, 3P and 4.5 Hz variations of mechanical power with defined amplitude (described in Sect. 5.2.6),
- drive-train model (described in Sect. 5.2.7),
- asynchronous generator model (described in Sect. 5.3.2),
- grid model (described in Sects. 5.7.5, 6.1).

The model is described in natural units. The WTGS and its model parameters are presented in Tables 6.8 and 6.9.

This type of WTGS is not equipped with wind-wheel and generator control systems[2], and therefore is cheaper and theoretically more reliable than other types of wind turbine [26, 89]. The power generated results directly from the wind velocity and the wind-wheel characteristics $P = f(v)$. The reactive power is usually partially compensated (no-load reactive power compensation) and therefore during the wind turbine operation, the reactive power is consumed according to the nonlinear function $Q = f(P, V)$.

For the considered WTGS, the function is presented in Fig. 6.3. The curves show that an increase of the real power generation causes an increase of the reactive power consumption. Also, the increase of the terminal voltage gives the same effect, but at the rated power the influence of the terminal voltage becomes small. This is a result of constant capacitor utilization.

[2] Of course, the supervisory control system does exist in the case of the WTGS.

**Table 6.8.** WTGS parameters [89]

| Object | Parameter | Symbol | Value | Unit |
|---|---|---|---|---|
| Operational parameters | Nominal output | $P_n$ | 900 | kW |
| | Power regulation | Stall | | |
| | Cut-in wind speed | $v_{cut\text{-}in}$ | 3.5 | m/s |
| | Rated wind speed | $v_n$ | 15 | m/s |
| | Cut-out wind speed | $v_{cut\text{-}out}$ | 25 | m/s |
| Rotor | Rotor diameter | $2R$ | 52.2 | m |
| | Number of blades | $B$ | 3 | - |
| | Rotor revolution | $n$ | 22.4/14.9 [3] | rpm |
| | Moment of inertia | $J_W$ | $1.6\times10^6$ | kg·m$^2$ |
| | Shaft stiffness | $K$ | *$6\times10^7$* [4] | N·m/rad |
| | Shaft damping coefficient | $D$ | *$10^6$* | N·m/rad$^2$ |
| | Gearbox ratio | $\upsilon$ | 67.5 | - |
| Generator (asynchronous with squirrel cage rotor) | Rated power | $P_n$ | 900/200 | kW |
| | Rated voltage | $V_n$ | 690/690 | V |
| | Rated current | $I_n$ | 840/207 | A |
| | Power factor | $\cos\varphi_n$ | 0.89/0.81 | - |
| | Rated speed | $n_n$ | 1510/1005 | rpm |
| | Losses at rated power | $\Delta P_n$ | 25.5/10.97 | kW |
| | No-load current | $I_0$ | 213/90 | A |
| | Start-up current factor | $K_{Istart}$ | 7.3/8.5 | - |
| | Start-up torque factor | $K_{\tau start}$ | 1.0/1.6 | - |
| | Maximum torque factor | $K_{\tau max}$ | 2.6/3.4 | - |
| | Moment of inertia | $J_G$ | 35.184 | kg·m$^2$ |
| Var compensation | Rated power | $Q_c$ | *150* | kvar |
| Wind wheel model | Coefficients of (5.16) | $c_1$ | *0.5* | - |
| | | $c_2$ | *67.56* | - |
| | | $c_3$ | *0* | - |
| | | $c_4$ | *0* | - |
| | | $c_5$ | *1.517* | - |
| | | $c_6$ | *16.286* | - |

**Table 6.9.** Asynchronous generator model data (Δ–connection)

| Parameter | Symbol | Value | Unit |
|---|---|---|---|
| Stator resistance | $R_S$ | 0.0034 | Ω |
| Rotor resistance | $R'_R$ | 0.003 | Ω |
| Stator leakage reactance | $X_{lS}$ | 0.055 | Ω |
| Rotor leakage reactance | $X'_{lR}$ | 0.042 | Ω |
| Magnetizing reactance | $X_m$ | 1.6 | Ω |

[3] Data for the delta (Δ) and star (Y) winding connection of asynchronous generator.

[4] Italic typeface indicates assumed data.

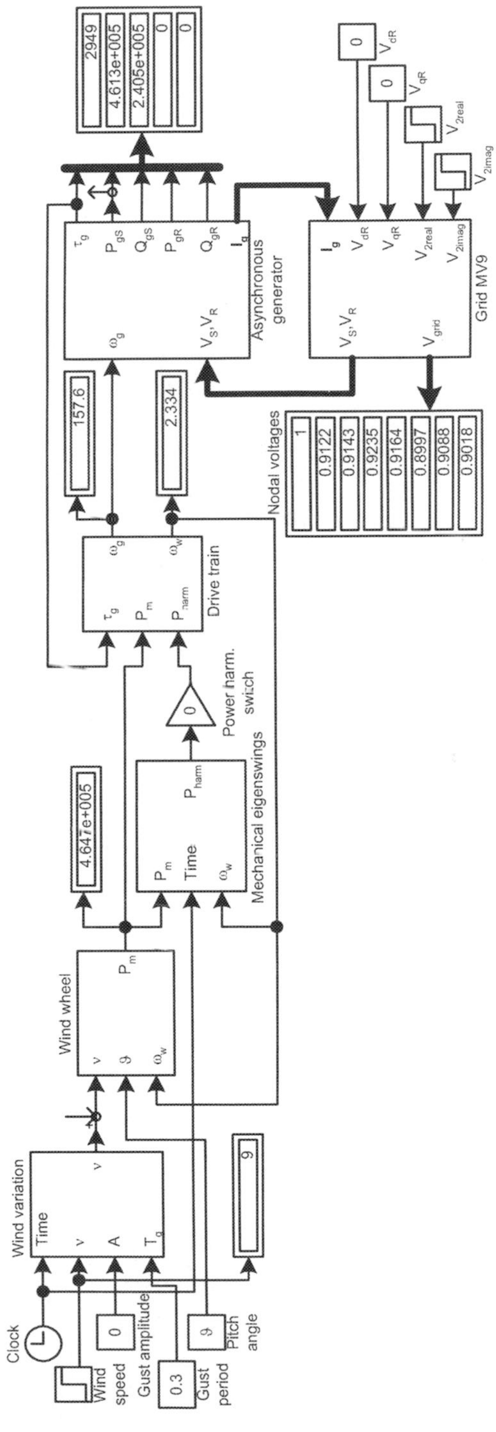


**Fig. 6.2.** Block diagram of the WTGS equipped with a squirrel-cage asynchronous generator model

When the terminal voltage changes, the power consumed by the capacitors changes as well $Q_c(V = 0.95) = 158$ kvar, $Q_c(V = 1.0) = 175$ kvar, $Q_c(V = 1.05) = 193$ kvar and compensates the increase of the real power consumed by the asynchronous generator (this effect has no general character but depends on the system data). At low real power generation, the change of reactive power generation by the capacitor (as a terminal voltage change) is not compensated by a change of the reactive power consumption.

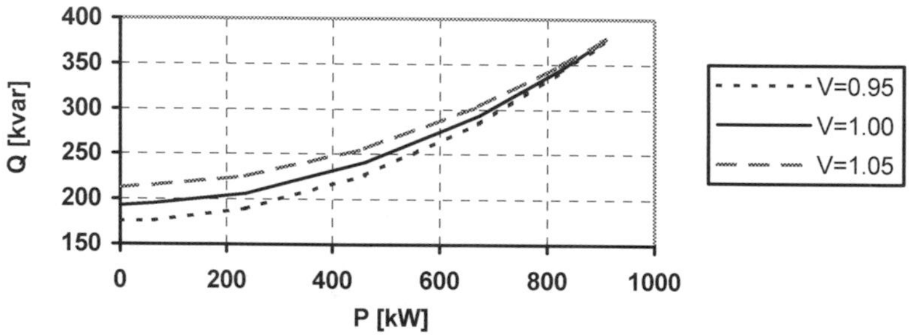


**Fig. 6.3.** Reactive power as a function of real power generated for various values of terminal voltage

Figure 6.4 presents the dependence of the wind turbine real power generation and reactive power consumption on the value of the wind velocity. The curves show that the terminal voltage practically does not influence the real power generation level but influences the real power consumption. The influence is of a similar character to the one presented in Fig. 6.3, i.e. at high wind velocity the reactive power consumption is almost independent of the terminal voltage, but at low wind velocity the reactive power consumption increases when the terminal voltage increases.

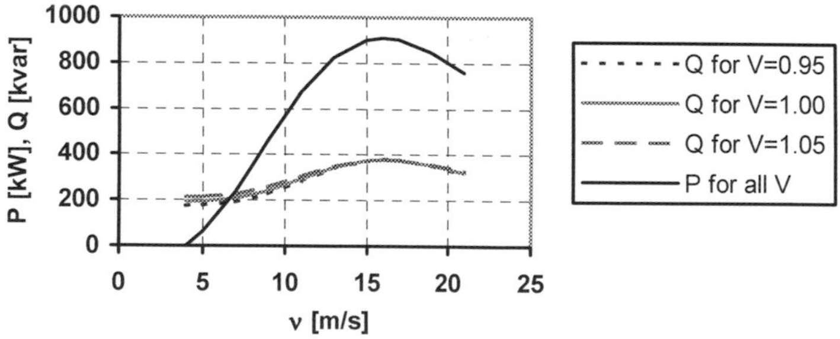


**Fig. 6.4.** Generator power as a function of wind velocity

The dynamic properties of the considered system are presented in the following figures. Most of the tests have been made for a WTGS operating at a wind velocity equal to 9 m/s (partial load with highest slope of WTGS characteristics, $P = f(v)$) and at wind speed equal to 15 m/s (full load). In general, the properties at partial load and at full load are quite different.

The response of the system to the step change of the wind velocity is presented in Fig. 6.5. The curves show the nodal voltages, real power and shaft speed change after the wind velocity increases by 2.7 m/s. The wind velocity harmonics and mechanical eigenswings are neglected.

At partial load, the wind velocity increase causes a high change of the operating point, which is related to the increase of the nodal voltages, power and shaft speed. The passage to the new operating point takes place with small oscillations, with a maximum change of voltage 1% at nodes 5, 8, 9, i.e. located at and to the right of the point of common coupling. The real power increase is about 250 kW, while the shaft speed increase is small and equal to 0.3 rad/s (typical for the constant-speed system resulting from the asynchronous generator mechanical characteristics). Torsional oscillations of the drive train are also visible.

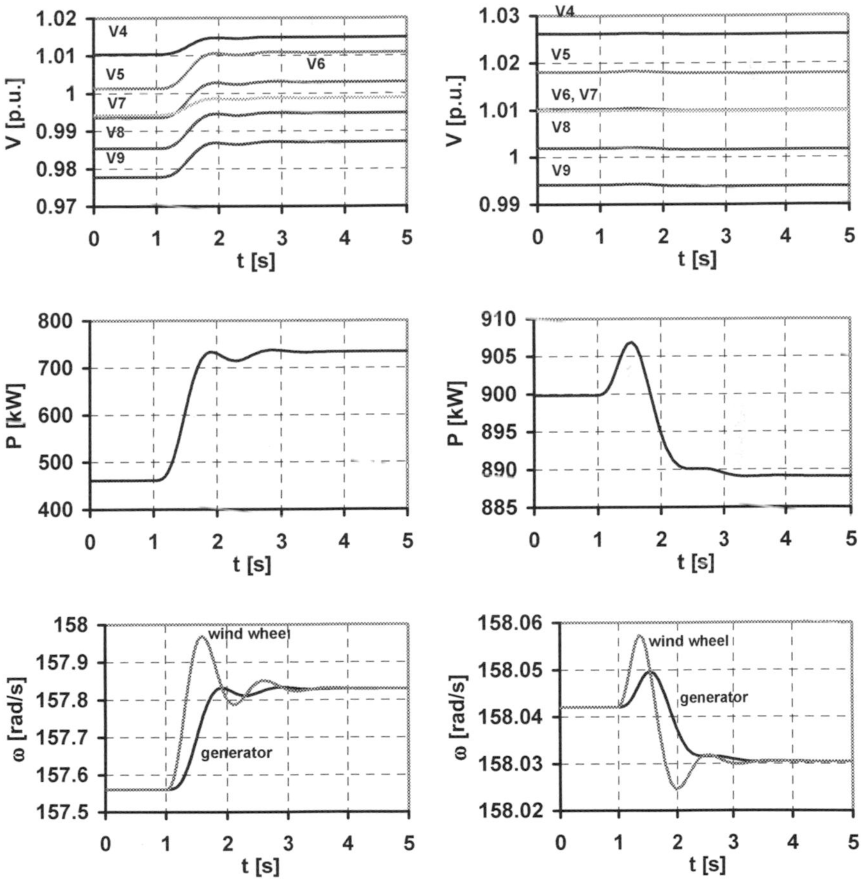


**Fig. 6.5.** WTGS response to a wind velocity step increase (left-hand graphs – partial load, wind velocity step from 9 to 11.7 m/s; right-hand graphs – full load, wind velocity step from 15 to 17.7 m/s; $V$ – nodal voltages, $P$ – real power, $\omega$ – generator and wind wheel shaft speed (on generator side))

At full load, the system reaction to wind velocity increase is much smaller, because of the relatively flat WTGS characteristic, $P = f(v)$, in this area. The nodal voltage change (and amplitude of oscillation) is below 0.1%. Also, the real power change is very small and equal to about 10 kW.

In general, the electromechanical swings in both cases are relatively well damped and last up to 2 s. This results mainly from the dynamic properties of the asynchronous machine.

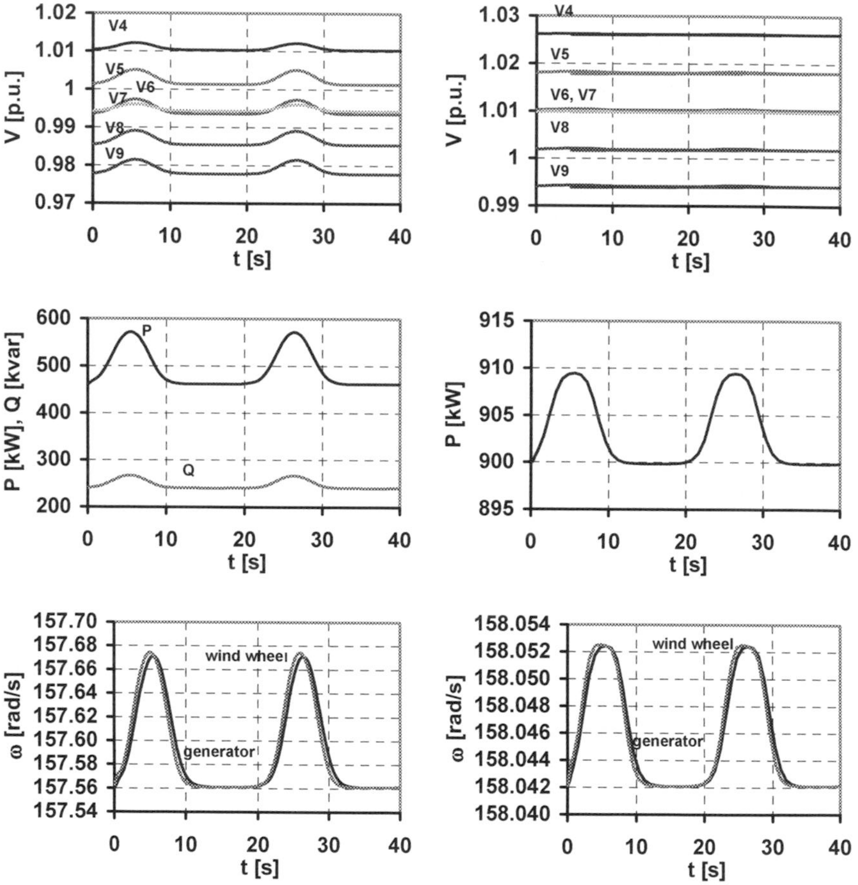


**Fig. 6.6.** WTGS operation during gusts (left-hand graphs – $v$ = 9 m/s; right-hand graphs – $v$= 15 m/s, Gust amplitude 1 m/s; $V$ – nodal voltages, $P$ – real power, $Q$ – reactive power, $\omega$ – generator and wind wheel shaft speed)

The next test is related to the response of the system to wind gusts with amplitude equal to 1 m/s. The response is presented in Fig. 6.6. In this case, because of the relatively small period of the gusts, the reaction of the WTGS can be called the follow-up type. The process runs without oscillations. Only a small delay is visible between the wind wheel and the generator rotor course (speed). As in the previous example the power and the voltage change are significantly less at full load operation than at partial load. At full load the amplitude of the real power change is up to 10 kW, while at partial load the amplitude reaches 100 kW. The voltage change amplitude reaches 0.5% at partial load, while at full load the amplitude is less than 0.03%.

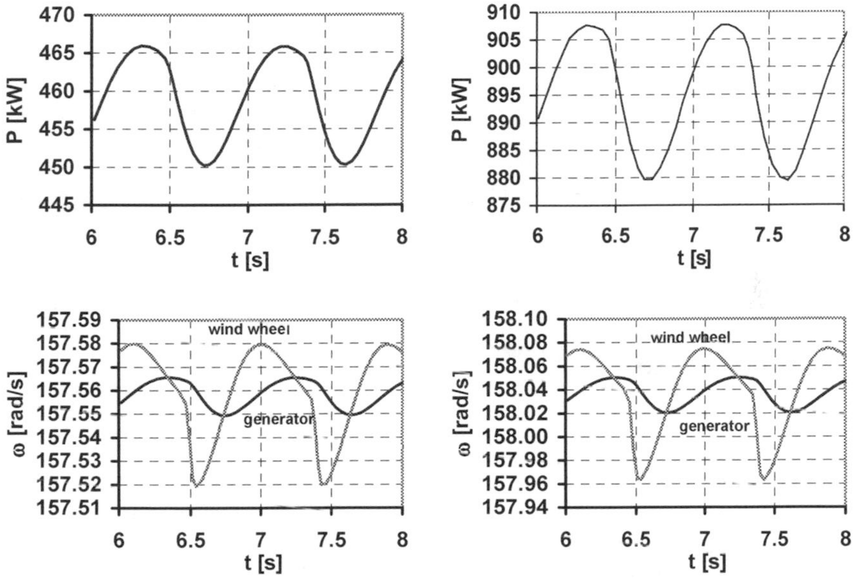


**Fig. 6.7.** Effect of blades passing in front of the WTGS tower (3P amplitude 7 %; left-hand graphs – $v$ = 9 m/s; right-hand graphs – $v$ = 15 m/s)

Unfortunately, during the WTGS operation there appear oscillations resulting from the wind velocity and the turbine construction. The oscillations proportional to the number of blades and to the wind velocity (3P oscillations in the case of the considered turbine) are the most important oscillations (because of magnitude). Figure 6.7 shows the operation of the system in the case when 3P oscillations of amplitude 7% exist. Because of the lack of control elements, damping of these oscillations depends on the WTGS characteristics. Contrary to the previous examples, the oscillations are better damped at partial load (amplitude 8 kW) while at full load the amplitude of oscillations reaches 15 kW.

In practice, other oscillations appear during the WTGS operation as well. Figure 6.8 shows the WTGS operation when the 1P, 3P oscillations and tower vibration effect (4.5 Hz) with amplitude 1%, 7% and 10%, respectively, exist. The figure shows the complex response of the turbine. The band of the real power oscillations is higher, as in the previous example, at the full load equal to 60 kW and at the partial load equal to 35 kW. The same is true of the voltage variation, visible in nodal voltages.

It is worth remembering that in a real system the amplitudes of a given oscillation (1P, 3P, etc.) depend on the WTGS operating point. Usually, the increase of the wind velocity and the power generated by the WTGS causes an increase of these amplitudes.

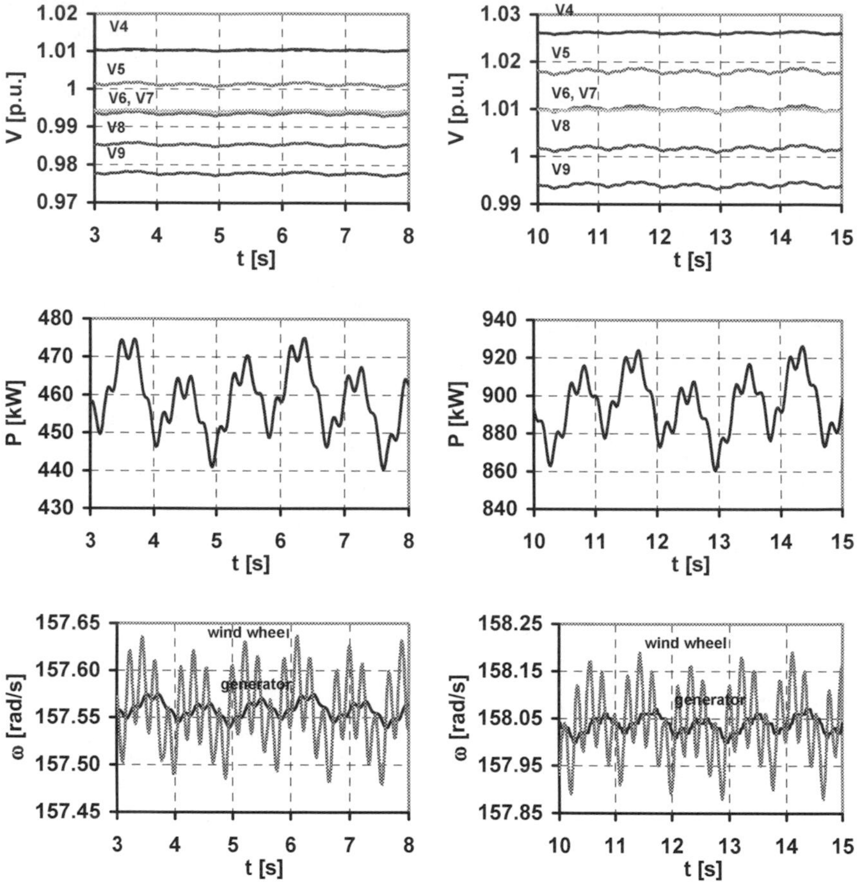


**Fig. 6.8.** 1P, 3P and tower vibration effects (1P amplitude 1%, 3P amplitude 7%, 4.5 Hz amplitude 7%; left-hand graphs – $v = 9$ m/s, right-hand graphs – $v = 15$ m/s)

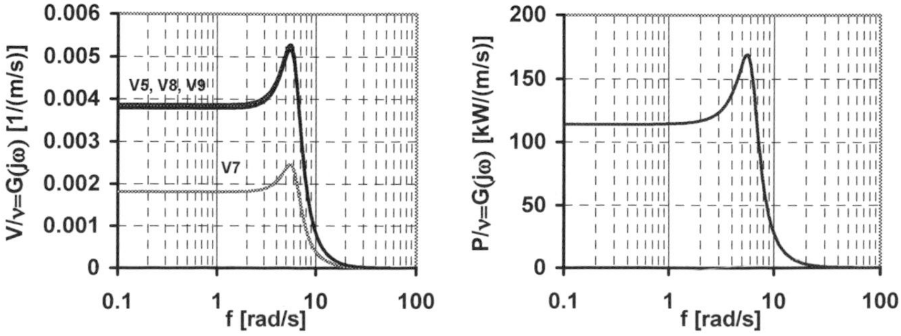


**Fig. 6.9.** Frequency characteristics of WTGS at partial load ($v = 9$ m/s)

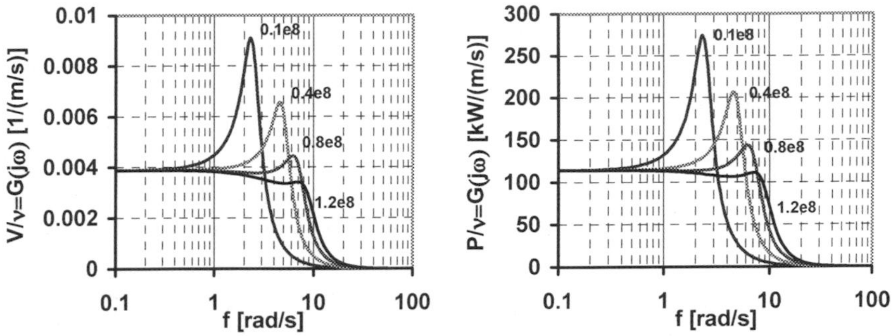


**Fig. 6.10.** Influence of the stiffness coefficient $K$ on the WTGS frequency characteristics ($v = 9$ m/s)

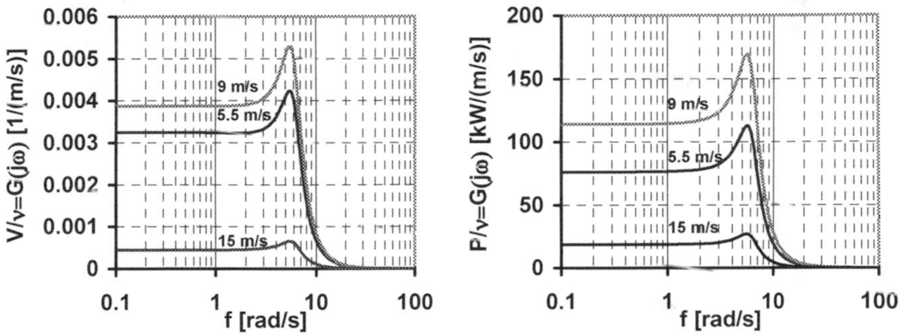


**Fig. 6.11.** Influence of the wind velocity (power generation) influence on the WTGS frequency characteristics

(a) (b)

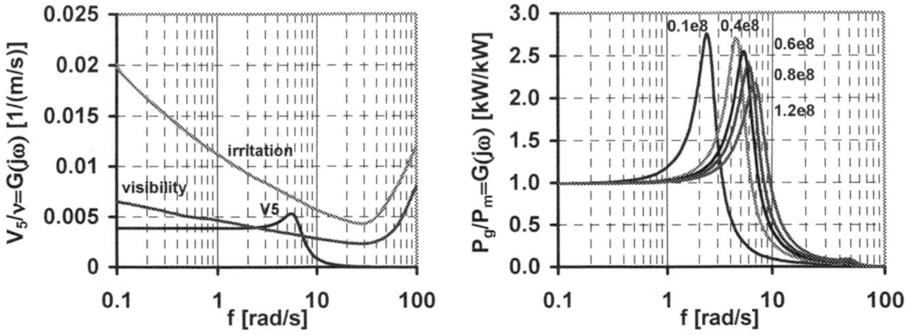


**Fig. 6.12.** Frequency characteristics: (a) comparison of the WTGS voltage characteristics with the borders of irritation and visibility (voltage dips effected by lights – see Fig. 3.4) defined by standard IEEE 141-1993; (b) drive train frequency characteristics for various values of the stiffness coefficient; $P_g$ – generator terminal power; $P_m$ – mechanical power)

The time-domain simulations allow us to analyze the performance of a dynamic system, but only at a given operating point and for a given disturbance. A more general view on the dynamic properties of the system can be achieved by utilizing other techniques, e.g. frequency analysis (but in this case, the linear system has to be used). Therefore, as a supplement to the time-domain simulations, the next Figs. 6.9–6.12 present various frequency characteristics of the analyzed system. Figure 6.9 shows the frequency characteristics $V_i/v = G(\mathrm{j}\omega)$ and $P/v = G(\mathrm{j}\omega)$ achieved for the system operating at partial load ($v = 9$ m/s). The left-hand graph shows the voltage variation as a response to wind velocity variation. Voltages at the low-voltage side of MV/LV transformers are presented because most customers consume electric energy at low voltage. The right-hand graph shows the real power variation as a response to wind velocity variation. Both characteristics, i.e. voltage and power, are similar to each other. The curves allow us to formulate the following remarks:

- The voltage variation in nodes located between the PCC and the feeding point (node 7) are much smaller than voltages in nodes located "behind" the WTGS (nodes 8, 9). The voltage variation in these nodes is similar to the voltage variation at the PCC (node 5). For a high range of frequency, the voltage variation is less than 0.4% when the wind speed variation is less than 1 m/s.
- The curves have a single eigenfrequency equal to about 5.5 rad/s. At that point, the voltage variation reaches 0.5% in the case of a wind velocity variation with amplitude 1 m/s (this is equivalent to a real power variation amplitude equal to 170 kW/(m/s) – right-hand graph in Fig. 6.9).
- Wind velocity variation with frequency higher than 11 rad/s is practically eliminated.

The resonance frequency of the system, shown in Fig. 6.9, should not be close to the mechanical eigenfrequencies appearing during the WTGS operation, e.g. 1P, 3P tower vibrations. To achieve this the drive-train shaft should be properly designed. For the considered example, this condition is fulfilled because $\omega_{1P} \approx 2.3$ rad/s, $\omega_{3P} \approx 7.0$ rad/s, while the resonance frequency is equal to 5.5 rad/s. Figure 6.10 shows the influence of the shaft stiffness coefficient (which is of essential significance here), which causes a decrease of the resonance amplitude and simultaneously shifts the resonance frequency to higher values (to the right in Fig. 6.10).

Figure 6.11 shows the influence of the operating point (wind velocity) on the frequency characteristics. The curves show that the higher response of the system takes place at a wind velocity equal to 9 m/s, i.e. where the WTGS slope of the characteristics $P = f(v)$ is the biggest. The lower response takes place when the slope of the characteristics is smaller, i.e. at the rated wind velocity (15 m/s) and at very low wind velocity (near the cut-in speed).

The last figure in this group shows a comparison of the voltage frequency characteristics (from Fig. 6.9) with the limits defined by the standards (Fig. 6.12a) and the drive train frequency characteristics (Fig. 6.12b). The first one shows that for frequencies between 2 and 7 rad/s the characteristics with higher amplitude

$V_5/v = G(j\omega)$ is located higher than the curve defining the border of visibility, which is due to the relatively long lines (high impedances) and weakly loaded power system.

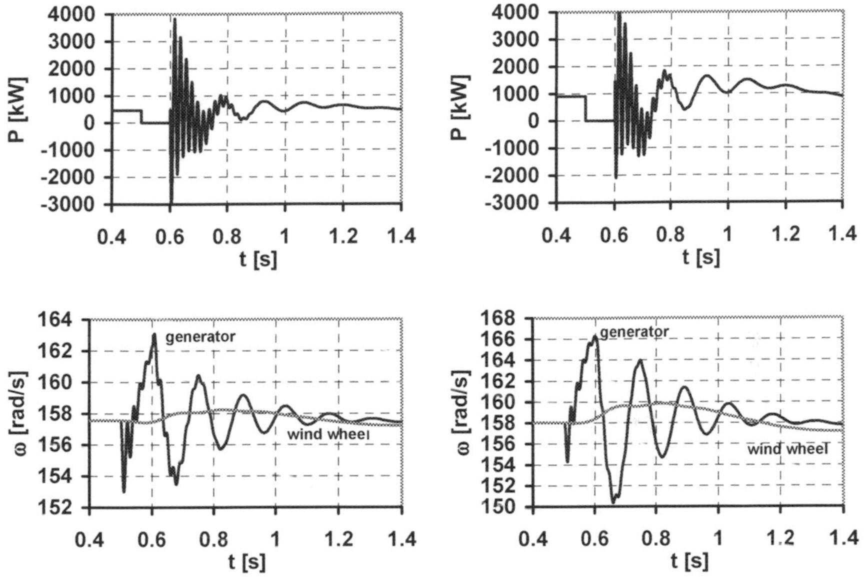


**Fig. 6.13.** WTGS response to a short circuit at the terminal bus lasting 100 ms (left-hand graphs – $v$ = 9 m/s; right-hand graphs – $v$ = 15 m/s)

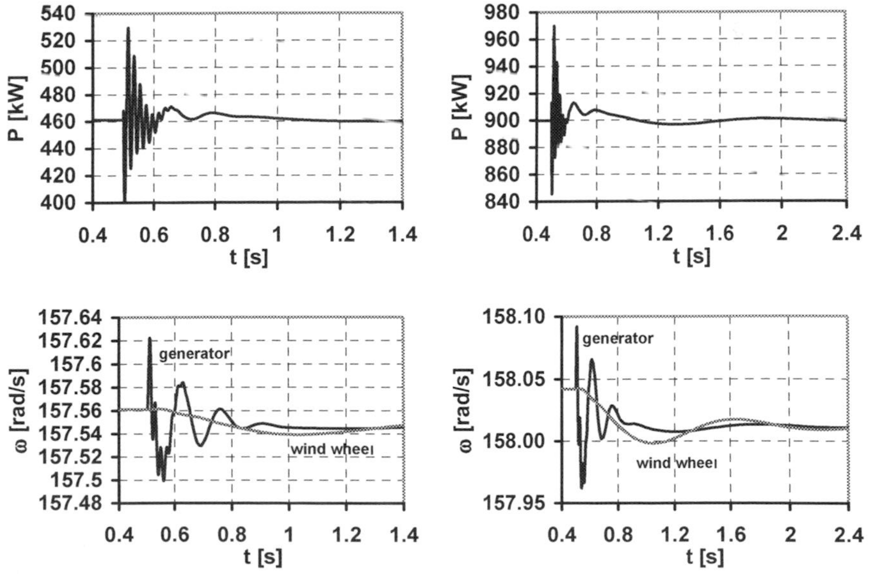


**Fig. 6.14.** WTGS response to a 1.5% step change of the terminal voltage (left-hand graphs – $v$ = 9 m/s; right-hand graphs – $v$ = 15 m/s)

But at the same time, for all the frequencies, the characteristic lies below the curve defining the border of irritation. The curve lies also below the border defined by the IEC standard (Fig. 3.4, $P_{st} = 1$). Of course, an increase of wind velocity variation amplitude above 1 m/s will cause the characteristics to move up and then the limits will not be fulfilled.

The last figures show again the time domain response of the system. Figure 6.13 shows the response of the system to a short-circuit at the terminal bus, while Fig. 6.14 shows the response of the system to a 1.5% step increase of the feeding bus voltage, which can be considered as a change of the feeding transformer tap. The curves indicate high damping of the electromechanical processes, which (in both cases) last up to 1.5 s. The real power surge as a result of the voltage change is less than 70 kW.

## 6.3 WTGS with Asynchronous Generator and Dynamic Slip Control

A model of the wind turbine generator system equipped with a slip-ring asynchronous generator and a dynamic slip control system operated in an MV system is presented in Fig. 6.15 and consists of the elements (i.e. generator, drive train, etc.) presented in Sect. 6.2. The model is described in natural units. The WTGS and its model parameters are shown in Tables 6.10 and 6.11.

The turbine and generator controllers are additional elements of the considered WTGS, in comparison to the system described in the previous chapter. The turbine controller acts on the pitch angle, while the generator controller acts on the resistor connected to the rotor winding of the asynchronous generator.

The turbine controller, presented in Fig. 6.16, consists of the following controllers[5]:

- Pitch angle controller (Fig. 6.17). The pitch angle controller structure consists of a control element, an actuator and an element modeling the dynamics of the object. The controller's aim is to keep constant the reference value of the pitch blade angle $\vartheta$ (named $J_{ref}$ in figures).
- Power controller (Fig. 6.18). The power controller's aim is to keep constant the reference value of the real power. The controller can operate at partial load and therefore allow us to keep constant, but lower than the WTGS capacity in the given windy conditions, the value of power generated by the WTGS. When at partial load $P_{gref} = P_n$, the controller keeps the blade pitch angle equal to the minimum value (e.g. zero). The controller, at partial load, can also be switched

[5] All the controllers considered in this and the following sections are PI-type controllers with parameters $K_R$, $T_I$. The utilization of controllers of this type are given as examples only and does not mean that these types of controller are or should be used in real WTGSs. The controller parameters are given in the figure captions. The parameters should not be treated as optimal either.

off. Then the blade pitch angle has to be kept equal to the set value (e.g. zero). In this case, the power extracted from the wind and transferred to the grid is not maximal but can be equal to near the optimum value.

- Speed controller (Fig. 6.19). The speed controller's aim is to keep constant the reference value of the shaft speed. The controller operates at the WTGS's full load.

**Table 6.10.** WTGS parameters[6] [24]

| Object | Parameter | Symbol | Value | Unit |
|---|---|---|---|---|
| Operational parameters | Nominal output | $P_n$ | 900 | kW |
| | Power regulation | Blade pitch control | | |
| | Cut-in wind speed | $v_{cut\text{-}in}$ | 3 | m/s |
| | Rated wind speed | $v_n$ | 15 | m/s |
| | Cut-out wind speed | $v_{cut\text{-}out}$ | 25 | m/s |
| Rotor | Rotor diameter | $2R$ | 57 | m |
| | Number of blades | $B$ | 3 | - |
| | Rotor revolution | $n$ | 15–28 | rpm |
| | Moment of inertia | $J_W$ | *1.6×10^6* | kg·m² |
| | Shaft stiffness | $K$ | *6×10^7* (7) | N·m/rad |
| | Shaft damping coefficient | $D$ | *10^6* | N·m/rad² |
| | Gearbox ratio | $\upsilon$ | 65.27 | - |
| Asynchronous generator with wounded rotor | Rated apparent power | $S_n$ | 1000 | kVA |
| | Rated power | $P_n$ | 900 | kW |
| | Rated stator voltage | $V_n$ | 690 | V |
| | Rated stator current ($\cos\varphi=1$) | $I_n$ | 670 | A |
| | No-load rotor voltage | $E_{20}$ | 241 | V |
| | Power factor | $\cos\varphi$ | $0.9_{lead}/0.9_{lag}$ | - |
| | Rated speed | $n_n$ | 1800 | rpm |
| | Synchronous speed | $n_s$ | 1500 | rpm |
| | Speed range | $n_{min}$–$n_{max}$ | 1000–2000 | rpm |
| Var compensation | Rated power | $Q_c$ | *200* | kvar |
| Wind-wheel model | Coefficients of (5.16) | $c_1$ | *0.5* | - |
| | | $c_2$ | *116* | - |
| | | $c_3$ | *0.4* | - |
| | | $c_4$ | *0* | - |
| | | $c_5$ | *5* | - |
| | | $c_6$ | *21* | - |

[6] For modeling purposes, the Enron EW 900 wind turbine data have been utilized here. This WTGS was originally a system equipped with a doubly-fed asynchronous generator (considered in the next section). For the considerations conducted here, the use of this wind turbine equipped with a variable resistor connected to the rotor winding does not introduce qualitative errors, but at the same time allows us to compare its dynamic properties to the system equipped with the doubly-fed asynchronous machine.

[7] Italic typeface indicates assumed data.

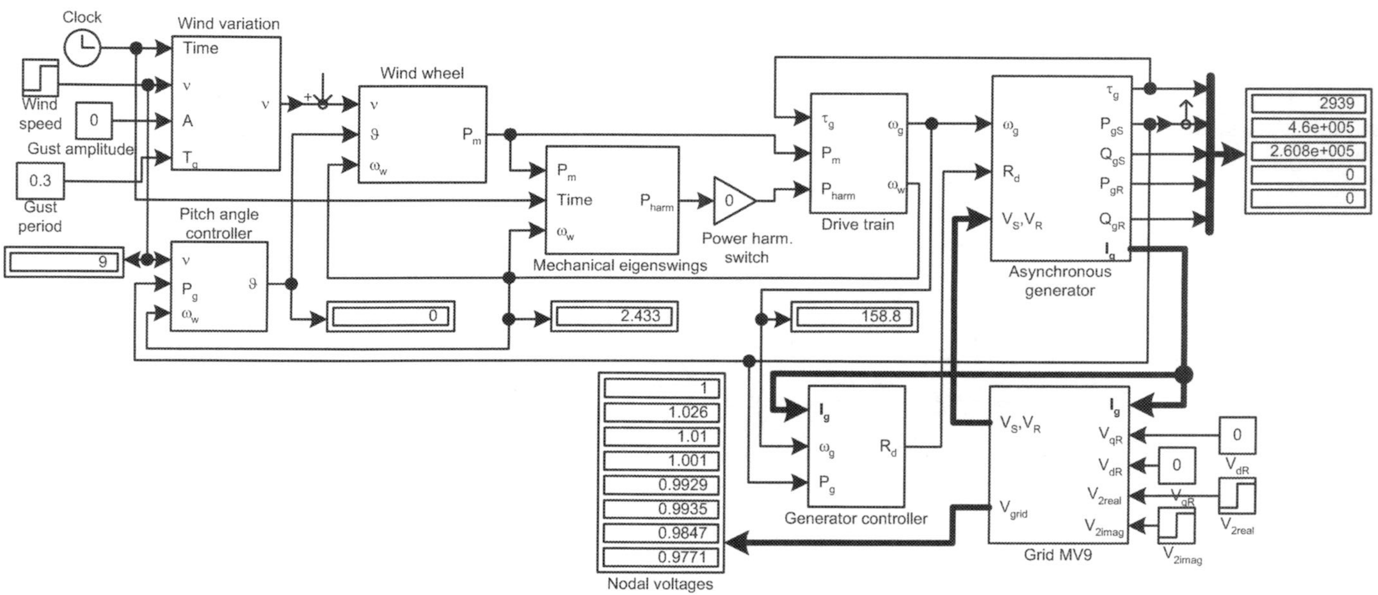


Fig. 6.15. Block diagram of the WTGS with asynchronous generator and dynamic slip control system model

**Table 6.11.** Asynchronous generator model data [24] ($\Delta$ – connection)

| Parameter | Symbol | Value | Unit |
|---|---|---|---|
| Stator resistance | $R_S$ | 0.0027 | Ω |
| Rotor resistance | $R'_R$ | 0.0022 | Ω |
| Stator leakage reactance | $X_{lS}$ | 0.025 | Ω |
| Rotor leakage reactance | $X'_{lR}$ | 0.046 | Ω |
| Magnetizing reactance | $X_m$ | 1.38 | Ω |

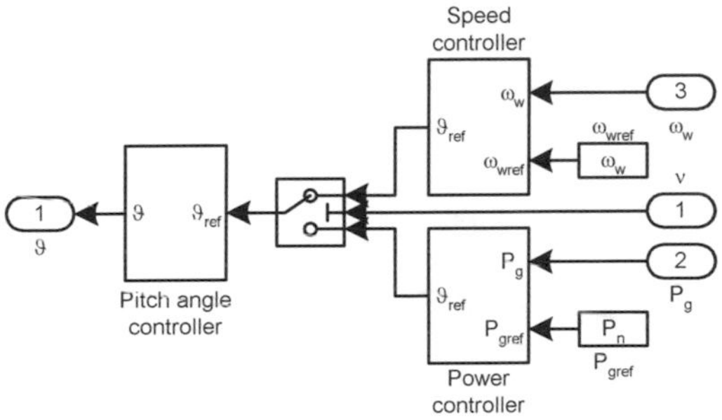


**Fig. 6.16.** Block diagram of turbine controller

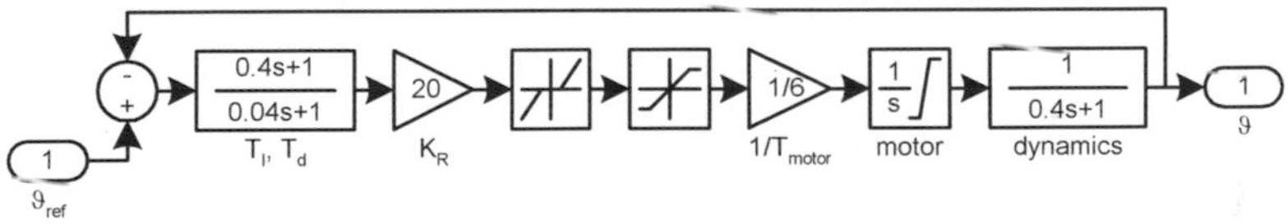


**Fig. 6.17.** Block diagram of pitch controller in turbine control system

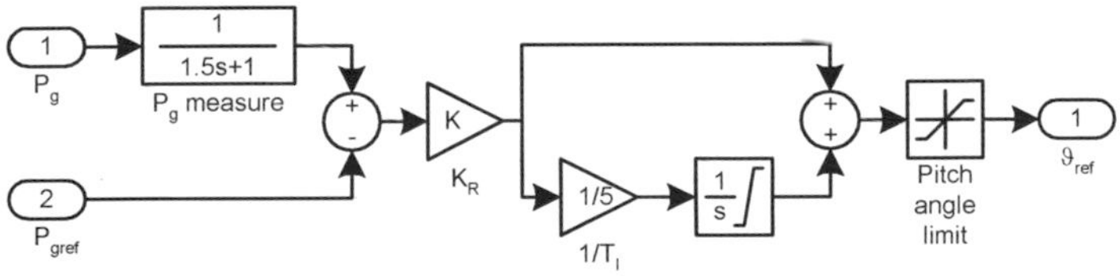


**Fig. 6.18.** Block diagram of power controller in turbine control system ($K_R$ = $80/P_n$, $T_I = 5$ s)

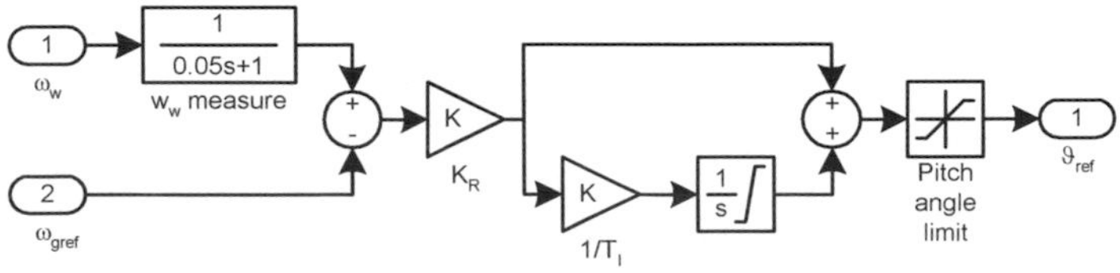


**Fig. 6.19.** Block diagram of speed controller in turbine control system ($K_R$ = 2.4, $T_I = 5.65$ s)

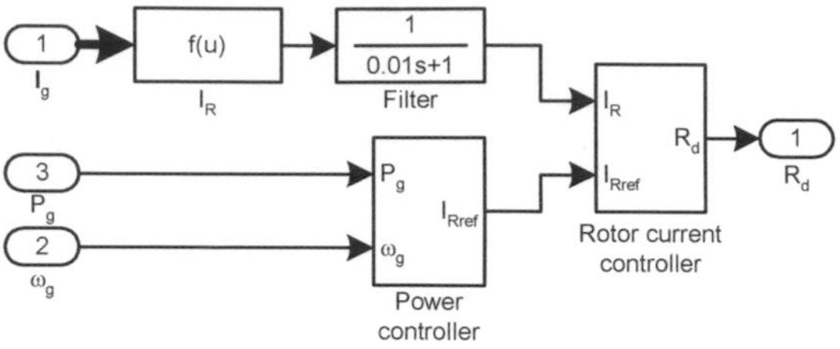


**Fig. 6.20.** Block diagram of generator controller

During WTGS operations switching between the power and the speed controllers is realized when the WTGS passes from partial to full load (and the opposite). In the modeled system, this operation is made according to the wind velocity threshold, i.e. it takes place when the wind velocity exceeds the rated value. This method of switching is given here only as an example.

The generator controller, presented in Fig. 6.20, consists of the following two controllers:

- Rotor current controller (Fig. 6.21). The aim of the controller is to keep constant the reference value of the rotor current. In real WTGSs (e.g. Vestas), the controller is located on the generator rotor.
- Power controller (Fig. 6.22). The aim of the power controller is to keep constant the reference value of the real power. The controller can operate both at partial and at full load. For the given structure of the control system, the controller parameters, especially the filter time constant (rotor speed measuring element), have to be changed when the operating point (partial load or full load) of the wind turbine changes. The real power reference is computed as a function of the shaft speed $P_{ref} = f(\omega)$.

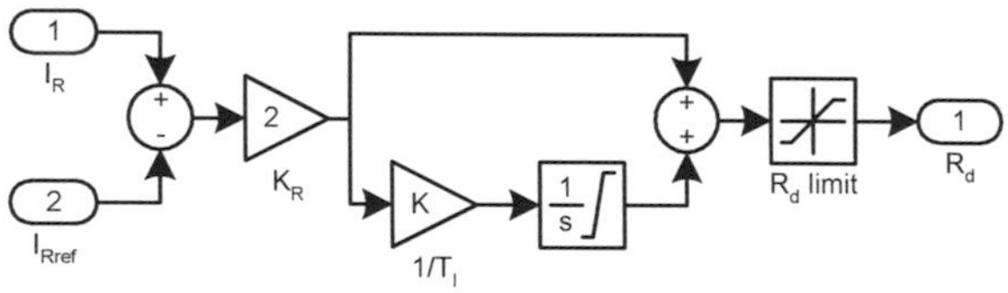


**Fig. 6.21.** Block diagram of the rotor current controller in the generator control system ($K_R = 2$, $T_I = 0.5$ s)

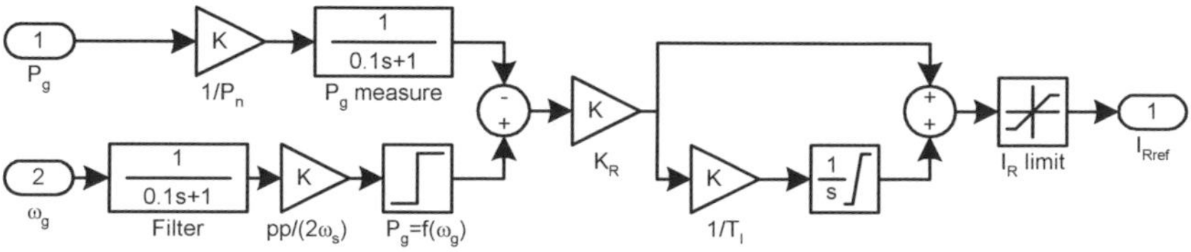


**Fig. 6.22.** Block diagram of the power controller in the generator control system ($K_R = 1000$, $T_I = 0.03$ s, filter time constant $T_f = 0.1$ s (partial load) or 10 s (full load))

In a WTGS with dynamic slip control system in the steady state, the rotor additional resistance $R_d$ (effective resistance connected to the rotor winding) is not equal to zero. The asynchronous machine is constructed in such a way that the rated slip is equal to about 1%. The resistance $R_d$ is chosen to achieve a rated speed equal to 2%. Then, for the considered machine, which is shown in Fig. 6.23, the additional (external) resistance should be about four times higher than the rotor resistance, i.e. $R_d = 4R_R$.

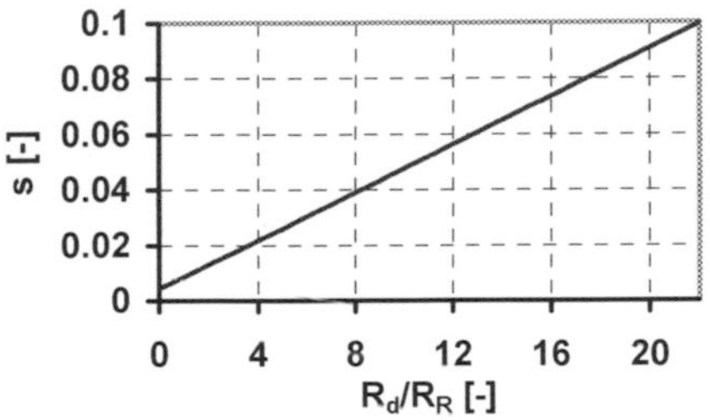


**Fig. 6.23.** Rotor slip versus resistances quotient $R_d/R_R$ characteristics ($P$ = 900 kW, $v$= 15 m/s)

The connection of an additional resistance to the rotor changes the asynchronous machine static characteristics and allows us to effectively change their dynamic properties. In the considered case, connection of the resistors ($R_d = 4R_R$ in each phase) to the rotor influences the power versus wind velocity characteristics relatively weakly, as shown in Fig. 6.24. This effect can be treated as positive in the steady-state because it shows that the additional power loss resulting from the $R_d$ connection is small. The difference is visible only near the WTGS rated power .

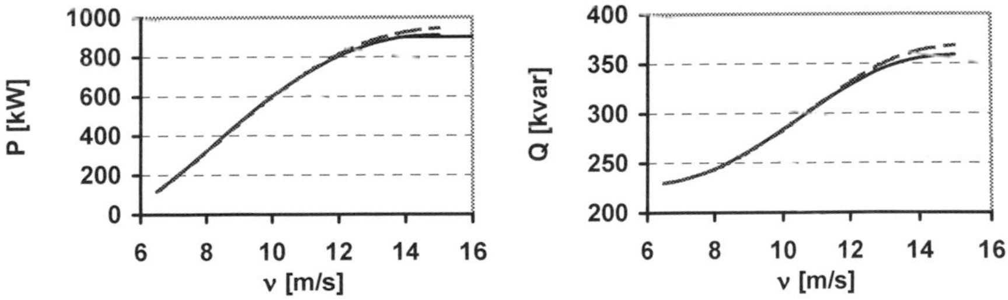


**Fig. 6.24.** Power versus wind velocity characteristics (lines: continuous – $R_d$= 0; dashed – $R_d = 4R_R$; horizontal – WTGS characteristic $P = f(v)$ at full load, i.e. $P$ = 900 kW)

The performance of a WTGS with dynamic slip control system has been tested in a similar way to that presented in the previous chapter. The results of the tests made are presented in Figs. 6.25–6.35.

The response of the system to a step change of wind velocity is presented in Fig. 6.25. The curves show the nodal voltages, real power, shaft speed change, resistance $R_d$ and blade pitch angle after a wind velocity increase of 2.7 m/s. Wind velocity harmonics and mechanical eigenswings are neglected.

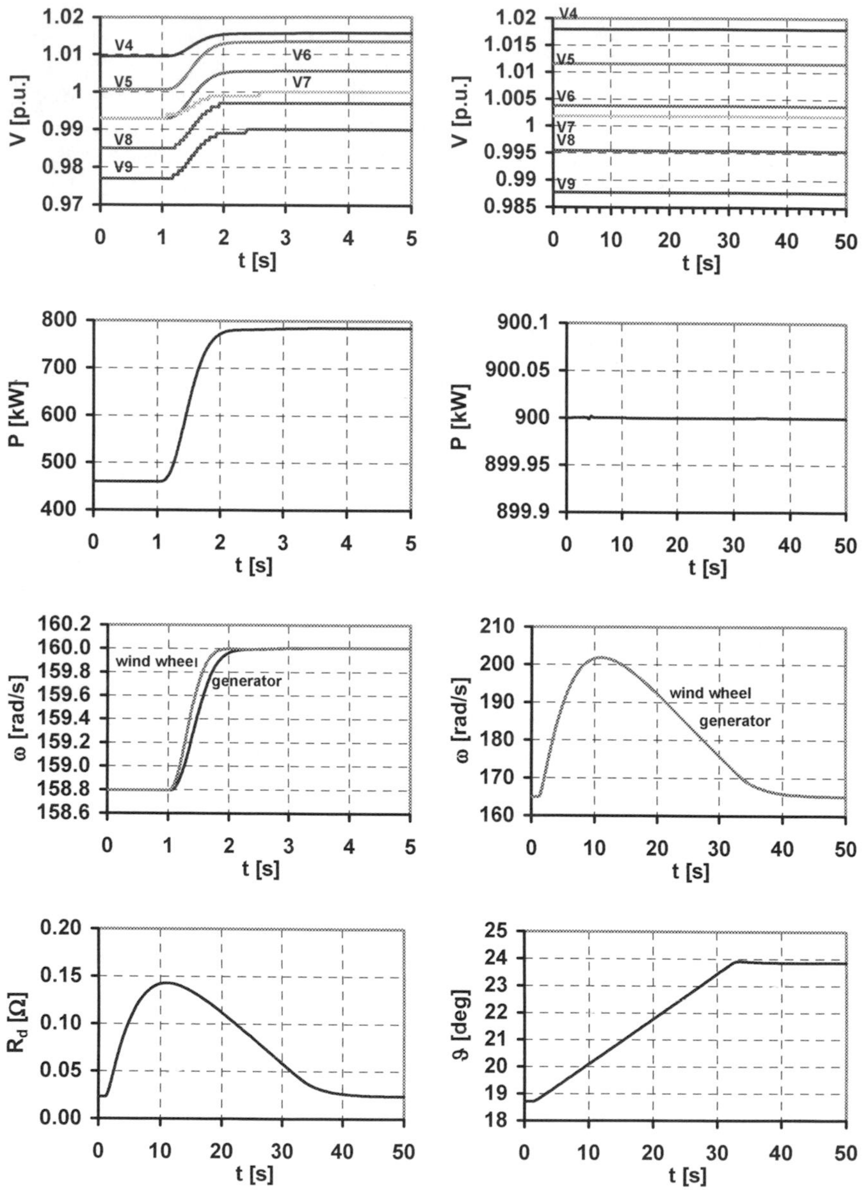


**Fig. 6.25.** WTGS response to step increase of a wind velocity (left-hand graphs – partial load, wind velocity step from 9 to 11.7 m/s; right-hand graphs and $R_d(t)$ – full load, wind velocity step from 16 to 18.7 m/s; $V$ – nodal voltages, $P$ – real power, $\omega$ – generator and wind-wheel shaft speed, $R_d$ – additional rotor resistance, $\vartheta$ – blade pitch angle)

At partial load, the wind velocity increase causes a high change of the operating point. The change to the new operating point has an aperiodic character. The maximum change of voltage is about 1.2% at nodes 5, 8, 9, i.e. located at and to the right of the point of common coupling.[8] The real power increase is equal to about 300 kW and the shaft speed increase is small and equal to 1.3 rad/s. The power and the shaft speed change are higher than in the case of the WTGS with squirrel-cage rotor considered in the previous section. This is the result of the higher slope of the power versus wind velocity characteristics of the WTGS and the lower slope of the generator's mechanical characteristics. There are no torsional oscillations of the drive train.

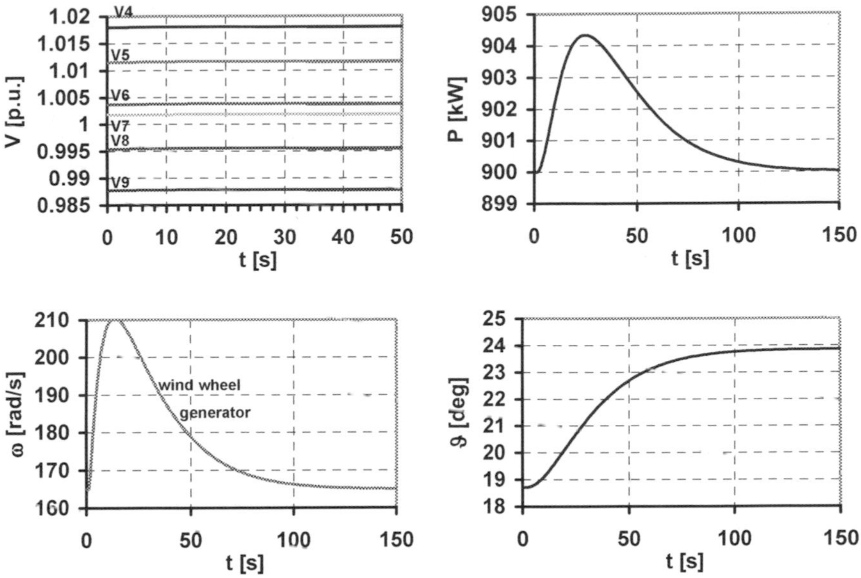


**Fig. 6.26.** WTGS response to a wind velocity step increase from 16 to 18.7 m/s in the system with the turbine power controller operating (at full load)

At full load, the system reaction to a wind velocity increase is completely different. For the given structure of the control system, the real power change (variation) is almost completely eliminated. The real power change is much less than 0.001 kW. The same effect is observed at the nodal voltages, where the voltage changes during the considered process are almost "invisible". Instead, the rotor shaft speed increases drastically and reaches 200 rad/s, when in the steady state it was equal to about 165 rad/s (which corresponds to a 22% speed increase). This is the result of the very fast and effective operation of the generator power controller. In this case, the surplus of energy is converted into the energy of the rotating shaft. The shaft speed decreases when the turbine controller changes the blade pitch angle and reduces the power extracted from the wind. The blade pitch angle and the resistance $R_d$ changes during the considered process are presented in the two lower figures.

---

[8] The change of the operating point depends on the power versus wind velocity $P = f(v)$ and motor mechanical $\tau = f(\omega)$ characteristics.

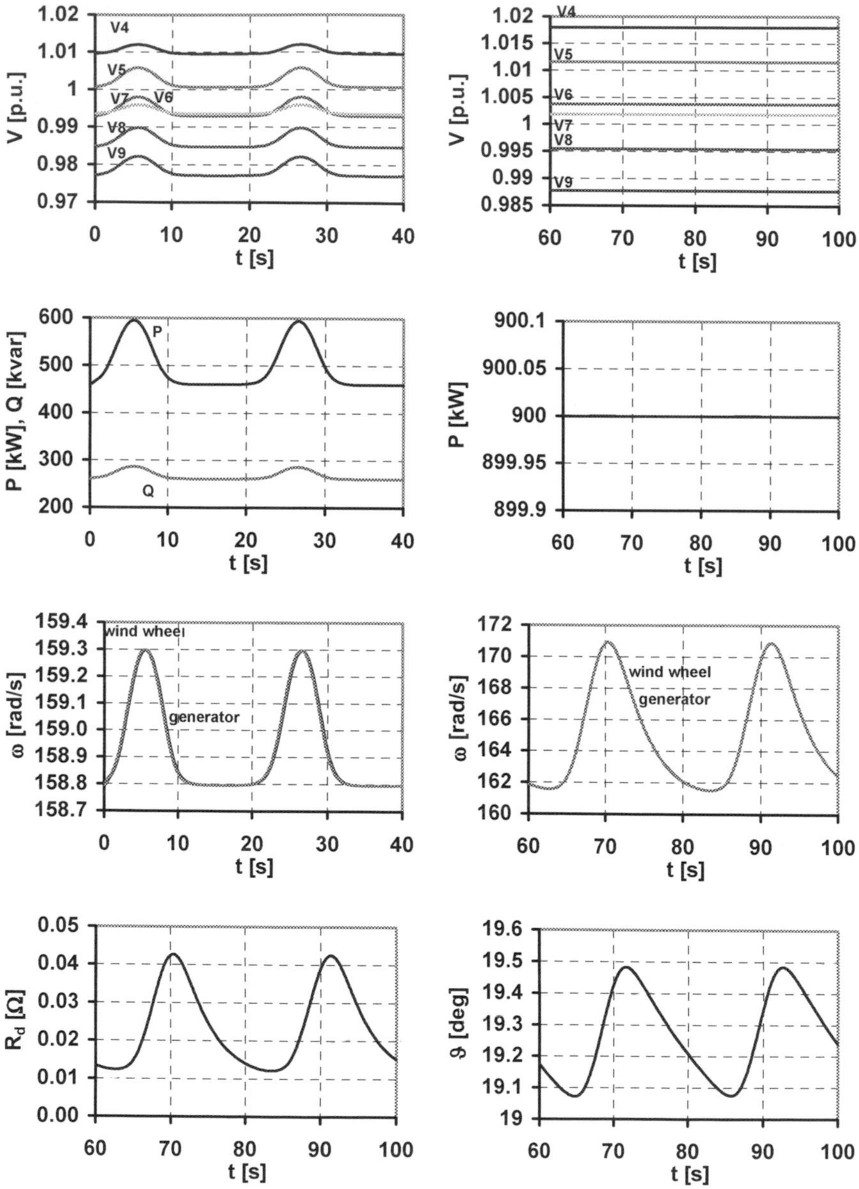


**Fig. 6.27.** WTGS operation during gusts (left-hand graphs – $v = 9$ m/s; right-hand graphs and $R_d(t)$ – $v = 16$ m/s; gust amplitude 1 m/s)

Generally, it is, of course, possible to decrease the effectiveness of the generator power controller. Then, as a result, we can achieve an increase of the real power variation but with the consequence of decreasing the rotor speed overshoot (which in some systems can be desirable).

Figure 6.26 shows the response of the WTGS equipped with an "inverse" control system to a step increase of the wind velocity. The inverse control system here consists of a turbine power controller and generator speed controller.[9] The use of the power controller in the turbine control system is possible when the $P_{ref} = f(\omega)$ characteristics (in the generator power controller, Fig. 6.22) has non-zero slope for rotor speed higher than rated (related to the wind velocity rated power).

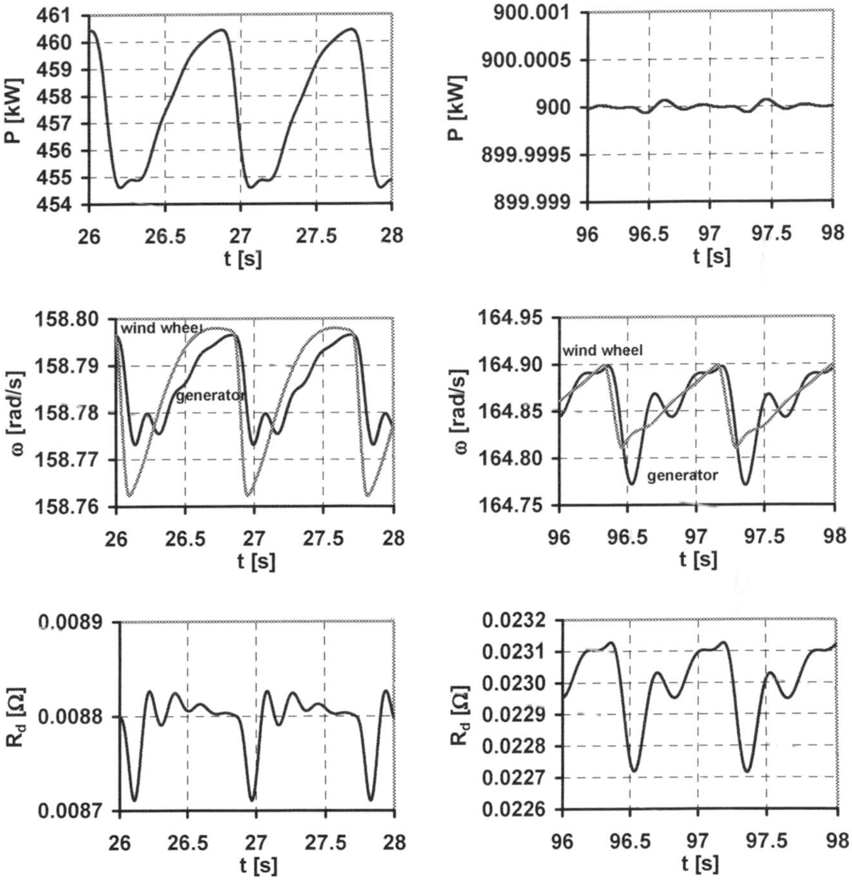


**Fig. 6.28.** Effect of the blades passing in front of the WTGS tower (3P amplitude 7%, left-hand graphs – $v = 9$ m/s; right-hand graphs – $v = 16$ m/s)

As seen in Fig. 6.15, this type of control is less effective (for a given structure of the controllers). The power oscillations (variations) in full load operation are less damped, but in fact are still small (4 kW). The control process is very slow. The turbine controller, in particular, reacts slowly because its input, which is the

[9] This is in contrast to the previously and further considered WTGS with turbine speed controller and generator power controller operating at full load.

power error, is small. The rotor speed overshoot is less than in the previous case (Fig. 6.25) because part of the power extracted from the wind is transferred to the grid (power overshoot).

The next test (carried out in the WTGS equipped with a standard control system) is related to the response of the system to wind gusts with amplitude equal to 1 m/s. The response is presented in Fig. 6.27. In this case, because of the relatively small period of the gusts, the reaction of the WTGS can be called a follow-up type. The process lasts without oscillations. Only a very small delay between the wind wheel and the generator rotor speed is visible. As above, the power and the voltage change are significantly less in full load operation. At full load, the amplitude of the real power change is below 0.001 kW, while at partial load, the amplitude reaches 130 kW. The voltage change amplitude reaches 0.6% at partial load, while at full load, the amplitude is less than 0.01%.

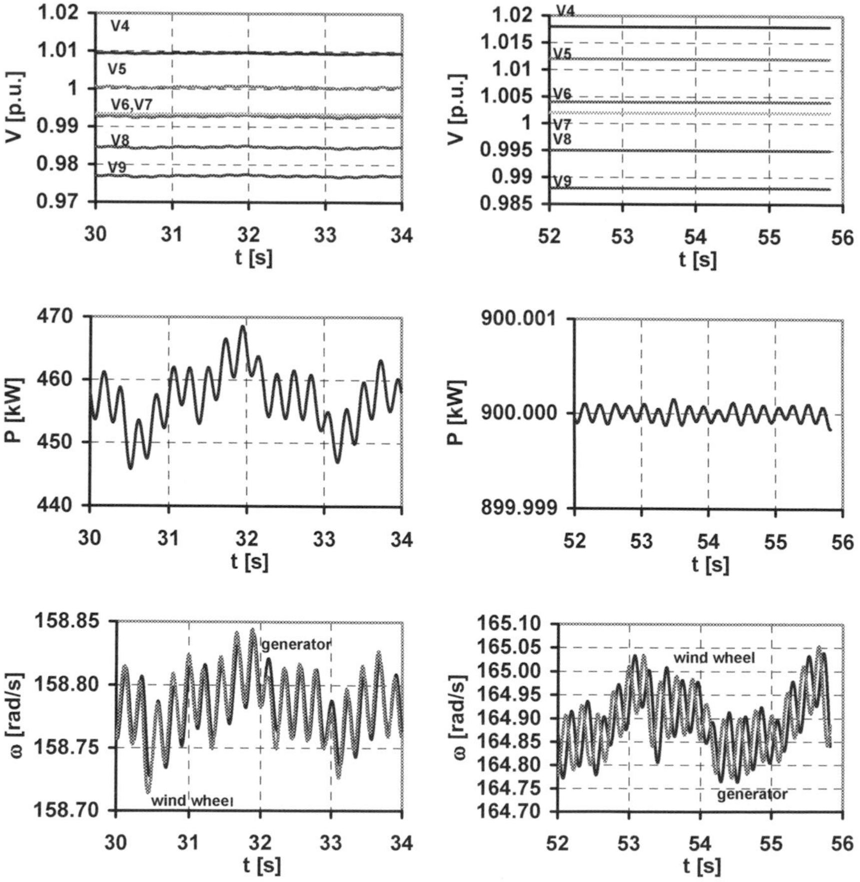


**Fig. 6.29.** 1P, 3P and tower vibration effects (1P amplitude 1%, 3P amplitude 7%, 4.5Hz amplitude 10%; left-hand graphs – $v = 9$ m/s; right-hand graphs – $v = 16$ m/s)

We can say that wind gusts of the given amplitude do not appear on the electric power system side – is not visible in the grid – when the WTGS operates at full load. The surplus of energy extracted from the wind is "consumed" by the rotor and resistor $R_d$ and it is "eliminated" by the action of the turbine control system.

Figure 6.28 shows the operation of the system in the case when 3P oscillations with amplitude 7% exist. In the case of the considered WTGS, the control systems also exist and their aim (especially of the generator controller) is to damp these types of oscillations. Contrary to the previously regarded wind turbine (Sect. 6.2), the oscillations are well damped at partial load (amplitude 6 kW), and very well damped at full load, where the amplitude is less than 1 W. This is the result of the active control of the resistance $R_d$, which is shown in the lower part of Fig. 6.28.

Figure 6.29 presents the WTGS operation when the 1P, 3P oscillations and the tower vibration effect (4.5 Hz) with amplitude 1%, 7% and 10%, respectively, exist. The curves show the complex response of the turbine. The band of the real power oscillations at full load is less than 1 W (which correspond practically to the elimination of the oscillations) and at partial load is equal to 25 kW. The same effect is visible in the nodal voltages. We can say that the dynamic slip control system eliminates the effects of mechanical eigenswings at full load and slightly increases their damping at partial load (in contrast to the WTGS equipped with a squirrel-cage rotor asynchronous machine).

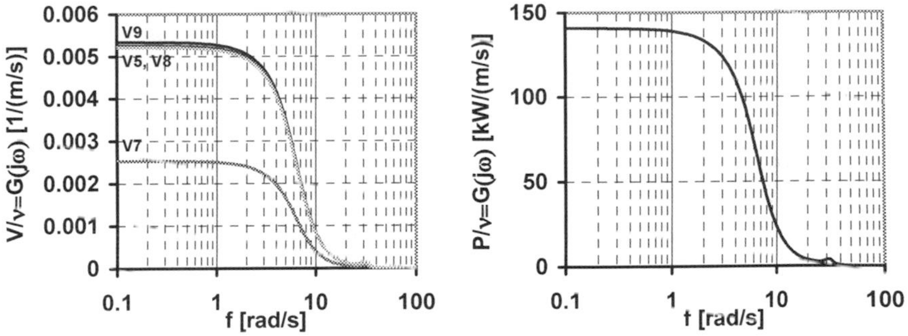


**Fig. 6.30.** Frequency characteristics of the WTGS at partial load ($v = 9$ m/s)

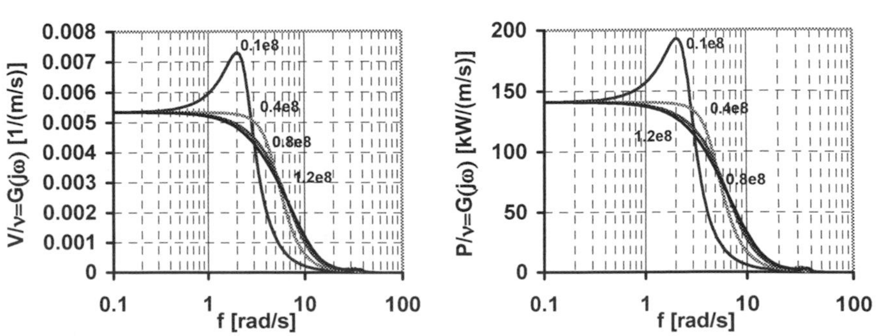


**Fig. 6.31.** Influence of the stiffness coefficient $K$ on the WTGS frequency characteristics ($v = 9$ m/s)

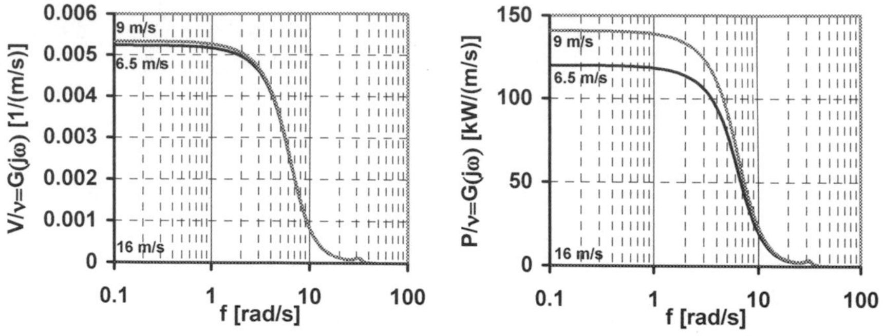


**Fig. 6.32.** Influence of the wind velocity (power generation) on the WTGS frequency characteristics

(a) (b)

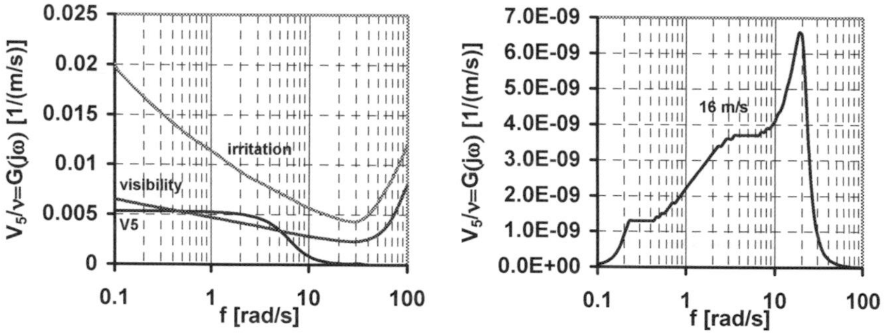


**Fig. 6.33.** Frequency characteristics: (a) comparison of the WTGS voltage characteristics with the borders of irritation and visibility (voltage dips effected by lights – see Fig. 3.4) defined by standard IEEE 141-1993; (b) frequency characteristics at wind speed 16 m/s

Figures 6.30–6.33 present various frequency characteristics of the analyzed system. Figure 6.30 shows the frequency characteristics $V_i/v = G(j\omega)$ and $P/v = G(j\omega)$ achieved for the system operating at partial load ($v = 9$ m/s). The first figure shows the voltage variation as a response to the wind velocity variation. The second figure shows the real power variation as a response to the wind velocity variation. The curves show that:

- The voltage variation in nodes located between the PCC (node 5) and the feeding point (node 2) are much smaller than the voltages in nodes located "behind" the WTGS (nodes 8, 9). The voltage variation in these nodes is similar to the voltage variation at the PCC. For the whole range of frequency, the voltage variation is less than 0.54%/(m/s), while the real power variation is less than 145 kW/(m/s).
- The system has no eigenfrequency.
- Wind velocity variation for a frequency higher than 11 rad/s is practically eliminated. There are no passes to the grid.

The lack of the resonance frequency in the system results mainly from the WTGS construction, parameters and the existence of control elements. Figure 6.31 (as in the previous section) shows the influence of the shaft stiffness coefficient on the frequency characteristics. The curves show that only for relatively small values of the stiffness coefficient ($< 3\times10^7$ Nm/rad), the resonance frequency appears on the frequency characteristics. For higher values of stiffness coefficient, the resonance does not appear. For the higher stiffness coefficients, the lower slope of the frequency characteristics is visible.

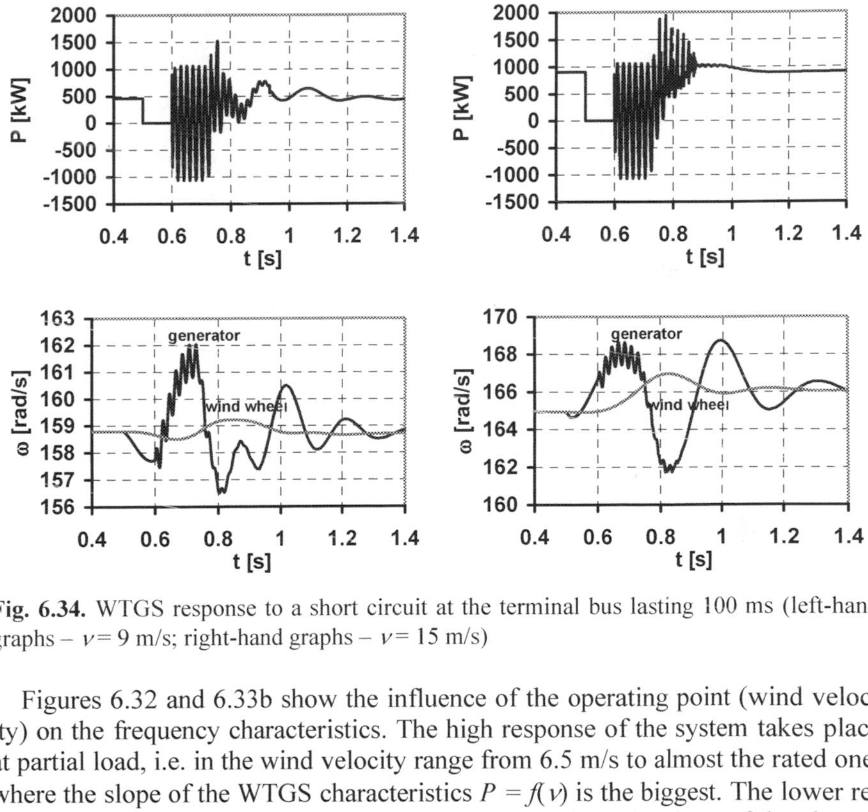


**Fig. 6.34.** WTGS response to a short circuit at the terminal bus lasting 100 ms (left-hand graphs – $v$ = 9 m/s; right-hand graphs – $v$ = 15 m/s)

Figures 6.32 and 6.33b show the influence of the operating point (wind velocity) on the frequency characteristics. The high response of the system takes place at partial load, i.e. in the wind velocity range from 6.5 m/s to almost the rated one, where the slope of the WTGS characteristics $P = f(v)$ is the biggest. The lower response takes place at WTGS full load (Fig. 6.33b), where the slope of the characteristics is small (near zero). For these characteristics, a resonance frequency equal to 11 rad/s but with very small amplitude equal to $6.6\times10^{-7}$%/(m/s) is visible. These characteristics confirm the results of time-domain simulations at full load presented above. Figure 6.33a in this group shows the comparison of the frequency characteristics with the limits defined by the standards. The curves show that for frequencies between 0.3 and 4 rad/s, the characteristic $V_5/v = G(j\omega)$ is located higher than the curve defining the border of visibility (in spite of the lack of a resonant frequency). It is so because the characteristics for low frequencies have greater values than the values achieved for the WTGS equipped with a squirrel-cage rotor asynchronous machine.

As in the case of the previously considered WTGS, for all frequencies the characteristics lie below the curve defining the border of irritation and below the border defined by the IEC standard (Fig. 3.4, $P_{st} = 1$). Of course, an increase of wind velocity variation amplitude above 1 m/s will cause the characteristics to move up and then the limits will not be fulfilled.

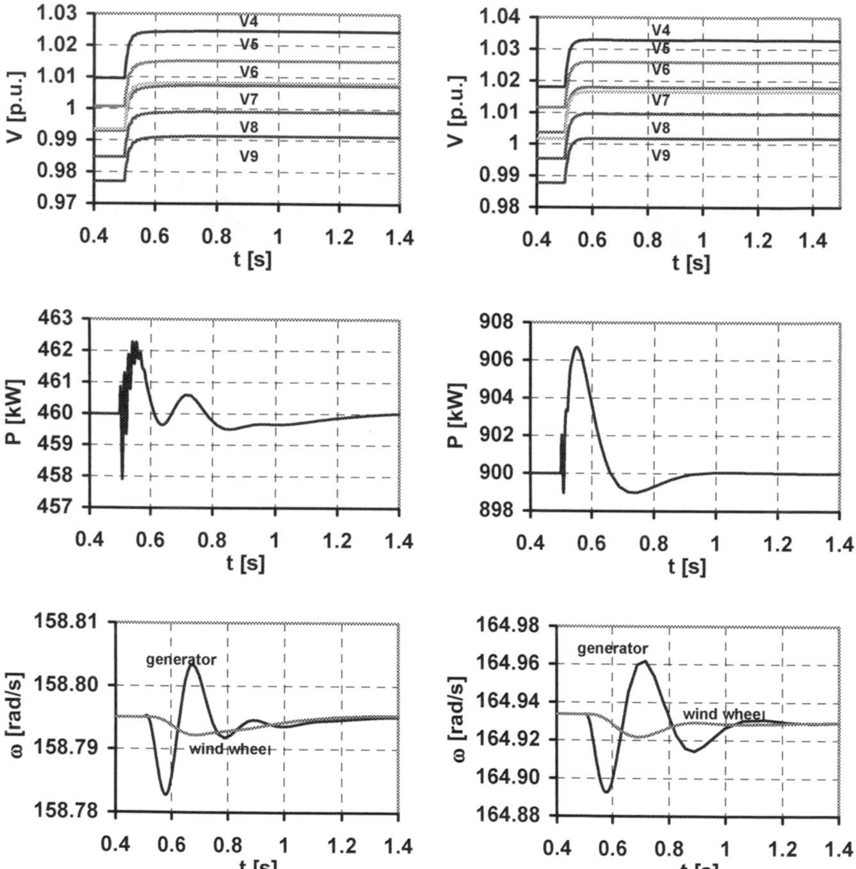


**Fig. 6.35.** WTGS response to a 1.5% step change of the terminal voltage (left-hand graphs – $v = 9$ m/s; right-hand graphs – $v = 15$ m/s)

The last figures show the time-domain response of the system. Figure 6.34 shows the response of the system to a short-circuit at the terminal bus, while Fig. 6.35 shows the response of the system to a 1.5% step increase of the feeding bus voltage, which can be considered as a change of the feeding transformer tap. The curves indicate high damping of the electromechanical processes, which (in both cases) last up to 1.5 s. The real power surge as a result of the voltage change is less than 7 kW (which is about ten times less than in the case of the WTGS considered in Sect. 6.2). The process of terminal voltage "regulation" lasts about 0.2 s and has an aperiodic character.

## 6.4 WTGS with Doubly-Fed Asynchronous Generator

A model of the wind turbine generator system equipped with a doubly-fed asynchronous generator operating in an MV system is presented in Fig. 6.38 and consists of the elements that are presented in Sect. 6.2. The model is described in natural units. The WTGS and its model parameters (ENRON wind turbine EW 900sL) are presented in Tables 6.10 and 6.11. The doubly-fed asynchronous generator operates without a compensating capacitor.

The WTGS equipped with a doubly-fed asynchronous generator is a fully controlled system, which means utilization of the turbine and generator controllers. The turbine controller acts on the blade pitch angle, while the generator controller acts on the generator rotor voltage through the power electronic converter, which is located between the machine terminal and the machine rotor. The voltage source inverter (VSI) is modeled here in a simplified form, where the generator controller creates rotor voltages $V_{dR}$, $V_{qR}$ in the *dq*-rotating frame directly.

The turbine controller, presented in Figure 6.36, consists of a pitch angle controller, power controller and speed controller, whose structures are presented in Figs. 6.17–6.19 and whose aim is described in Sect. 6.3.

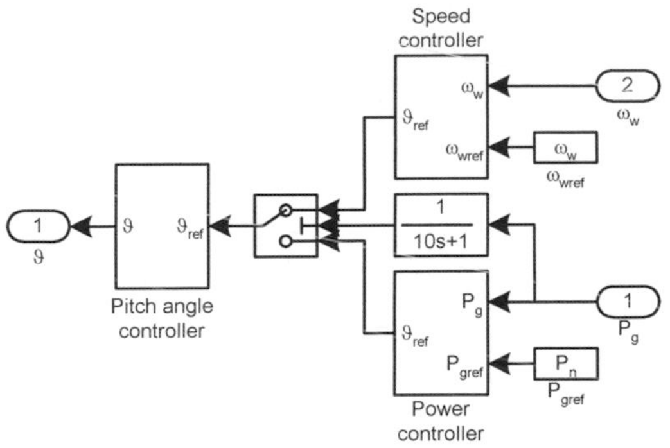


**Fig. 6.36.** Block diagram of turbine controller (power controller – $T_f = 1.5$ s, $K_R = 1.1/P_n$, $T_I = 1.5$ s, speed controller – $T_f = 0.05$ s, $K_R = 10$, $T_I = 0.32$ s)

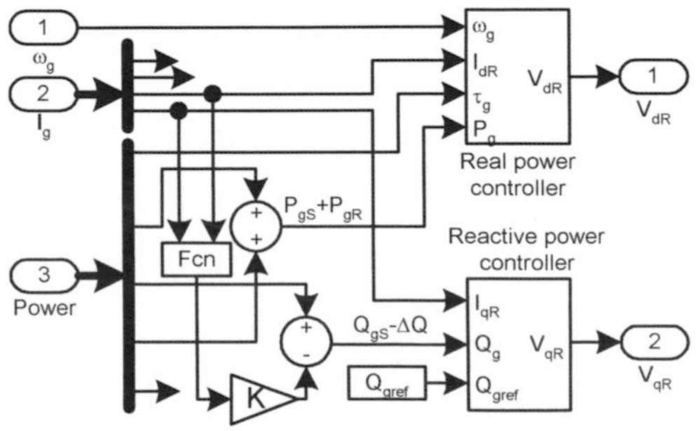


**Fig. 6.37.** Block diagram of generator controller

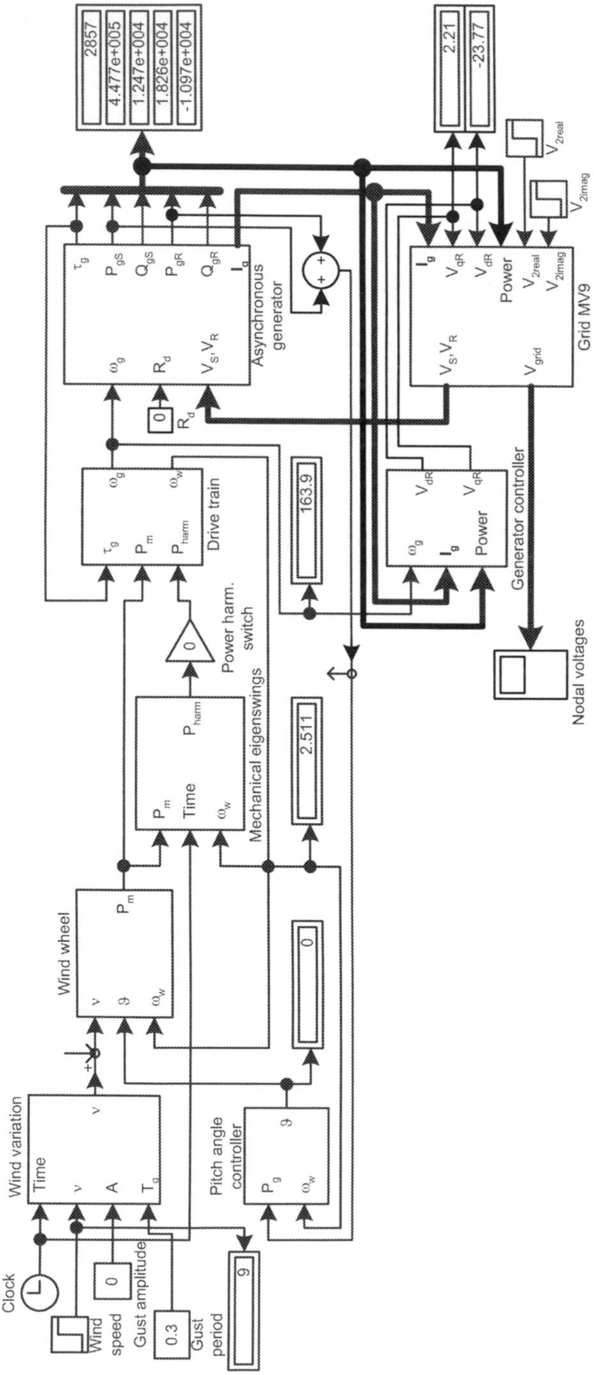


**Fig. 6.38.** Block diagram of WTGS with a doubly-fed asynchronous generator model

The generator controller, presented in Fig. 6.37, consists of two basic controllers: a reactive power controller and a real power controller, which consist of adequate sub-controllers.

The reactive power controller (Fig. 6.39) forms the $q$-axis rotor voltage and at the same time controls the WTGS reactive power (or power factor). The controller consists of a $q$-axis rotor current controller and a reactive power controller ($Q_g$ controller). Both controllers are of the PI-type with a measuring element (filter) and output limiter, as presented in Fig. 6.18. The rotor current controller's aim is to keep constant the reference value of the rotor current in the $q$-axis, while the reactive power controller keeps constant the reference value of the reactive power. Its output is the reference value of the $q$-axis rotor current. The controller can operate both at partial and full load. But for the given structure of the control system, the controller parameters have to be changed when the operating point (partial load or full load) of the wind turbine changes.

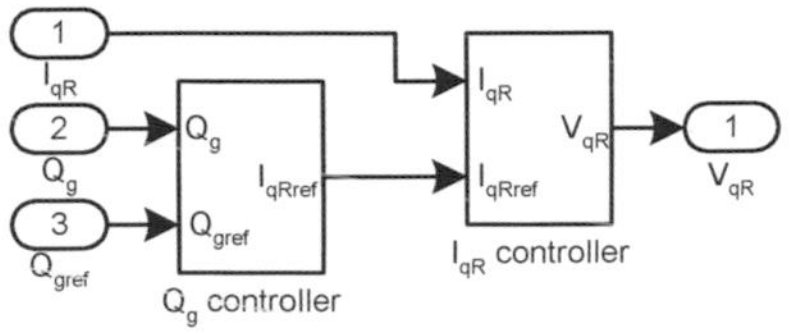


**Fig. 6.39.** Block diagram of reactive power controller ($Q_g$ controller – $T_f$ = 0.3 s, $K_R = 4.3/P_n$, $T_I$ = 0.2 s, $I_{qR}$ controller – $T_f$ = 0.01 s, $K_R$ = 1000, $T_I$ = 0.02 s)

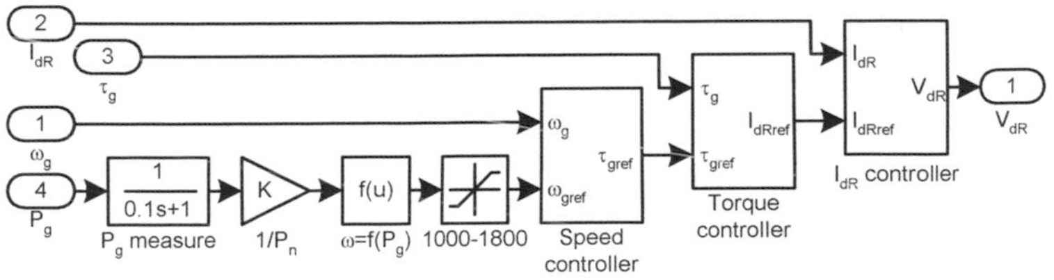


**Fig. 6.40.** Block diagram of a real power controller (speed controller – $T_f$ = 0.1 s, $K_R = 5/\omega_n$, $T_I$ = 0.149, torque controller – $T_f$ = 0.1 s, $K_R$ = 1000, $T_I$ = 0.1 s, $I_{dR}$ controller – $T_f$ = 0.01 s, $K_R$ = 1000, $T_I$ = 0.02 s)

The real power controller (Fig. 6.40) forms the $d$-axis rotor voltage and at the same time controls the WTGS real power or rotor speed. The controller consists of a $d$-axis rotor current controller, torque controller and real power (or rotor speed) controller. All the controllers are of the PI type with a measuring element (filter) and output limiter. The rotor current controller's aim is to keep constant the reference value of the rotor current in the $d$-axis, while the torque controller keeps constant the reference value of the torque. Its output is the reference value of the $d$-axis rotor current. And the real power/rotor speed controller keeps the reference value of the real power or the rotor speed. The controllers form the reference value of the torque. The utilization of the real power or rotor speed controller depends on the WTGS operating point. At partial load, the speed controller is used. Then its input, i.e. the shaft speed reference, is computed as a function of the WTGS real power $\omega_{ref} = f(P)$; this is function *Opt* in Fig. 6.41. In full load operation, the speed and torque controllers are replaced by a power controller (PI-type) with parameters $T_f = 0.1$ s, $K_R = 5000$, $T_I = 0.1$ s. The power controller output is then the reference value of the *d*-axis rotor current.

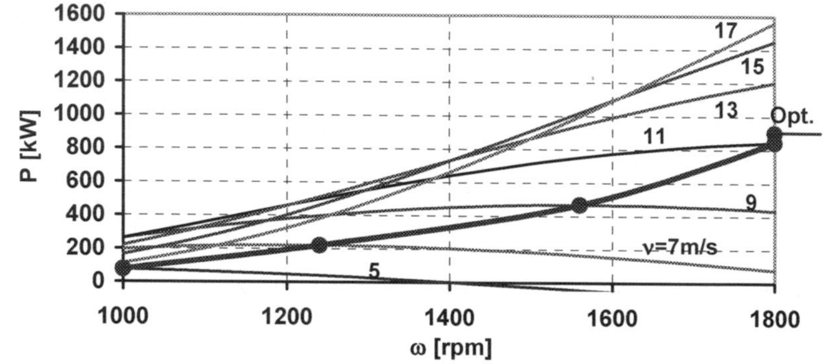


**Fig. 6.41.** Power versus shaft speed characteristics for the considered WTGS with a doubly-fed asynchronous generator for various values of wind velocity

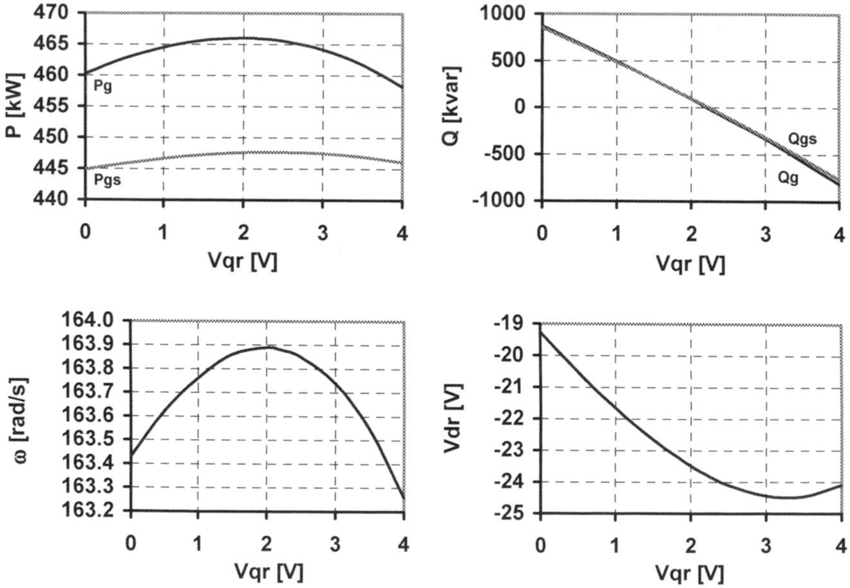


**Fig. 6.42.** Influence of rotor voltage $V_{qR}$ on the WTGS operating point at a wind speed equal to $v = 9$ m/s

The WTGS equipped with a doubly-fed asynchronous generator allows its rotor speed to be changed over a wide range and at the same time allows the achievement of the control algorithm described in Sect. 5.6.1 (optimal tip speed ratio $\lambda_{opt}$ scheme). For the considered model, its power versus rotor speed characteristic is presented in Fig. 6.41. The WTGS characteristic (marked as a bold curve called *Opt*) consists of points located on the maxima of the power versus wind velocity curves. For wind velocity higher than $v = 11$ m/s, the shaft speed is kept constant and equal to 1800 rpm. For the wind velocity higher than about 12 m/s, the turbine power is additionally limited to the rated value $P_n = 900$ kW.

The system with the doubly-fed asynchronous machine allows us to influence the system state by controlling the rotor voltage (here the *dq*-reference frame voltages). For the considered system, a few characteristics that show the possibility of influencing the wind turbine operating point by changing the $V_{qR}$ voltage[10] are presented in Fig. 6.42. The characteristics are made in the steady-state at a wind velocity equal to $v = 9$ m/s (partial load at the operating point with the maximum slope of the $P = f(\omega)$ characteristics).

The curves show that by changing the $V_{qR}$ voltage over a small range (0–4 V), it is possible to change the reactive power transferred to the grid over a wide range (±800 kvar). The power transferred to the grid[11] $Q_g$ is close to the stator power $Q_{gs}$ when the difference is equal to the power loss in the transformer feeding the machine rotor and the power converter. At the same time, the $V_{qR}$ voltage change causes a relatively small change of the real power transferred to the grid $P_g$. In fact, this change is mainly related to the shaft speed $\omega$ and $V_{dR}$ voltage change (lower part of Fig. 6.42), which are necessary to keep the WTGS in the steady-state for a given wind velocity, and which are directly correlated.

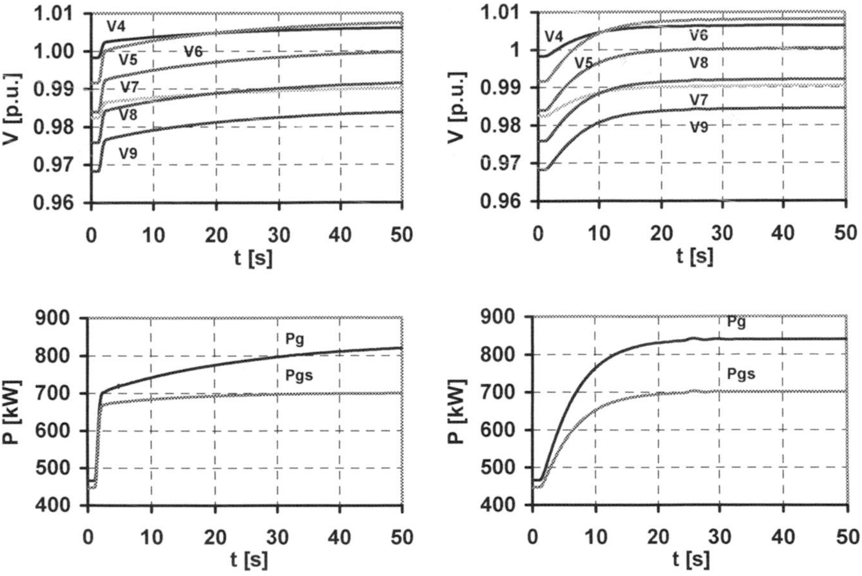


---

[10] Voltages $V_{dR}$, $V_{qR}$ are recalculated on the stator side.

[11] The real power transferred to the grid is equal to the sum of the stator and rotor power, i.e. $P_g = P_{gS} + P_{gR}$. The reactive power is equal to the difference of the stator power and the reactive power losses in a grid-side of converter and transformer, i.e. $Q_g = Q_{gS} - \Delta Q$.

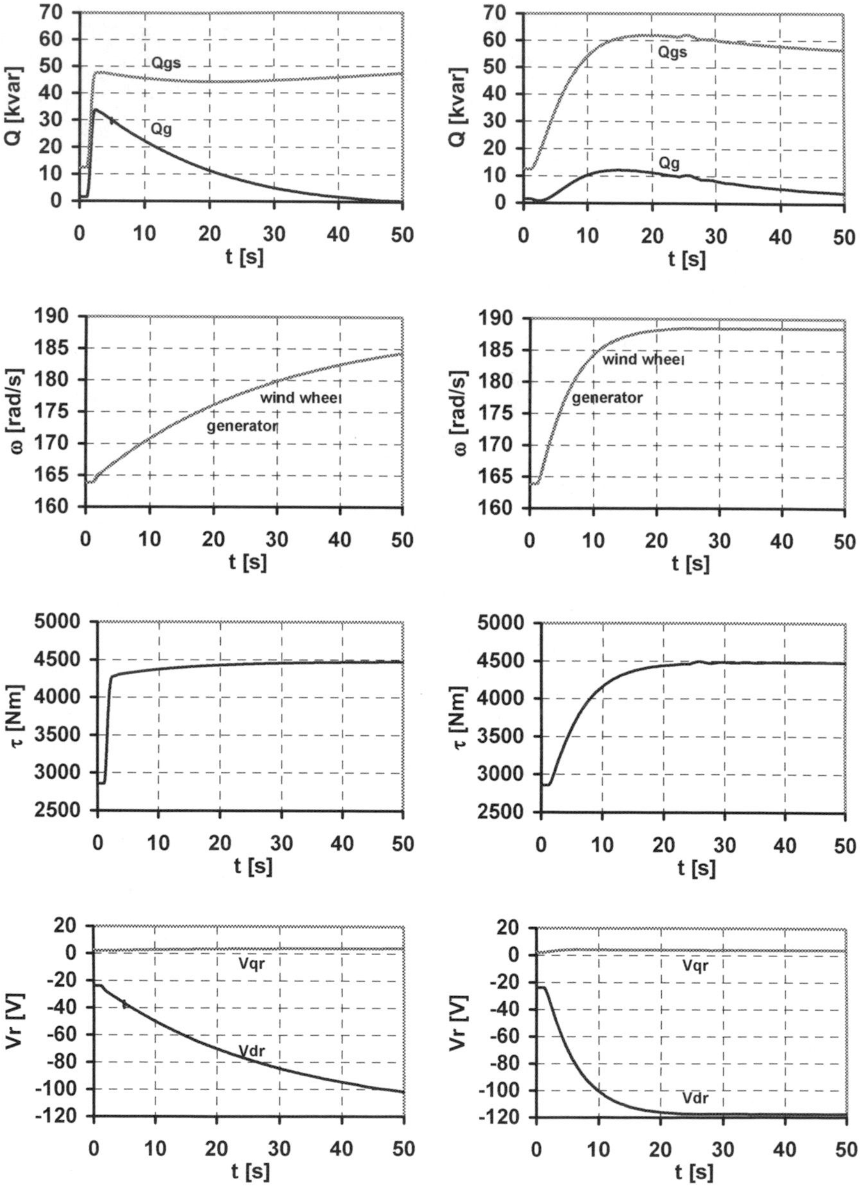


**Fig. 6.43.** WTGS response to a wind velocity step increase from 9 to 11 m/s (left-hand graphs – generator speed controller $K_R = 50$, $T_I = 14.9$ s, $T_f = 21$ s; right-hand graphs – generator speed controller $K_R = 5$, $T_I = 0.15$ s, $T_f = 0.1$ s; $V$ – nodal voltages, $P_g$ – WTGS real power, $P_{gs}$ – stator real power, $Q_q$ – WTGS reactive power, $Q_{gs}$ – stator reactive power, $\omega$ – generator and wind wheel shaft speed, $\tau$ – torque, $V_{dr}$ and $V_{qr}$ – $dq$-axis rotor voltages)

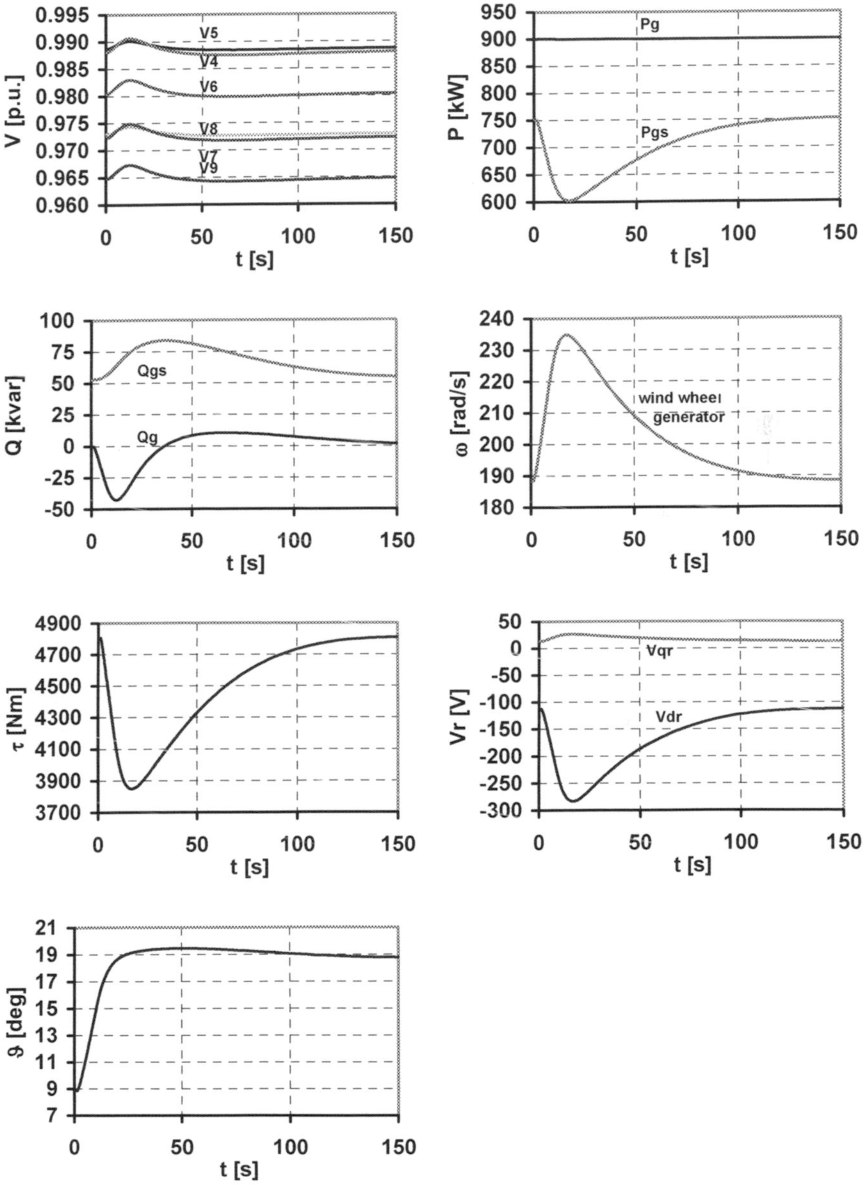


**Fig. 6.44.** WTGS response to a wind velocity step increase from 15 to 17 m/s (generator real power controller $T_f = 0.1$ s, $K_R = 5000$, $T_I = 0.1$ s, $V$ – nodal voltages, $P_g$ – WTGS real power, $P_{gs}$ – stator real power, $Q_q$ – WTGS reactive power, $Q_{gs}$ – stator reactive power, $\omega$ – generator and wind wheel shaft speed, $\tau$ - torque, $V_{dr}$ and $V_{qr}$ – $dq$-axis rotor voltages, $\vartheta$ – pitch angle)

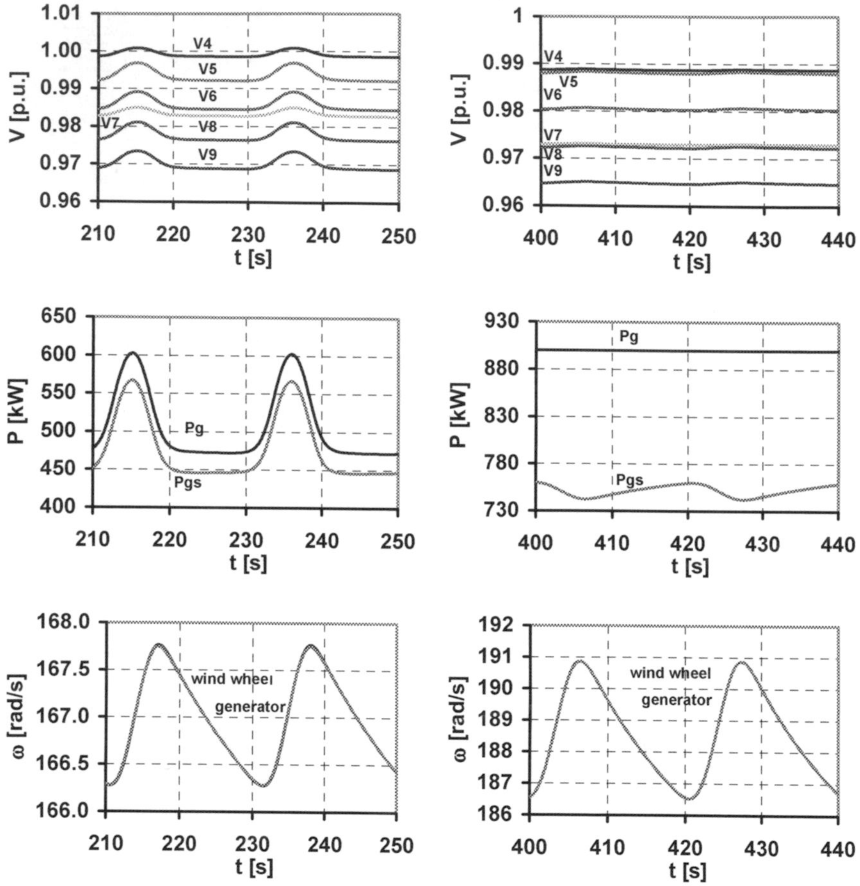


**Fig. 6.45.** WTGS operation during gusts (gust amplitude 1 m/s; left-hand graphs – $v$= 9 m/s; right-hand graphs – $v$=15 m/s; generator speed controller $K_R$ =5, $T_I$ =0.15 s, $T_f$=0.1 s)

The next figures present the dynamic properties of the considered WTGS in the form presented in previous sections.

The response of the system to the step change of wind velocity is presented in Figs. 6.43 and 6.44. The curves show the nodal voltages, real and reactive power, shaft speed, rotor voltages and torque change after a wind velocity increase of 2.7 m/s. The wind velocity harmonics and mechanical eigenswings are neglected.

Figure 6.43 presents the response of the WTGS operating at partial load for two (taking into account the values of the parameters) generator speed controllers. Speed controllers are located in the real power controller of the generator. For the two sets of controller parameters, the control process is slow. The process presented in the left-hand graphs is characterized by two steps. The first step is related to fast change of nodal voltages, real power and torque while the rotor speed and $V_{dr}$ voltage change slowly. Next, in the second step, the process of passing to the new equilibrium point becomes slow (with a time constant of about 10 s). The process presented in the right-hand graphs is generally faster, but the response has an aperiodic character, with a time constant of about 6 s.

The pass to the new operating point is related to a change of nodal voltages of up to 1.5% at nodes 5, 8, 9 and with an increase of the real power equal to about 370 kW. The shaft speed increase is equal to 25 rad/s (15%). There are no torsional oscillations of the drive train.

At full load, the system reaction to a wind velocity increase (Fig. 6.44) is much smaller. The nodal voltage change (and amplitude of oscillation) is less than 0.3%. Also, the WTGS real power change is relatively small and equal to about 0.35 kW when the stator (and at the same time the rotor) real power change (amplitude) reaches 150 kW. The increase of the rotor speed with a speed increment of 45 rad/s (23%) is reduced by the turbine control system in 100 s, after the blade pitch angle $\vartheta$ changes, which takes about 20 s.

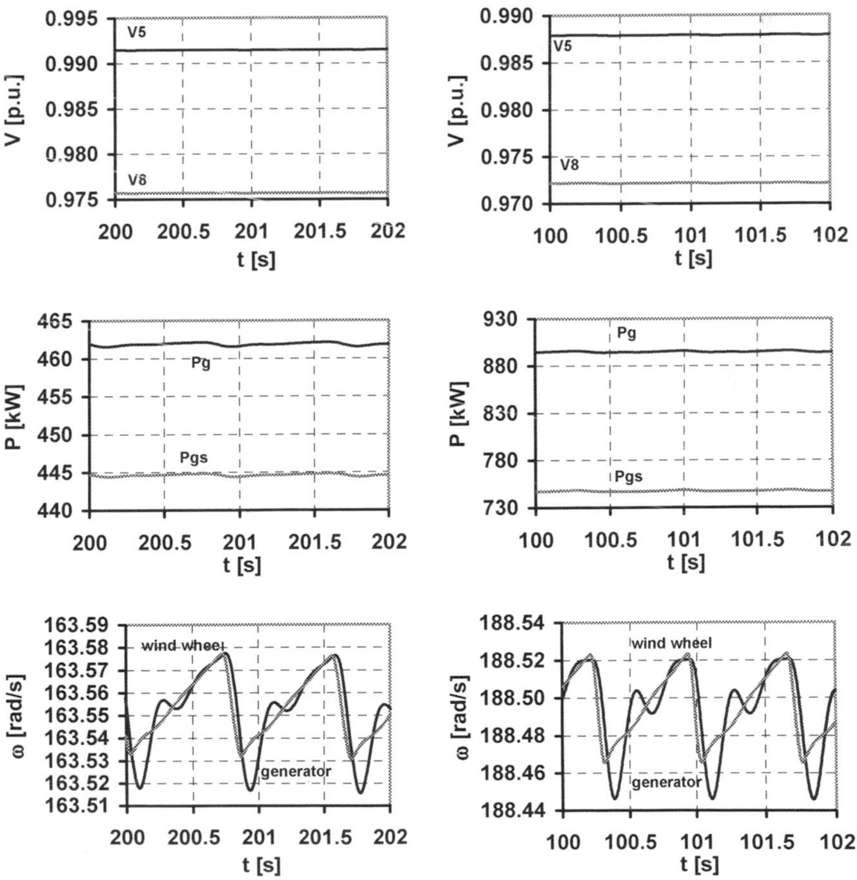


**Fig. 6.46.** Effect of blades passing in front of the WTGS tower (3P amplitude 7 %, left-hand graphs – $v = 9$ m/s; right-hand graphs – $v = 15$ m/s; generator speed controller $K_R = 5$, $T_I = 0.15$ s, $T_f = 0.1$ s)

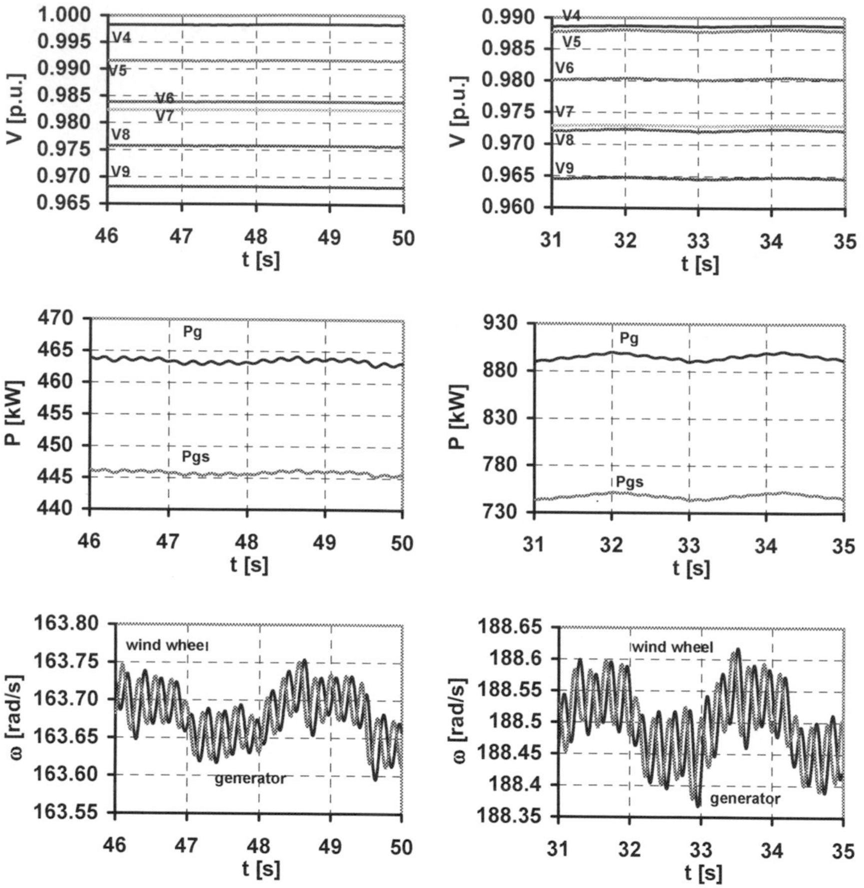


**Fig. 6.47.** 1P, 3P and tower vibration effects (1P amplitude 1%, 3P amplitude 7%, 4.5 Hz effect 10%; left-hand graphs – $v = 9$ m/s; right-hand graphs – $v = 15$ m/s; generator speed controller $K_R = 5$, $T_I = 0.15$ s, $T_f = 0.1$ s)

The next test is related to the response of the system to wind gusts with amplitude equal to 1 m/s. The response is presented in Fig. 6.45. As in the case of the previously considered wind turbine, because of relatively the small period of the gusts, the reaction of the WTGS is a follow-up type. The process occurs without oscillations. There is no visible delay between the wind-wheel and the generator rotor speed. Also, as in the previous example, the power and the voltage change are significantly smaller in full load operation. At full load, the amplitude of the real power change is equal to 1 kW, while at partial load, the amplitude reaches 130 kW. The voltage change amplitude reaches 0.5% at partial load, while at full load the amplitude is less than 0.04%.

Figure 6.46 shows the operation of the system in the case when the 3P oscillation with amplitude 7% exists. In the considered case, at both operating points (partial and full load), the oscillations are well damped. At partial load, the band of the real power variation is equal to 0.6 kW, while at full load, the band reaches 1.6 kW. The nodal voltage variation amplitudes are less than 0.01% in both cases.

Figure 6.47 presents the WTGS operations when the 1P, 3P oscillations and tower vibration effect (4.5 Hz) with amplitude 1%, 7% and 10%, respectively, exist. The figure shows the complex response of the wind turbine. The WTGS response at partial and full load has slightly different characteristics. The oscillations at partial load are more damped. The band of the real power oscillations, with the frequency of about 0.2 Hz, is equal to 1.6 kW, while the amplitudes of the nodal voltage oscillation are less than 0.1%. At full load, the real power oscillations with a band equal to about 11 kW have a frequency of about 2 Hz. There are also visible nodal voltage oscillations smaller than at partial load.

It is worth remembering that in real system, the amplitude of a given oscillation depends on the WTGS operating point. Usually, an increase of a wind velocity causes an increase of these amplitudes.

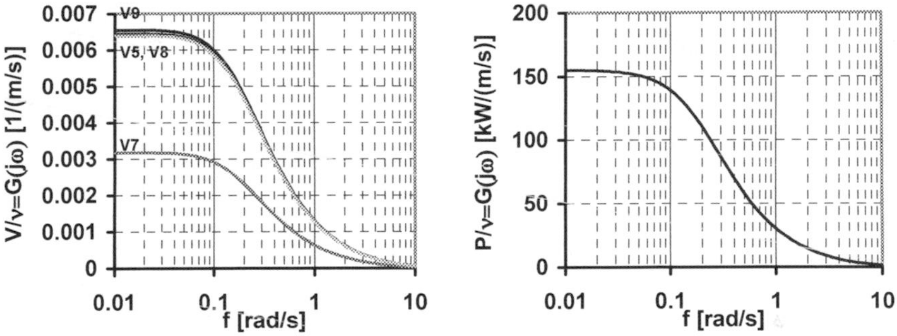


**Fig. 6.48.** Frequency characteristics of the WTGS at partial load ( $v$= 9 m/s)

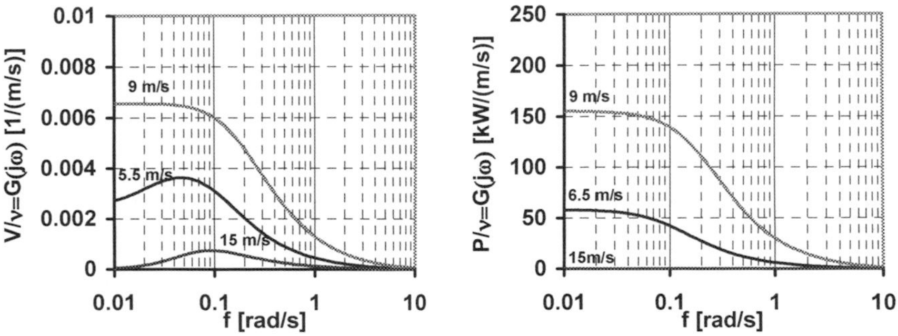


**Fig. 6.49.** Power generation (wind velocity) influence on the WTGS frequency characteristics

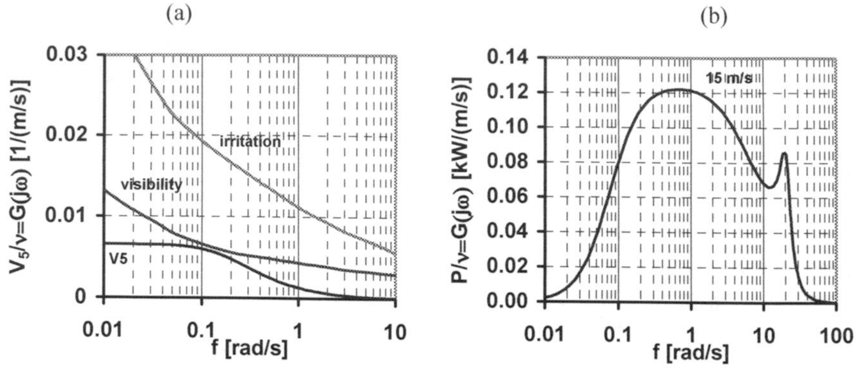


**Fig. 6.50.** Frequency characteristics: (a) comparison of WTGS voltage characteristics with the borders of irritation and visibility (voltage dips effected by lights – see Fig. 3.4) defined by standard IEEE 141-1993; (b) frequency characteristics at wind speed 15 m/s (lower curve on the right-hand side of Fig. 6.49)

Figures 6.48–6.50 present various frequency characteristics of the analyzed system. Figure 6.48 shows the frequency characteristics $V_i/v = G(j\omega)$ and $P/v = G(j\omega)$ achieved in the system operating at partial load ($v = 9$ m/s). The curves allow us to formulate the following conclusions:

- The voltage variation in the nodes located between the PCC and the feeding point is much smaller than the voltages at the nodes located "behind" the WTGS (nodes 8, 9). The voltage variation in these nodes is similar to the voltage variation at the PCC. Over a high range of frequency, the voltage variation is less than 0.65% when the wind speed variation is less than 1 m/s. Then the WTGS can be considered as a low-pass filter. The cut-off frequency of the "filter", equal to about 0.2 rad/s, is 10 times lower than the one achieved in the case of the previously considered wind turbines.
- The real power variation amplitude is close to 160 kW/(m/s) for the frequency of the wind velocity variations $\omega < 0.1$ rad/s and decreases for higher frequency.
- Wind velocity variation with a frequency higher than 10 rad/s is practically eliminated.

In the case of the considered WTGS, and contrary to the previously considered WTGSs, the stiffness coefficient value practically does not influence the frequency characteristics (module). The influence is insignificant and therefore the characteristics are not presented.

Figure 6.49 shows the influence of the operating point (wind velocity) on the frequency characteristics. The higher response of the system, as in the previous cases, takes place at a wind velocity of 9 m/s, where the WTGS characteristics $P = f(v)$ slope is the biggest. The lower response takes place when the slope of the characteristics is smaller, i.e. at rated and higher wind velocity ($v \geq 15$ m/s) – see also Fig. 6.50b.

The last figure in this group shows the comparison of the frequency characteristics with the limits defined by the standards (Fig. 6.50). The frequency characteristic $V_5/v = G(j\omega)$, with the highest amplitude, for all the frequencies, is located below the borders of irritation and visibility. The curve, of course, lies also below the border defined by the IEC standard (Figure 3.4, $P_{st} = 1$). It is still worth remembering that an increase of the wind velocity variation amplitude above 1 m/s will cause the characteristics to move up and then the limits will not be fulfilled.

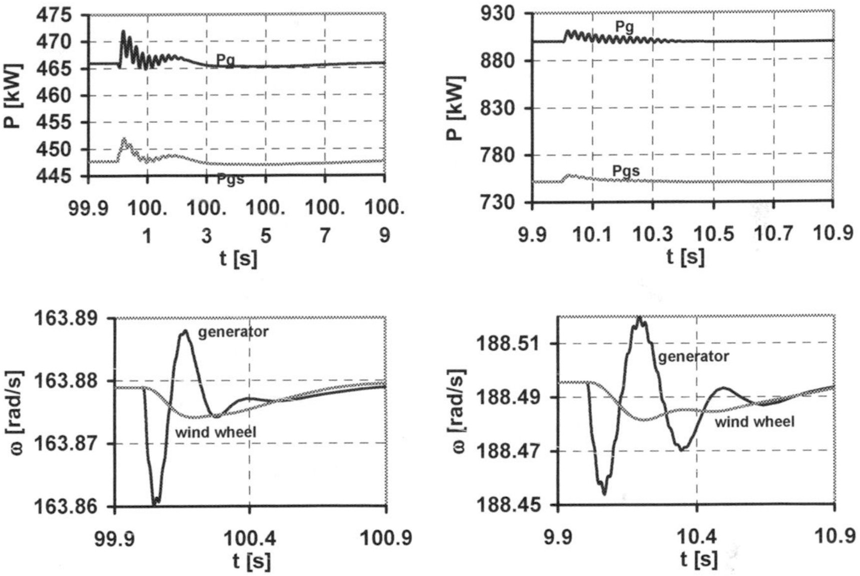


**Fig. 6.51.** WTGS response to a 1.5% step change of the feeding bus voltage (left-hand graphs – $v$= 9 m/s; right-hand graphs – $v$= 15 m/s)

The last figure shows again the time domain response of the system. Figure 6.51 shows the response of the system to a 1.5% step increase of the feeding bus voltage. The curves indicate high damping of the electromechanical processes, which (in both cases) pass in about 1 s. The real power surge as a result of the voltage change is less than 6 kW at partial load and 12 kW at full load.

## 6.5 WTGS with Synchronous Generator

### 6.5.1 Synchronous Generator with Current-Source Inverter

A model of the wind turbine generator system equipped with a synchronous generator operated in an MV system is presented in Fig. 6.52 and consists of:

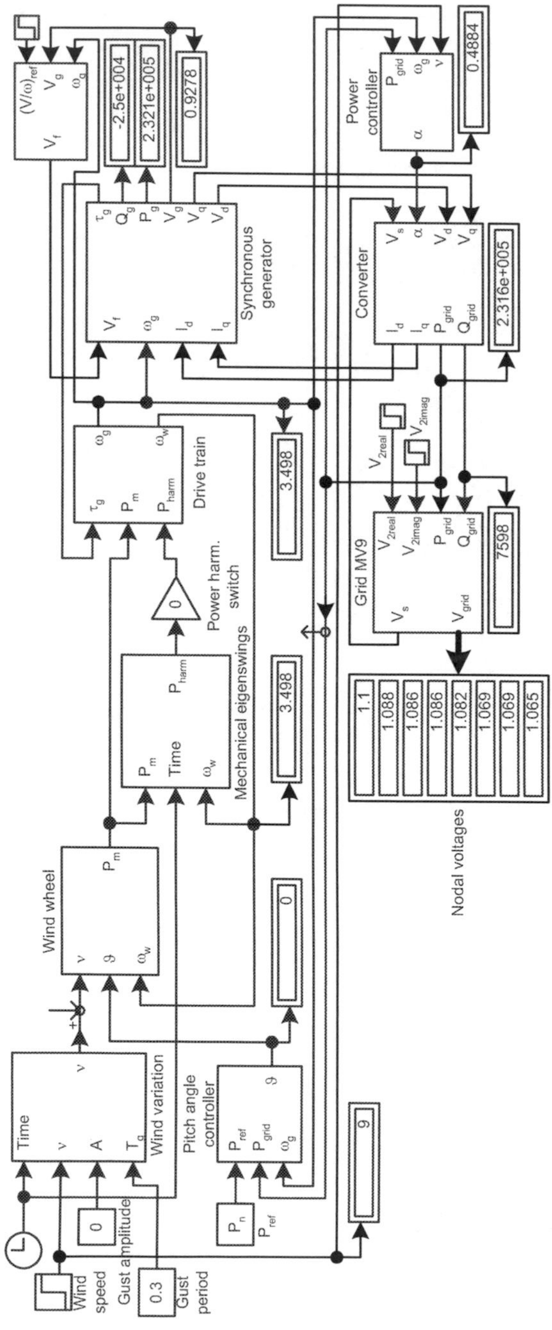


**Fig. 6.52.** Block diagram of the WTGS with a synchronous generator model

- wind-variation model,
- wind-wheel model (Sect. 5.2.5, 5.2.8),
- mechanical-eigenswing model (Sect. 5.2.6),
- drive-train model (Sect. 5.2.7),
- synchronous-generator model (Sect. 5.4.2) – a 5-th order model of a synchronous generator with terminal voltages ($V_d$, $V_q$) as the output and terminal currents ($I_d$, $I_q$) as the input is used here. Then in the differential equations defining the model, the transformation voltages $d\Psi_d/dt$ and $d\Psi_q/dt$ are neglected.
- converter model (Sect. 5.5.2) – the functional models of the uncontrolled rectifier and the controlled inverter are utilized. The intermediate element is modeled by the first order differential equation:

$$L_d \frac{di_d}{dt} + R_d i_d = v_{drec} - v_{dinv}\,, \tag{6.6}$$

where $L_d$, $R_d$ – intermediate element (reactor) parameters,
$i_d$ – intermediate element current,
$v_{drec}$, $v_{dinv}$ – DC side voltage of rectifier and inverter.

- grid model (Sects. 5.7.5, 6.1).

The WTGS and its model parameters are presented in Tables 6.12 and 6.13. The WTGS is equipped with a turbine, a generator and inverter controllers. The turbine controller acts on the pitch angle, the generator controller acts on the generator field voltage and the inverter controller acts on the firing angle of the thyristors.

The turbine controller model is presented in Fig. 6.53 and consists of a pitch angle controller, power controller and speed controller. The structures and functions of the controllers are described in Sect. 6.3 (Figs. 6.17–6.19).

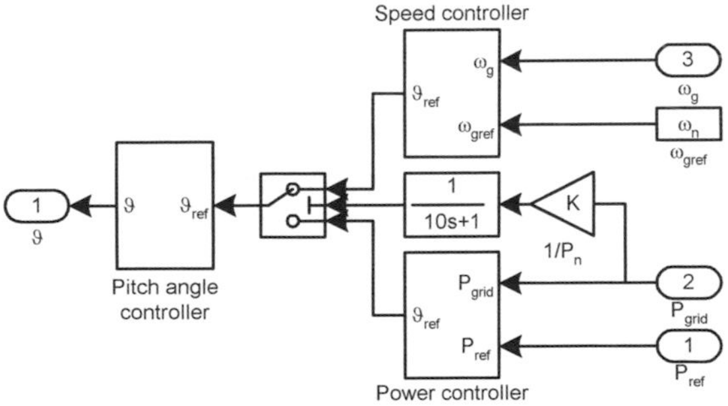


**Fig. 6.53.** Block diagram of the turbine controller (power controller – $T_f = 3$ s, $K_R = 1.2/P_n$, $T_I = 5$ s; speed controller – $T_f = 3$ s, $K_R = 0.24$, $T_I = 0.91$ s)

The voltage controller structure is presented in Fig. 6.54. In the presented form, the controller in fact controls the flux defined by the $V_G/\omega$ quotient. Using this quantity, in the case when the rotor speed varies over a high range, allows us avoids over-excitation of the generator rotor for low values of the rotor speed. In the modeled system, the reference value of $V_G/\omega$ is set equal to 1 (in the per unit).

**Table 6.12.** WTGS parameters (ENERCON E-40 wind turbine)

| Object | Parameter | Symbol | Value | Unit |
|---|---|---|---|---|
| Operational parameters | Nominal output | $P_n$ | 500 | kW |
| | Nominal voltage | $V_n$ | 400 | V |
| | Nominal current | $I_n$ | 720 | A |
| | Power regulation | Blade pitch control | | |
| | Cut-in wind speed | $v_{cut\text{-}in}$ | 3.5 | m/s |
| | Rated wind speed | $v_n$ | 12 | m/s |
| | Cut-out wind speed | $v_{cut\text{-}out}$ | 25 | m/s |
| Rotor | Rotor diameter | $2R$ | 52 | m |
| | Number of blades | $B$ | 3 | - |
| | Rotor revolution | $n$ | 18-36 | rpm |
| | Moment of inertia | $J_W$ | *1.6×10⁶* [12] | $kg{\cdot}m^2$ |
| Synchronous generator | Rated apparent power | $S_n$ | 510 | kVA |
| | Rated power | $P_n$ | 500 | kW |
| | Rated voltage | $V_n$ | 440 | V |
| | Power factor | $\cos\varphi_n$ | 0.98 | - |
| | Synchronous speed | $n_s$ | 36 | rpm |
| | Moment of inertia | $J_G$ | *3800* | $kg{\cdot}m^2$ |
| Wind wheel model | Coefficients of (5.16) | $c_1$ | *0.5* | - |
| | | $c_2$ | *116* | - |
| | | $c_3$ | *0.4* | - |
| | | $c_4$ | *0* | - |
| | | $c_5$ | *5* | - |
| | | $c_6$ | *21* | - |

**Table 6.13.** Synchronous generator model data

| Parameter | Symbol | Value | Unit |
|---|---|---|---|
| d-axis synchronous reactance | $X_d$ | *1.0* | p.u. |
| d-axis transient reactance | $X'_d$ | *0.22* | p.u. |
| d-axis subtransient reactance | $X''_d$ | *0.13* | p.u. |
| q-axis synchronous reactance | $X_q$ | *0.54* | p.u. |
| Armature leakage reactance | $X_l$ | *0.07* | p.u. |
| Zero-sequence reactance | $X_0$ | *0.05* | p.u. |
| Negative-sequence reactance | $X_2$ | *0.141* | p.u. |
| Short-circuit d-axis transient time constant | $T'_d$ | *0.25* | s |
| Short-circuit d-axis subtransient time constant | $T''_d$ | *0.06* | s |
| Short-circuit q-axis subtransient time constant | $T''_q$ | *0.06* | s |
| Armature winding time constant | $T_a$ | *0.025* | s |

The inverter controller structure is presented in Fig. 6.55. The controller operates here as the real power transferred to the grid controller (therefore, in Fig. 6.52 it is labelled as the power controller). The controller's aim is to keep constant the reference value of the real power, which is defined by the function $P_{ref} = f(\omega)$ described in Fig. 6.56 as *Char* (bold line). For the shaft speed close to the rated one

[12] The data in Italic typeface indicate assumed data.

the characteristics are defined by the function $P_{\text{ref}} = K\omega$ with a high value of coefficient $K$.

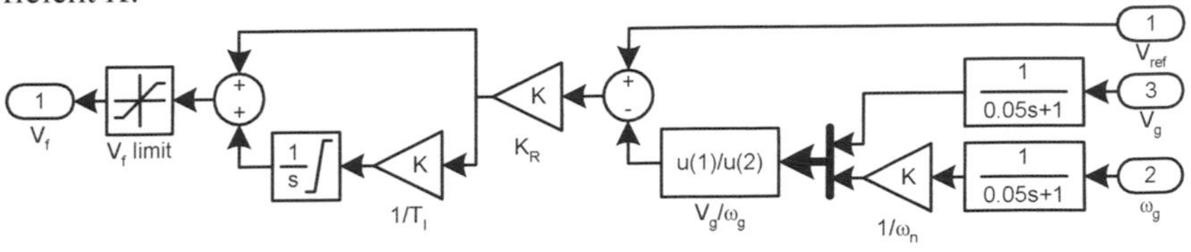


**Fig. 6.54.** Block diagram of the synchronous generator voltage controller ($K_R = 1$, $T_I = 0.4$s)

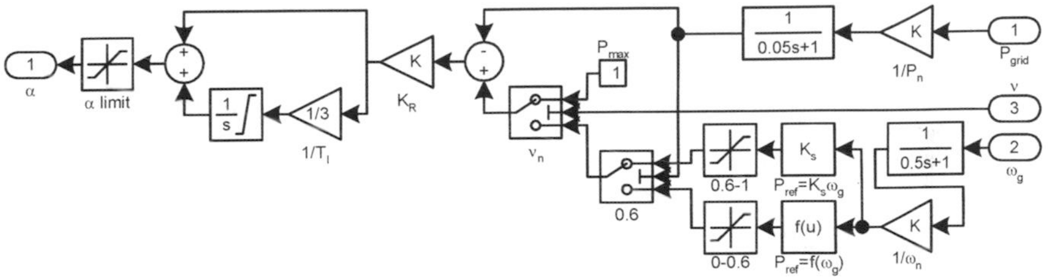


**Fig. 6.55.** Block diagram of the inverter power controller ($K_R = 0.001$, $T_I = 3$ s; at partial load $K_R$ can be higher, e.g. 0.1)

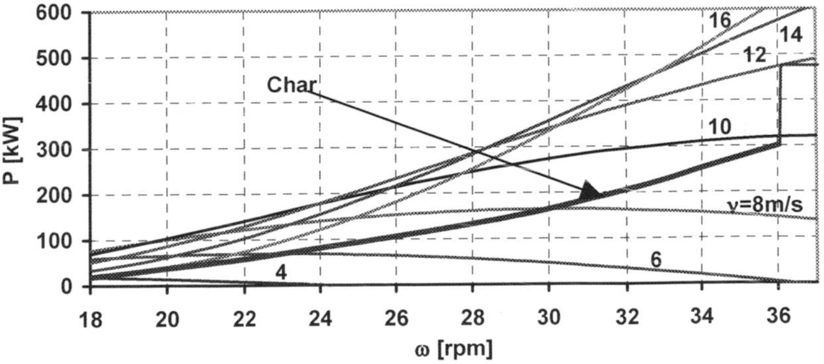


**Fig. 6.56.** Power versus rotor speed characteristics of the considered WTGS

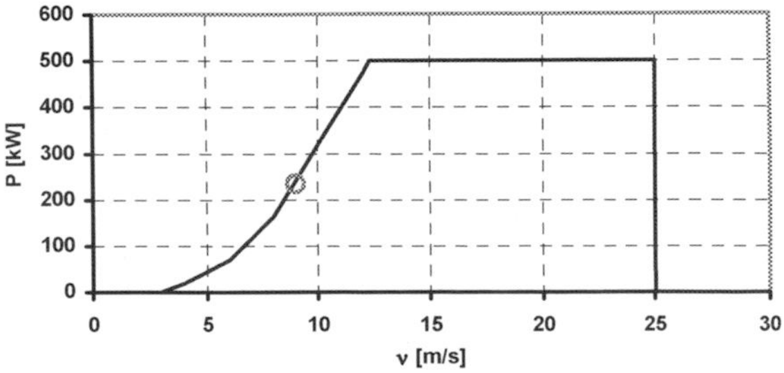


**Fig. 6.57.** Power versus wind velocity characteristics of the considered WTGS

The latter function in the form of the WTGS real power versus wind velocity characteristic is also presented in Fig. 6.57. There the operating point (partial load) for which further analysis is made is marked by a circle. The characteristic gives a good fit to the ENERCON E-40 wind turbine real characteristic.

The controller operates both at partial and full load. Its control loop (structure) changes when the real power exceeds 300 kW (0.6 in per unit) and when the real power reaches the rated value.[13] At the first operating point, the turbine speed controller is switched on to keep the rotor speed equal to 36 rpm.

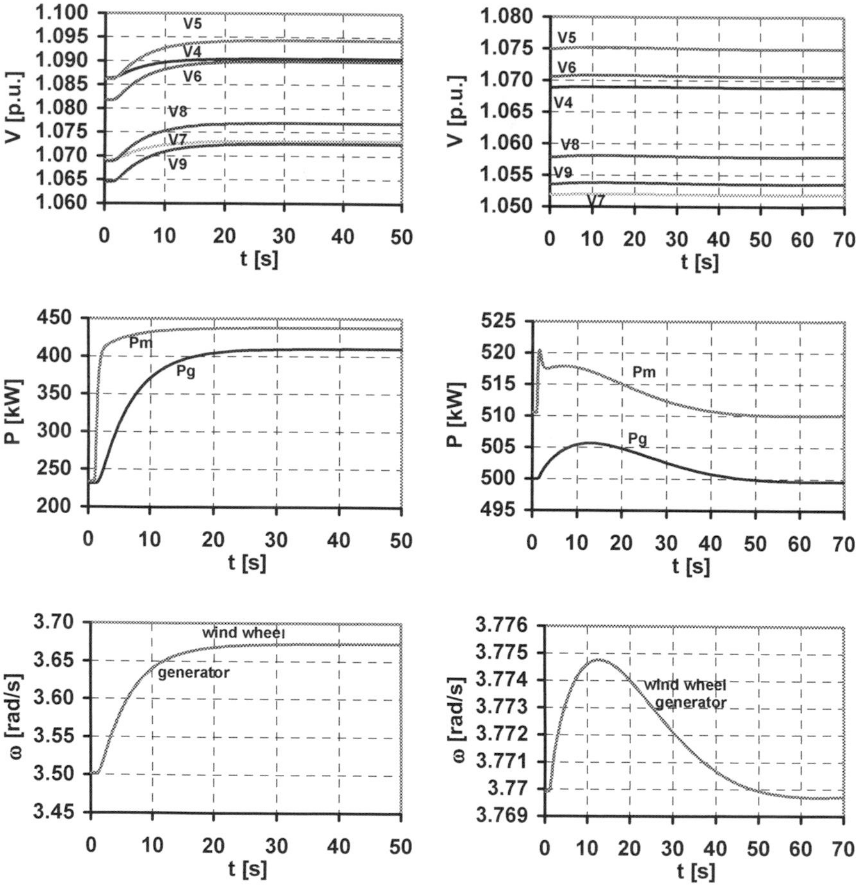


**Fig. 6.58.** WTGS response to a wind velocity step increase (left-hand graphs – partial load, wind velocity step from 9 to 11.7 m/s; right-hand graphs – full load, wind velocity step from 15 to 17.7 m/s; $V$ – nodal voltages, $P_g$ – WTGS real power, $P_m$ – mechanical power, $\omega$ – generator and wind-wheel shaft speed, $V_g$ – generator terminal voltage, $\vartheta$ – blade pitch angle)

[13] The first state takes place at wind velocity slightly below 10 m/s, the second one at a wind velocity equal to 12 m/s.

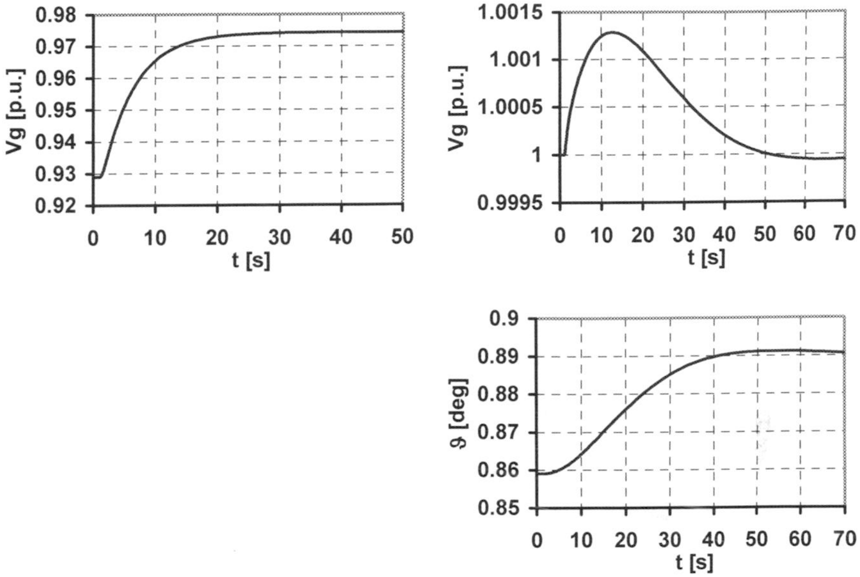


**Fig. 6.58.** (cont.) WTGS response to a wind velocity step increase

From the second operating point, the power (inverter) controller keeps constant the rated value of the real power transferred to the grid. The controller, as for all the controllers presented above, is a PI-type controller, which, as mentioned before, has to be considered as only an example of a structure.

As an element of dynamic system analysis, the response of the system to the step change of wind velocity is presented in Fig. 6.58. At partial load, the wind velocity increase of 2.7 m/s causes an aperiodic change of the operating point with a time constant equal to about 7 s. The change to the new operating point is related to a change of the nodal voltages by up to 0.8% at nodes 5, 8, 9. The real power increase is equal to about 170 kW, the shaft speed increase is equal to 0.17 rad/s (4.9 %), and the generator voltage increase is equal to 0.042 (in per units), which is directly related to the shaft speed change.

At full load, the system reaction to the wind velocity increase is much smaller, which is typical for all the considered wind turbines and results from the flat WTGS characteristic $P = f(v)$. The nodal voltage change (and the amplitude of oscillations) is less than 0.01%. Also the real power change is very small and equal to about 6 kW. The rotor speed increase is small and equal to 0.13% and is reduced to the initial value as a result of the turbine controller small reaction (pitch angle change) during 50 s. In general, the electromechanical swings in both cases are relatively well damped.

It is worth realizing that in the considered example a WTGS with a rated power of 500 kW is tested, while in the previous examples, the wind turbines with a rated power equal to 900 kW were considered, which undoubtedly influences the values of the amplitudes and changes of the various presented quantities.

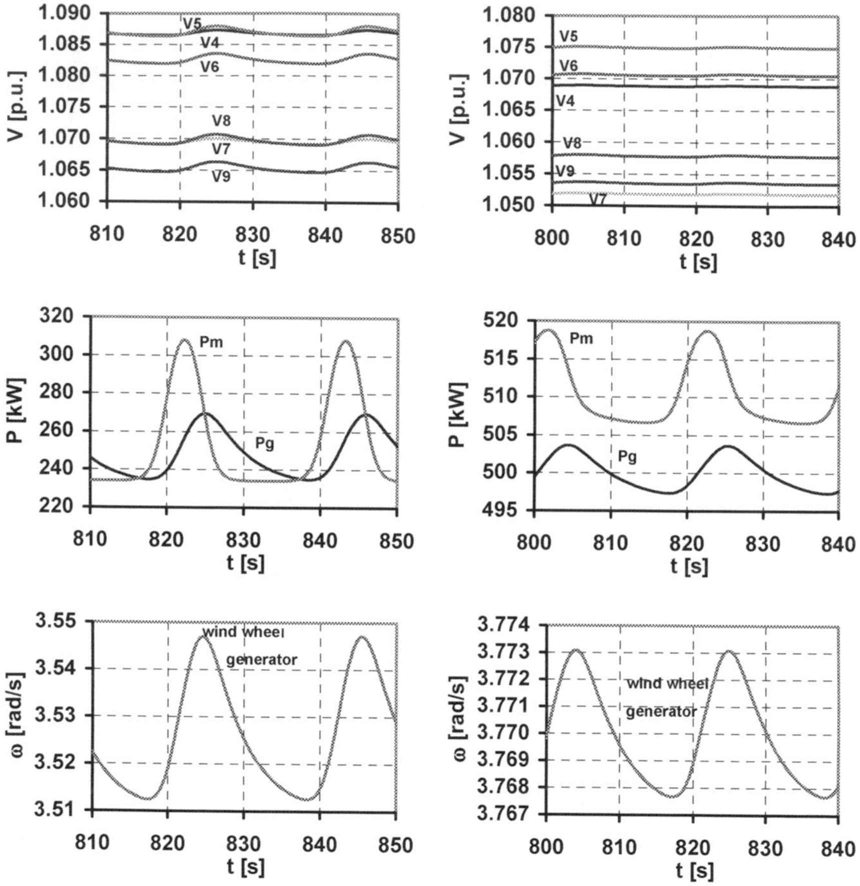


**Fig. 6.59.** WTGS operation during gusts (left-hand graphs – $v = 9$ m/s; right-hand graphs – $v = 15$ m/s; gust amplitude 1 m/s)

The next test is related to the response of the system to wind gusts with amplitude equal to 1 m/s. The response is presented in Fig. 6.59. The process runs without oscillations. As in the case of the step increase of wind velocity, in the considered test there are no oscillations between the wind wheel and generator rotors. Also, as in the previous examples, the power and the voltage changes are significantly smaller in full load operation. At full load, the amplitude of the real power change is equal to 6 kW, while at partial load the amplitude reaches 35 kW. The figure also shows the mechanical power, which in comparison to the WTGS power $P_g$ allows us to evaluate the level of power variation damping. The quotient of the power transferred to the grid and into mechanical power can be used as a measure of the power oscillation damping. The voltage change amplitude reaches 0.3% at partial load, while at full load the amplitude is less than 0.01%.

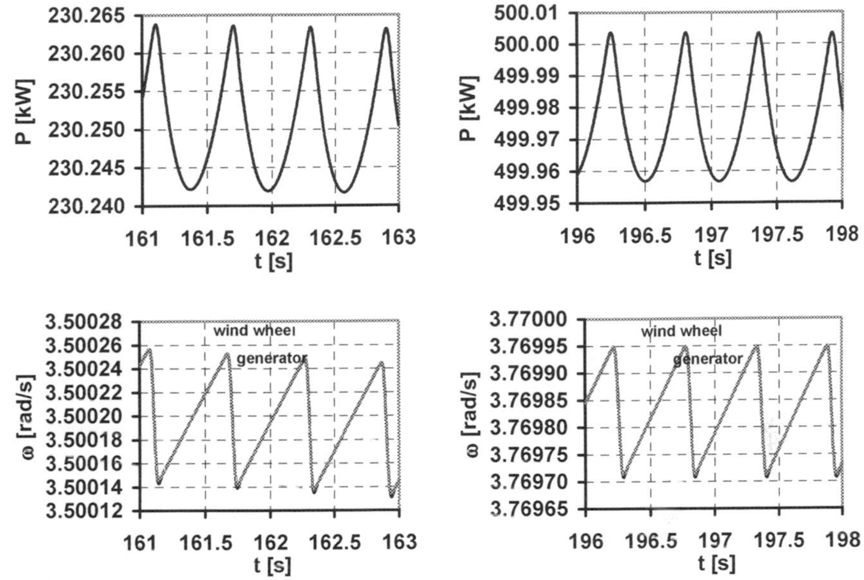


**Fig. 6.60.** Effect of blades passing in front of the WTGS tower (3P amplitude 7 %, left-hand graphs – $v = 9$ m/s; right-hand graphs – $v = 15$ m/s)

Figure 6.60 shows the operation of the system in the case when the 3P oscillations with amplitude 7% exist. The oscillations are very well damped and at partial load the real power oscillation band is equal to 0.025 kW, while at full load the band is equal to 0.05 kW. Both values of power oscillations are related to small nodal voltage variations.

Figure 6.61 presents the WTGS operations when the 1P, 3P oscillations and tower vibration effect (4.5 Hz) with amplitudes of 1%, 7% and 10%, respectively, exist. In this case, the band of the real power oscillations is higher, as in the previous example: at full load it is equal to 0.4 kW, while at partial load it is equal to 0.2 kW. The same, that is oscillations in both operating states with amplitude less than 0.01%, is visible in the nodal voltages.

Of course, an increase of the mechanical eigenswing amplitude will cause a matching increase of power and voltage variation.

Figures 6.62–6.64 present various frequency characteristics of the analyzed system. Figure 6.62 shows the frequency characteristics $V_l / v = G(\mathrm{j}\omega)$ and $P / v = G(\mathrm{j}\omega)$ achieved for the system operating at partial load ($v = 9$ m/s). The curves show that:

- The voltage variation at the nodes located between the PCC and the feeding point is much smaller than the voltages at the nodes located "behind" the WTGS (nodes 8, 9). The voltage variation at these nodes is similar to the voltage variation at the PCC. For a high range of frequency, the voltage variation is less than 0.33 % when the wind speed variation amplitude is less than 1 m/s. The real power variation amplitude is less than 70 kW/(m/s).

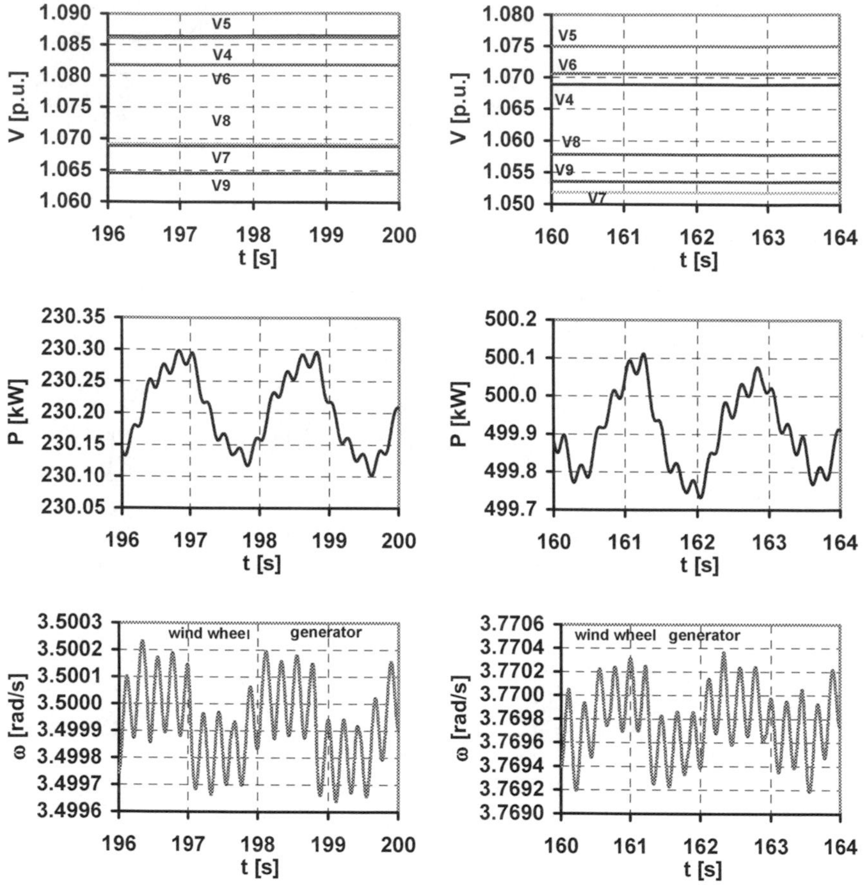


**Fig. 6.61.** 1P, 3P and tower vibration effects (1P amplitude 1%, 3P amplitude 7%, 4.5 Hz effect 10%; left-hand graphs – $v = 9$ m/s; right-hand graphs – $v = 15$ m/s)

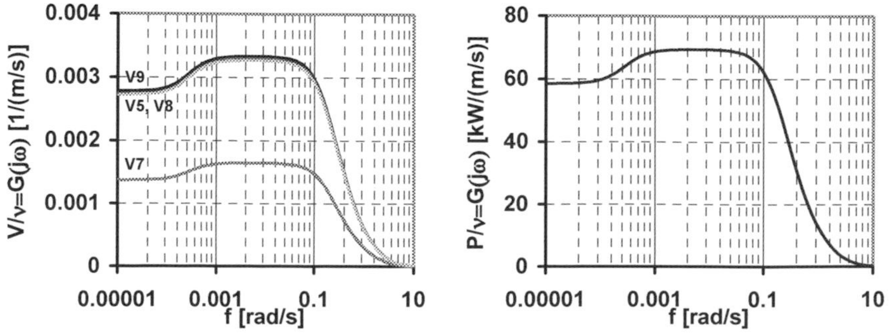


**Fig. 6.62.** Frequency characteristics of the WTGS at partial load ($v = 9$ m/s)

- The system has no resonance frequency.
- Wind velocity variation with frequency higher than 10 rad/s is practically eliminated.
- The shaft stiffness coefficient within the range $0.1\times10^8$–$1.2\times10^8$ N·m/rad has no meaningful influence on the frequency characteristics presented in Figs. 6.62–6.64.

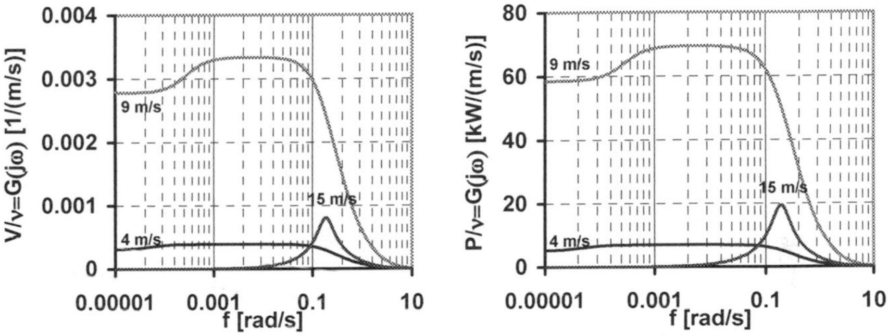


**Fig. 6.63.** Influence of the wind velocity (power generation) on the WTGS frequency characteristics

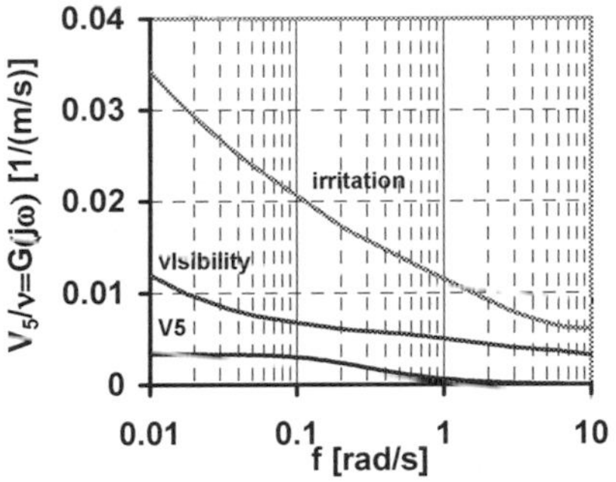


**Fig. 6.64.** Comparison of the WTGS voltage characteristics with the borders of irritation and visibility (voltage dips affected by lights – see Fig. 3.4) defined by standard IEEE 141-1993

Figure 6.63 shows the influence of the operating point (wind velocity) on the frequency characteristics. The higher response of the system takes place at a wind velocity equal to 9 m/s, where the slope of the WTGS characteristic $P = f(v)$ is greatest. The lower response takes place when the slope of the characteristic is smaller, i.e. at the rated wind velocity (15 m/s). But at this operating point, there appears an eigenfrequency with value equal to 1.5 rad/s and a real power amplitude of 20 kW/(m/s). This frequency is more than twice lower than the rotor speed and mechanical oscillations related to the wind-wheel rotation. It results from the structure and parameters of the WTGS control systems and therefore can be eliminated or modified by proper design of the controllers.

And finally Fig. 6.64 shows the comparison of the frequency characteristics with the limits defined by the standards. The curves show that the characteristic $V_5/v = f(\omega)$ (with higher amplitude, i.e. for $v = 9$ m/s) and all the other characteristics lie below the standard curves defining the border of visibility and the border of irritation. The characteristics also lie below the border defined by the IEC standard (Fig. 3.4, $P_{st} = 1$). As mentioned before, an increase of the wind velocity variation amplitude above 1 m/s will cause the characteristics to move up and then the limits will not be fulfilled.

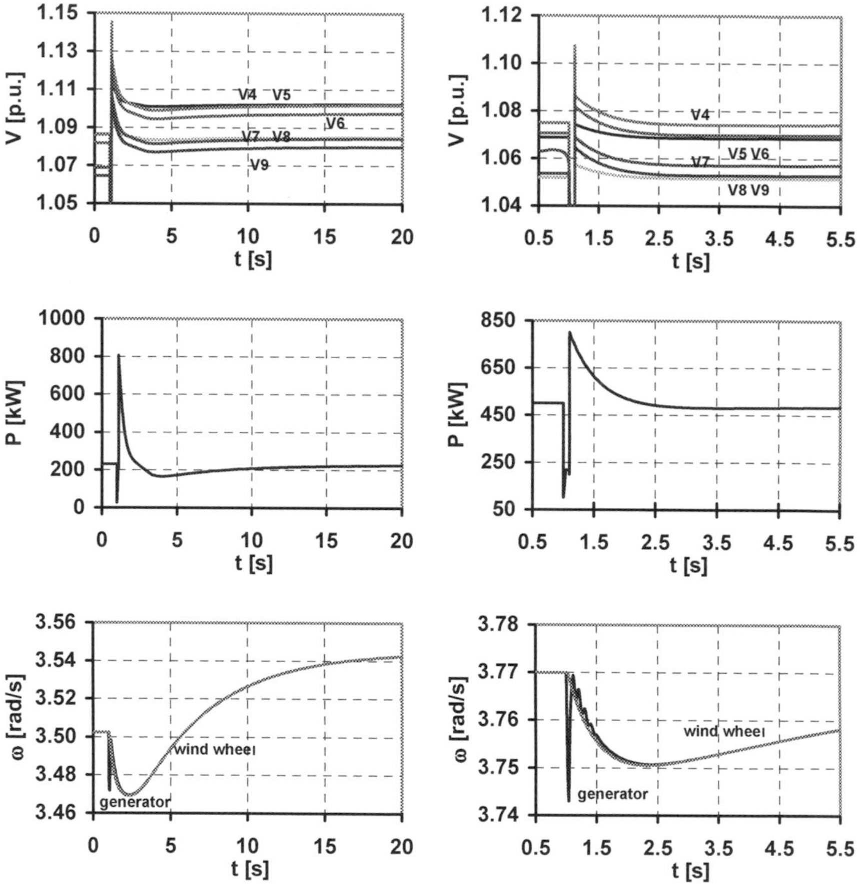


**Fig. 6.65.** WTGS response to a short circuit at the terminal bus lasting 100 ms (left-hand graphs – $v = 9$ m/s; right-hand graphs – $v = 15$ m/s)

The last figures show again the time-domain response of the system. Figure 6.65 shows the response of the system to a short-circuit at the terminal bus, while Fig. 6.66 shows the response of the system to a 1.5% step increase of the feeding bus voltage. The curves indicate that the transient processes caused by a short-circuit last about 5 s when the WTGS operates at partial load and last 1.5 s when the WTGS operates at full load. It is mainly related to the slow process of the rotor speed coming back to the equilibrium point. In the case of the feeding voltage change, the transient processes last much longer, i.e. 25 s and over 500 s, respectively. This results directly from the inverter type (current-source inverter here) and the inverter power controller structure and parameters.

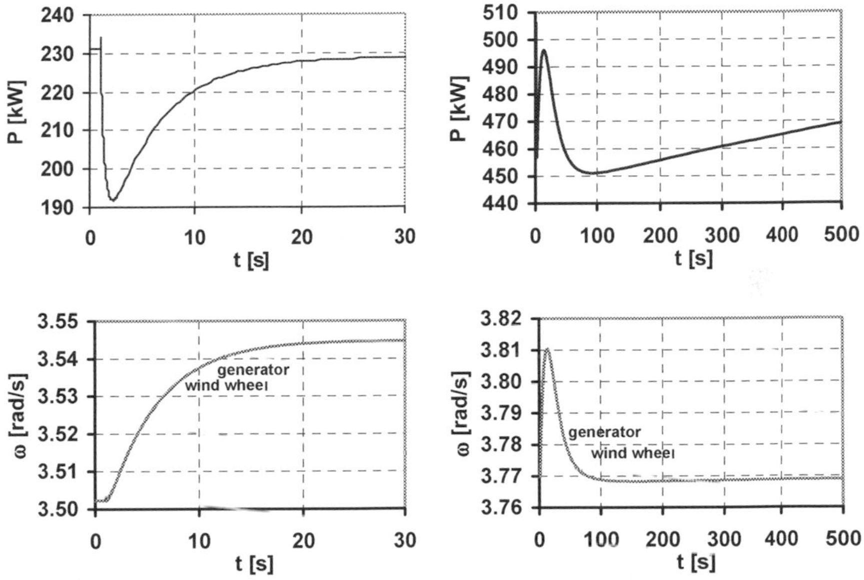


**Fig. 6.66.** WTGS response to a 1.5% step change of the feeding bus voltage (left-hand graphs – $v = 9$ m/s; right-hand graphs – $v = 15$ m/s)

## 6.5.2 Synchronous Generator with Voltage-Source Inverter

A model of the wind turbine generator system equipped with a synchronous generator and converter with voltage-source inverter (VSI) has a structure similar to the one presented in Fig. 6.52. The model components, i.e. the wind variation model, wind wheel, mechanical eigenswings, drive train, synchronous generator and grid, are modeled as in the case of the WTGS equipped with a current-source inverter (CSI) (Sect. 6.5.1). The converter is also modeled as a functional one equipped with an uncontrolled rectifier and controlled inverter. The intermediate element (capacitor) is modeled by the first-order differential equation:

$$C_{\mathrm{d}} \frac{\mathrm{d}v_{\mathrm{d}}}{\mathrm{d}t} = i_{\mathrm{drec}} - i_{\mathrm{dinv}} \tag{6.7}$$

where $C_{\mathrm{d}}$ – intermediate element (capacitor) parameter,
$v_{\mathrm{d}}$ – intermediate element voltage,
$i_{\mathrm{drec}}$, $i_{\mathrm{dinv}}$ – DC-side current of rectifier and inverter.

The WTGS and its model parameters are presented in Tables 6.12 and 6.13.

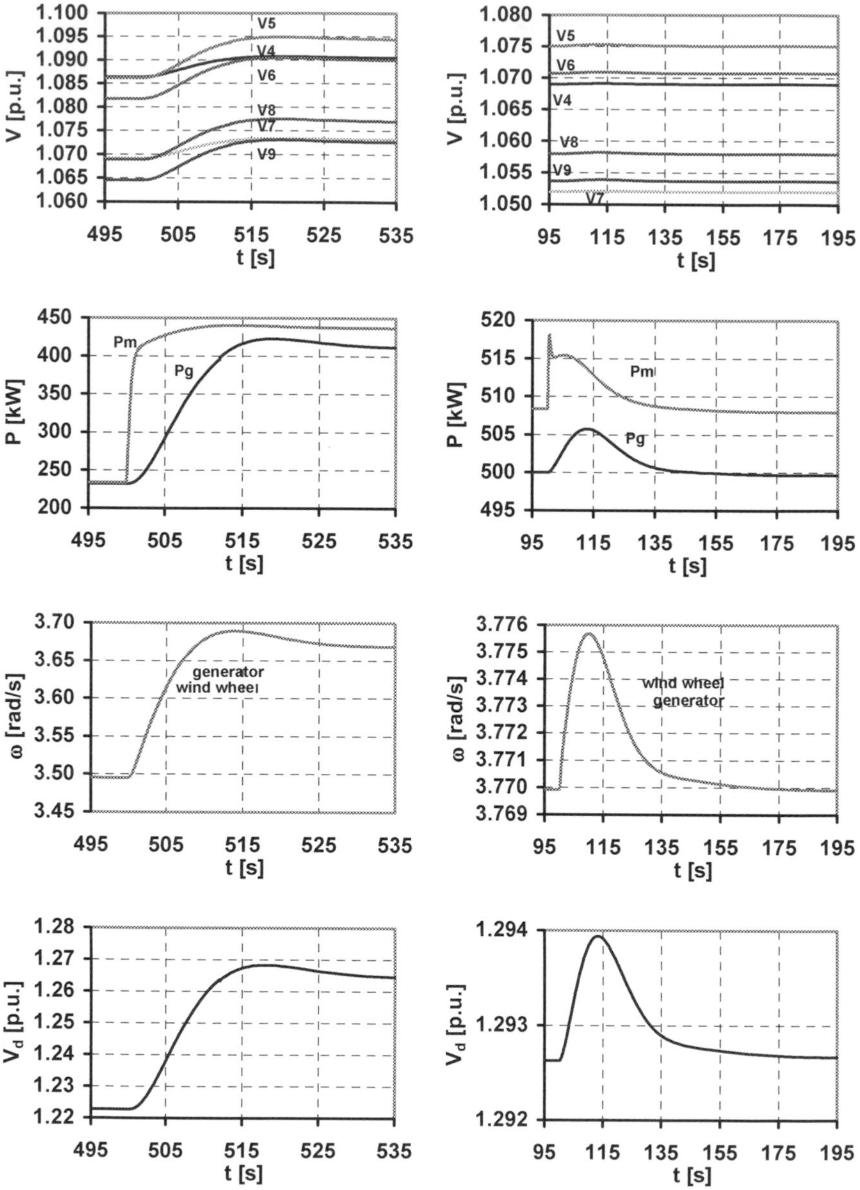


**Fig. 6.67.** WTGS response to a wind velocity step increase (left-hand graphs – partial load, wind velocity step from 9 to 11.7 m/s; right-hand graphs – full load, wind velocity step from 15 to 17.7 m/s; $V$ – nodal voltages, $P_g$ – WTGS real power, $P_m$ – mechanical power, $\omega$ – generator and wind wheel shaft speed, $V_d$ – VSI capacitor voltage; VSI controller $K_R = 0.03$, $T_I = 27$ s, $T_{f\omega} = 3$ s)

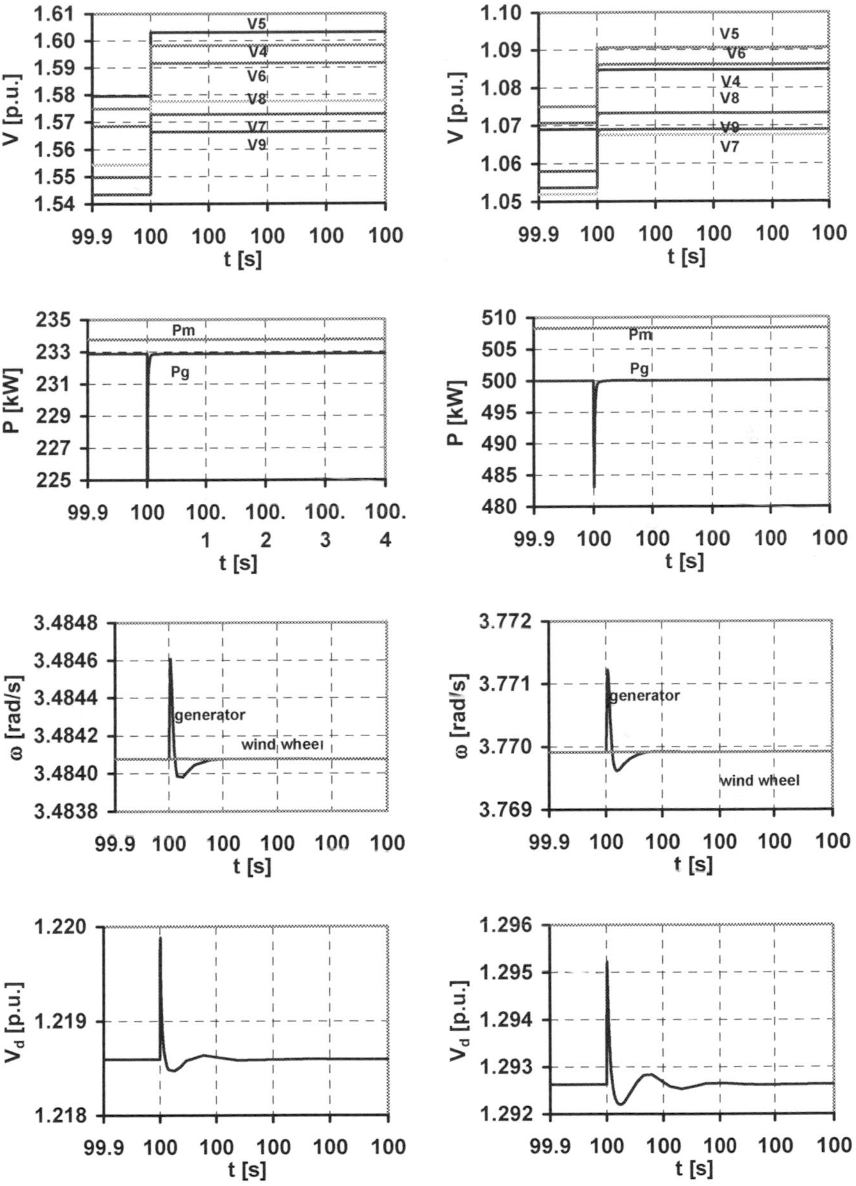


**Fig. 6.68.** WTGS response to a 1.5% step change of the feeding voltage (left-hand graphs – $v = 9$ m/s; right-hand graphs – $v = 15$ m/s; VSI controller $K_R = 100$, $T_I = 0.051$ s, $T_{f\omega} = 3$ s)

In the case of the considered system, control systems analogous (including the controller parameters) to the ones presented in the previous section are used. But because of the different dynamic properties of the converter, the synchronous generator voltage controller and the inverter controller parameters are changed. The parameters used in the case of the generator voltage controller (Fig. 6.54) are to $K_R = 1$, $T_I = 0.4$ s, $T_{f\omega} = 3$ s ($T_{f\omega}$ time constant of rotor speed measuring element), while the inverter power controller (Fig. 6.55) parameters are listed in the figure captions.

The considered WTGS has been tested in the same domain as the previously presented wind turbines. Figures 6.67 and 6.68 present the system response to a step change of the wind velocity and the response to a feeding voltage change.

The curves show that an increase of wind velocity (Fig. 6.67) at partial load causes a pass to a new operating point with slight overshoot. The process runs for about 25 s (which is similar to the response time in the system with the current-source inverter CSI). The nodal voltages, real power and rotor speed change are similar in qualitative and quantitative manner to those achieved in the system with the CSI.

Also, the system response to a wind velocity increase at full load behaves in a similar way to the one achieved in the WTGS with the current source inverter.

But in the considered system, in contrast to the system equipped with the CSI, its reaction to the feeding voltage change is much faster. The transient processes at partial and at full load last for up to 0.2 s (in comparison to 500 s in the previous case). The processes are related to the step decrease of the real power – 8 kW at partial load and 10 kW at full load. The process is simultaneously related to the step increase of the VSI capacitor voltage, which is relatively small (less than 0.3%).

Unfortunately, such a fast run of the transient process is possible for inverter power controller parameters different from the ones used in the system whose response is presented in Fig. 6.67. The VSI controller from Fig. 6.68 is characterized by a higher value of the gain $K_R$ and a smaller value of the time constant $T_I$ than the one from Fig. 6.67. Of course, it is possible to decrease the gain and increase the time constant, which slows down the system response and simultaneously makes the control system more effective in the case of disturbances located on the wind side. But, in fact, it is not a solution for the controller design. The controller effective (optimal) for grid-side disturbances is not effective for wind-side disturbance (and vice versa). In general, one can say that the converter controller of the presented structure cannot be at the same time effective in the case of disturbances located on the grid and on the wind side.

The above shows that the controller structure should be more complex than the one used in the model.

## 6.6 Concluding Remarks

The WTGS models and the analysis presented in this chapter allow us to formulate the following conclusions related to the dynamic properties of the modeled systems and related to the objects modeling:

- The dynamic properties of the WTGS depend on its construction and on the control system structure and parameters. Ability to damp the real power variation is the one of the dynamic properties. Table 6.14 presents the ability to keep the WTGS output constant after a wind velocity change in the short term and the long term. In the case of modern WTGSs, a permanent change of the WTGS real power is possible only by blade pitch control, while a temporary effect can be additionally achieved by using resistor $R_d$ (in the dynamic slip control system) and the rotor speed control. These two types of control are (can be) very fast and then can be used for fast reaction to the appearance of power imbalance (e.g. resulting from wind velocity change, torsional oscillations, etc.). With the exception of the wind turbine equipped with the dynamic slip control, the power imbalance can appear in the WTGS as a rotor speed change or in the real power transferred to the grid change. Therefore, if we want to eliminate power variations, we have to permit large (sometimes also fast) changes of the shaft speed, which unfortunately can lead to high changes of torques and to high mechanical stresses in the drive train. Therefore, while constructing the WTGS control system it is necessary to take account of the mechanical limitations of the turbine, i.e. speed, torque, etc., which can make it impossible to develop control systems that could eliminate the power variations.[14]

- In a radial network, the higher voltage variation caused by the WTGS operation appears at the network end opposite to the feeding point. In the network nodes located between PCC and the network end, the voltage variations amplitude is similar.
- WTGSs equipped with generator control systems allow damping power variations, which are especially effective at full load. In the case of the considered systems, the WTGS equipped with the dynamic slip control is best for eliminating the gust and mechanical eigenswing effects (Table 6.15). But at partial load, the best results, i.e. real power variation damping, are achieved by the WTGS with synchronous generator and with doubly-fed asynchronous generator.[15] The power variation at the WTGS terminals as a result of wind velocity variation is at least ten times lower at full load than at partial load. The same effect is visible in the network nodes, e.g. for nodal voltages.
- The utilization of control systems allows us to eliminate eigenfrequency that results from the drive-train features (the relatively soft and long shaft and the high masses of the wind wheel and generator rotor). In the considered examples (for the WTGS equipped with a doubly-fed asynchronous generator and with synchronous generator) a shift of the frequency characteristics to the left, with the cut-off frequency equal to about 0.1 rad/s, is achieved. The frequency of 0.1 rad/s is located below the operational frequencies of both WTGSs. The "pass to the left" of the frequency characteristic causes one (even for the worst operating case – at partial load at a wind velocity of about 9 m/s) is located be-

[14] This remark is related to the WTGS operation at full load. At partial load, elimination of real power variation is impossible for other reasons considered in this monograph.

[15] The conclusions related to the results of the simulations presented in this chapter do not have general validity and cannot be considered as valid for the all wind turbines equipped with a given type of control system. The control system structures, algorithms, parameters and strategies in various types of WTGSs vary considerably.

low the limits defined by the standards.

**Table 6.14.** Method of keeping constant output (real power) after a wind velocity change

| | Temporary action/effect | | |
|---|---|---|---|
| WTGS type |  $R_d$ = var | 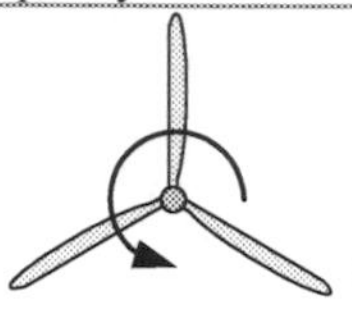 $\omega$ = var | 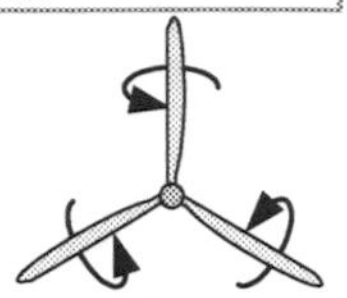 $\vartheta$ = var |
| Squirrel-cage rotor asynchronous generator + stall control | There is no way to keep the real power constant | | |
| Asynchronous generator with dynamic slip control | +[a] | +[b] | +[c] |
| Doubly-fed asynchronous generator, oversynchronous cascade | – | +[b] | +[c] |
| Synchronous generator connected to grid via power electronic converter | – | +[b] | +[c] |
| | **Permanent effect** | | |
| WTGS type |  $R_d$ = var | 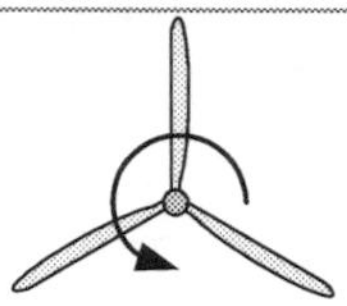 $\omega$ = var | 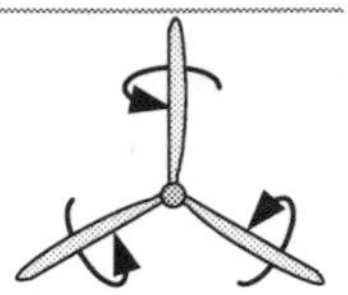 $\vartheta$ = var |
| Squirrel-cage rotor asynchronous generator + stall control | There is no way to keep the real power constant | | |
| Asynchronous generator with dynamic slip control | – | – | +[e] |
| Doubly-fed asynchronous generator, oversynchronous cascade | – | +[d] | +[e] |
| Synchronous generator connected to grid via power electronic converter | – | +[d] | +[e] |

[a] fast control, limited to a few percent of rated power
[b] can be fast but is generally limited by the maximum allowable torque (value and surge) in the drive train
[c] slow control
[d] at partial load
[e] at full load

**Table 6.15.** Comparison of dynamic properties of the considered WTGSs

| Parameter | Unit | Squirrel-cage rotor asynchronous generator | | Asynchronous generator with dynamic slip control | | Doubly-fed asynchronous generator | | Synchronous generator connected to grid via converter[16] | |
|---|---|---|---|---|---|---|---|---|---|
| | | Partial load | Full load | Partial load | Full load | Partial load | Full load | Partial load | Full load |
| $t_{\Delta v}$ | s | 2 | | 1 | | 50 | | 20 / 0.2 | |
| $\Delta P_{gust}$ | kW | 100 | 10 | 130 | 0.001 | 130 | 1 | 65 | 11 |
| $\Delta V_{gmax}$ | % | 0.5 | 0.03 | 0.6 | 0.01 | 0.5 | 0.04 | 0.55 | 0.02 |
| $\Delta P_{3P}$ | kW | 8 | 15 | 6 | 0.001 | 0.6 | 1.6 | 0.05 | 0.09 |
| $\Delta P_{1P,3P..}$ | kW | 35 | 63 | 25 | 0.001 | 1.5 | 11 | 0.36 | 0.75 |
| $\Delta V_{max}/v$ | %/(m/s) | 0.4 | | 0.54 | | 0.65 | | 0.6 | |
| $\Delta P_{max}/v$ | kW/(m/s) | 170 | | 145 | | 155 | | 125 | |
| $f_{eig}/f_{lim}$ Std. | rad/s | $f_{eig} = 5.5$ not fulfilled | | $f_{lim} = 3$ not fulfilled | | $f_{lim} = 0.1$ fulfilled | | $f_{lim} = 0.1$ fulfilled | |

$t_{\Delta v}$ – settling time after step change of wind velocity
$\Delta P_{gust}$ – real power amplitude during gust
$\Delta V_{gmax}$ – maximum nodal voltage change during gust
$\Delta P_{3P}$ – band of power variation during 3P mechanical eigenswings existence
$\Delta P_{1P,3P..}$ – band of power variation during 1P, 3P and 4.5 Hz mechanical eigenswings existence
$\Delta V_{max}/v$ – maximum nodal voltage variation as a result of wind velocity variation (from frequency characteristics)
$\Delta P_{max}/v$ – maximum real power variation as a result of wind velocity variation (from frequency characteristics)
$f_{eig}/f_{lim}$ – eigenfrequency or cut-off frequency
*Std.* – marker showing if the IEEE standard is fulfilled (the IEC standard is fulfilled for all the WTGSs when $\Delta v = 1$ m/s)

[16] The values printed in the Table and related to the WTGS equipped with a synchronous generator ($P_n = 500$ kW) are multiplied by factor 900/500 to allow comparison with other WTGSs ($P_n = 900$ kW).

# 7 Conclusions

The utilization of energy extracted from the wind is getting more and more popular today. It is due to European Union policy, which results directly from the benefits of the energy from the wind conversion process.[1] The environmental aspects play a significant role here as well.

Unfortunately, the utilization of modern wind energy converters (wind turbine generator systems) brings both benefits and problems for the existing electric power systems. The main problems that occur as a result of WTGS coupling to the electric power system are the power quality problems and the stability problems. The problems will increase as the number of WTGSs operating in the power system increases.

Therefore, in the case of each connection of a WTGS (or wind farm) to the power system, a detailed analysis is made. Adequate tools, i.e. mathematical models of the considered problem, object or phenomena, have to be used to handle the problems. Therefore, models of the WTGS components and the electric power system components as analytical tools are presented and considered in the monograph.

When considering power system models, the standard approach has been assumed here, in which the network and loads are modeled as a set of algebraic equations (steady-state models), while sources of energy (generators, turbines and their controllers) are modeled as dynamic objects. In the standard approach, unlike the generators, the turbines are usually more simplified, which has been assumed in the monograph as well.

The monograph, of course, does not answer all the questions and problems that can be formulated in area of WTGS operation in the electric power system. But it points out the problems and provides some insight related to modeling WTGS operation in the electric power system, which is the author's intention and hope. In particular, the monograph provides the basis for creating one's own (the researcher's) models, which is the first and basic step for solving real system problems.

Of course, when modeling the WTGS as an element of the electric power system, we can use "own" models (as presented in the monograph) or we can utilize commercial software designed specially for the analysis of power systems or power electronic systems. Each approach has both advantages and disadvantages.

The first approach (the use of "own" models) needs some effort in terms of system modeling – as a set of differential and algebraic equations. But in return, full

[1] We should not neglect the lobbying wind turbine manufacturers.

"control" over the mathematical model can be achieved here. The researcher has access to each equation, quantity, parameter, computing procedure, etc. It is also possible to make any type of analysis but, unfortunately, adequate software needs to be obtained or developed.

The second approach (utilization of commercial software) faces the following problems. The electric power system analysis software has not been not equipped (so far) with WTGS models[2], which means that in any particular case the WTGS has to be modeled by the user (when the software allows). Moreover, some commercial software does not permit some types of analysis, e.g. small-signal stability analysis when the inputs or outputs are located in the user (non-standard) model. The power electronic systems (i.e. converters) have to be modeled here in a simplified form.

The situation is quite similar when power electronic analysis software is used. The software allows us to model the WTGS in a detailed form. Unfortunately, the researcher has to do the modeling of the system. Additionally, unlike for power system analysis software, the size of the electric power system model has to be limited here. The researcher does not have access to the equations and computing procedures. Sometimes there is no information about it in the manuals, and then the researcher does not know exactly how it works.

However, commercial software has the following advantages:

- it allows the modeling of very large power systems,
- the models of the power system elements (with the exception of WTGS) are well defined,
- the mathematical procedures for various types of analysis are enclosed.

The conclusion is that it is not easy to give advice on what type of modeling should be chosen for the analysis. It depends on many factors, e.g. analysis domain, researcher's experience and habits, the software used in the company, etc. But in any case, the modeling basics are always worth knowing, and the dissemination of this knowledge is the intention of this monograph.

[2] WTGS models will be added to the commercial software probably fairly soon.

# References


1. Amora MAB, Bezerra UH (2001) Assessment of the effects of wind farms connected in a power system. In: Proc IEEE Porto Power Tech Conf, Sept. 10–13, Porto, Portugal, DRT3-260
2. Anderson PM, Fouad AA (1980) Power system control and stability. The Iowa State University Press, Ames (Iowa)
3. Argyrys JH, Braun KA, Kirchgasner B, Walther R (1980) Static and dynamic investigations for the model of a wind rotor. ISD Report no 272, Universitat Stuttgart
4. Akhmatow V, Knudsen H (1999) Modelling of windmill induction generators in dynamic simulation programs. In: Proc IEEE Power Tech Conf, Aug. 29–Sept. 2, Budapest, Hungary, BPT99-243-12
5. Akhmatow V, Nielsen AH, Knudsen H (2000) Electromechanical interactions and stability of power grids with windmills. In: Proc IASTED Conf on Power and Energy Systems, Sept. 19–22, Marbella, Spain, pp 398–405
6. Akhmatow V, Knudsen H, Nielsen AH (2000) Advanced simulation of windmills in the electric power supply. J Electrical Power & Energy Systems 22, pp 421–434
7. Akhmatow V, Knudsen H, Bruntt M, Nielsen AH, Pedersen JK, Poulsen NK (2000) The dynamic stability limit of grid-connected induction generators. In: Proc IASTED Int Conf on Power and Energy Systems, Sept. 19–22, Marbella, Spain, pp 235–244
8. Anderson PM, Bose A (1983) Stability simulation of wind turbine systems. IEEE Trans Power Appar and Syst, Vol PAS-102, No 12, pp 3791–3795
9. Amlang B, Arsurdis D, Leonhard W, Vollstedt W, Wefelmeier K (1992) Elektrische Energie-versorgung mit Windkraftanlagen. Abschlusbericht BMFT-Forschungsvorhaben 032-8265-B, Brauschweig
10. Barlik R, Nowak M (1997) Thyristor technique (in Polish). WNT, Warsaw
11. Bogalecka E (1997) Control of the doubly-fed asynchronous machine operating in the electric power system (in Polish). Monograph, ISSN 0860-4827, Maritime Academy, Gdynia
12. Bollen MHJ (2000) Understanding power quality problems. Voltage sags and interruptions. IEEE Press, Series on Power Engineering, New York
13. Bongers P (1990) DUWECS Reference Guide v1.0. TU Delft, MEMT 5
14. Bongers P, Bierbooms W, Dijkstra S, van Holten T (1990) An integrated dynamic model of a flexible wind turbine. TU Delft, MEMT 6
15. Canay IM (1983) Determination of model parameters of synchronous machines. IEE Proc B, Vol 130, No 2, pp 86–94
16. Christiansen P, Jorgensen KK, Sorensen AG (2000) Grid connection and remote control for the Horns Rev 150 MW offshore wind farm in Denmark. EP11746.03 Notat 00/N228pc, Elsamprojekt A/S, Fredericia
17. Committee report DEFU 111-E (1998) October, 2-edition. Connection of wind turbines to low and medium voltage network. DEFU a.m.b.a , Frederiksberg

18. Criado R, Soto J, Rodriguez JM, Martin L, Fernandez JL, Molina J, Tapia A, Saenz JR (2000) Analysis and control strategies of wind energy in the Spanish power system. In: Proc CIGRE Session, Paris, France, 38-103
19. de Mello FP, Hannet LH (1981) Validation of synchronous machine models and derivation of model parameters from tests. IEEE Trans Power Appar and Syst, Vol 100, No 2, pp 662–672
20. Dugan RC, McGranaghan MF, Beaty HW (1996) Electrical power systems quality. McGraw-Hill, New York
21. DEWI German Wind Energy Institute (2001), Technical Data for Utilities Planning Wind Turbine Connection, No 00 0910-4e, March, DEWI, Wilhelmshaven
22. Eigenerzeugungsanlagen am Mittelspannungsnetz (1998) Richtlinie fur Anschluβ und Parallelbetrieb von Eigenerzeugungsanlagen am Mittelspannungsnetz. Verlags- und Wirtschafttsgesellschaft der Elektrizitatswerke m.b.H. – VWEW 2
23. EN 50160: 1999 Std. Voltage characteristics of electricity supplied by public distribution systems. CENELEC, Bruxelles
24. Enron Wind 900 Series, 50 Hz Wind turbine generator system. Technical description and general turbine specification. 900serie_GD_allComp_50Hzxxxx, ENRON WIND
25. Feijoo A, Cidras A (2000) Modelling of wind farms in the load flow analysis. IEEE Trans Power Syst, Vol 15, No 1, pp 110–115
26. General Specification V80-2.0 MW, OptiSpeed™ -Wind Turbine. (2001) May, Vestas Wind Systems A/S, Ringkobing
27. Grid Connection of Wind Farms (2001) Danish - Polish seminar EWISEE '01, Nov. 22–23, Gdansk University of Technology, Gdansk, Poland
28. Hansen MH, Thomsen K, Petersen JT (2001) Rotor whirling modes and the relation to their aerodynamic damping. In: Proc European Wind Energy Conf, Copenhagen, Denmark, July 2–6, pp 422–425
29. Harley RG, Limebeer DJN, Chirricozzi E (1980) Comparative study of saturation methods in synchronous machine models. IEE Proc B, Vol 127, No 1, pp 1–7
30. Hau E (2000) Windturbines: fundamentals, technologies, application, economics. Springer Verlag, Berlin
31. Heier S (1998) Grid integration of wind energy conversion systems. John Wiley & Sons, Chichester
32. Hellmann W, Szczerba Z (1978) Frequency and voltage control in the electric power system (in Polish). WNT, Warsaw
33. IEC 61000-3-6: 1996, Section 6 Assessment of emission limits for distorting loads in MV and HV power system
34. IEC 61000-3-6: 1996, Section 7 Assessment of emission limits for fluctuating loads in MV and HV power systems
35. IEC 61400-21 (Draft) Power quality requirements for grid-connected wind turbines
36. IEEE Committee Report (1973) Dynamical models for steam and hydroturbines in power system studies. IEEE Trans Power Appar and Syst, Vol 92, No 6, pp 1904–1915
37. IEEE Committee Report (1968) Computer representation of excitation systems. IEEE Trans Power Appar and Syst, Vol 87, No 6, pp 1460–1464
38. IEEE Committee Report (1973) Excitation system dynamic characteristic. IEEE Trans Power Appar and Syst, Vol 92, No 1, pp 64–75
39. IEEE Committee Report (1981) Excitation system models for power system stability studies. IEEE Trans Power Appar and Syst, Vol 100, No 2, pp 494–509

40. IEEE, Std 421A (1978) IEEE Committee Report: Guide for identification, testing and evaluation of the dynamic performance of excitation control systems
41. IEEE, Std 421.5 (1992) IEEE Recommended Practice for Excitation System models for power system stability studies
42. IEEE, Std 141 (1993) Recommended practice for electric power distribution for industrial plants
43. IEEE, Std 519 (1992) Recommended practice and requirements for harmonic control in electric power system
44. IEEE, Std 1346 (1998) Recommended practice for evaluating electric power system compatibility with electronics process equipment
45. Jarzyna W, Rozycki M, Toborek K (1994) Forming of wind power station electric generator characteristics with the help of microprocessor control system. In: Proc PEMC'94 Conference, Sept. 20–22, Warsaw, Poland, pp 790–794
46. Jockel S, Hagenkort B, Hartkopf T, Schneider H, Ruckh AM (2001) Direct-drive synchronous generator system for offshore wind farms with active drive train damping by blade pitching. In: Proc European Wind Energy Conf, Copenhagen, Denmark, July 2–6, pp 991–998
47. Kacejko P, Machowski J (1993) Short-circuits in the power network. Basics of computation (in Polish). WNT, Warsaw
48. Kaczorek T (1998) Vectors and matrices in automatic and electrotechnics (in Polish). PWN, Warsaw
49. Kaczorek T (1999) Control and systems theory (in Polish). PWN, Warsaw
50. Kremens Z, Sobierajski M (1996) Analysis of power systems (in Polish). WNT, Warsaw
51. Kundhur P (1994) Power System Stability and Control. McGraw-Hill, New York
52. Kuehn M (2001) Dynamics and design optimization of offshore wind energy conversion systems. Wind Energy Research Institute, Delft University
53. Kruger T, Andresen B (2001) Vestas Optispeed™ - Advanced control strategy for variable speed wind turbines. In: Proc European Wind Energy Conf, Copenhagen, Denmark, July 2–6, pp 983–986
54. Krzeminski Z (2001) Digital control of asynchronous machines (in Polish). Monograph, Development of power drives series 45, Gdansk University of Technology Press, Gdansk
55. Latek W (1973) Turbogenerators (in Polish). WNT, Warsaw
56. Ledesma P, Usaola J (2001) Minimum Voltage protection in variable speed wind farms. In: Proc IEEE Porto Power Tech Conf, Sept. 10–13, Porto, Portugal, DRT2-245
57. Lekou DJ, Vionis PS, Philippidis TP (2001) Damping characterization of rotor blades by means of full-scale testing. In: Proc European Wind Energy Conf, Copenhagen, Denmark, July 2–6, pp 287–290
58. Lubosny Z, Szczerba Z (1993) Mathematical model of the multi-machine power system for analysing electromechanical transient processes. In: Proc Mathematical Methods in Power Systems Conf, Nov. 10–12, Zakopane, Poland, pp 34–38
59. Lubosny Z (1999) Self-organising controllers of generating unit in electric power system. Monograph No 4. Gdansk University of Technology, Gdansk
60. Machowski J, Bernas S (1989) Transient states and stability of an electric power system (in Polish). WNT, Warsaw
61. Machowski J, Bialek J, Bumby JR (1997) Power system dynamics and stability. John Wiley & Sons Ltd, Chichester

62. Malick IA, Umans SD, Wilson GL (1978) Modelling of solid rotor turbogenerators. Part I: Theory and techniques. Part II: Example of derivation and use in digital simulation. IEEE Trans Power Appar and Syst, Vol 97, No 1, pp 269–291
63. Manning CD, Halim MAA (1988) New dynamic inductance concept and its application to synchronous machine modelling. IEE Proc B, Vol 135, No 5, pp 231–239
64. Manwell JF, McGowan JG, Rogers AL. (2002) Wind energy explained. Theory, Design and Application. John Wiley & Sons Ltd, Chichester
65. Matsuzaka T, Kodama N, Inomata N (2001) Control strategy of a wind generator to reduce power variation using probabilistic optimal control, combining feed forward control from wind speed with feedback control. In: Proc European Wind Energy Conf, Copenhagen, Denmark, July 2–6, pp 1178–1181
66. Mitkowski W (1991) Stabilization of dynamic systems (in Polish). WNT, Warsaw
67. Meyer M, Hermann GM (2001) A model for the simulation of the aeroelastic response of large wind turbines. In: Proc European Wind Energy Conf, Copenhagen, Denmark, July 2–6, pp 390–393
68. Nijssen RPL, Zaaijer MB, Bierbooms WAAM, van Kuik GAM, van Delft DRV, van Holten Th (2001) The application of scaling rules in up-scaling and marinisation of a wind turbine. In: Proc European Wind Energy Conf, Copenhagen, Denmark, July 2–6, pp 619–626
69. Nowak M, Barlik (1998) Power electronics engineer's guide (in Polish). WNT, Warsaw
70. Noroozian M, Knudsen H, Bruntt M (2000) Improving a wind farm performance by reactive power compensation. In: Proc IAESTED Int Conf on Power and Energy Systems, Sept. 19–22, Marbella, Spain, pp 437–442
71. Paszek W (1998) Dynamics of AC electrical machines (in Polish). Helion, Gliwice
72. Patel MR (1999) Wind and solar power systems. CRC Press, New York
73. Pena RS, Clare JC, Asher GM (1995) Implementation of vector control strategies for a variable speed doubly-fed induction machine for wind generation system. In: Proc EPE'95 Conf, Sevilla, Spain, pp 3075–3080
74. Pierce K, Fingersh LJ (1998) Wind turbine control system modelling capabilities. In: Proc American Control Conf, Philadelphia, June 24–26, pp 456–461
75. Roeper R (1985) Short-circuit currents in three-phase systems. Siemens Aktiengesellschaft, John Wiley & Sons Ltd, Berlin und Munchen
76. Rodriguez-Amendo JL, Burgos S.A. (2001) Design criteria of variable speed wind turbines with doubly fed induction generator. In: Proc European Wind Energy Conf, Copenhagen, Denmark, July 2–6, pp 1116–1119
77. Ronkowski M (1995) Circuit-oriented models of electrical machines for simulation of converter systems. Monograph, Gdansk University of Technology Press, Gdansk. Elektryka LXXVIII, No 523
78. Rozycki M (1990) Wind turbine with wounded rotor induction machine (in Polish). Przegląd Elektrotechniczny, LXVI, No 4–5, pp 74–76
79. Slootweg JG, Polinder H, Kling WL (2001) Dynamic modelling of a wind turbine with direct drived synchronous generator and back to back voltage source converter and its control. In: Proc European Wind Energy Conf, Copenhagen, Denmark, July 2–6, pp 1014–1017
80. Sorensen JN, Mikkelsen R (2001) On the validity of the blade element momentum method. In: Proc European Wind Energy Conf, Copenhagen, Denmark, July 2–6, pp 287–290

81. Sorensen P, Hansen AD, Janosi L, Bech J, Blaabjerg F, Bak-Jensen B (2001) Simulation of wind farm interaction with grid. In: Proc European Wind Energy Conf, Copenhagen, Denmark, July 2–6, pp 1003–1006
82. Spooner E, Williamson AC (1996) Direct coupled, permanent magnet generators for wind turbine application. IEE Proc Electr Power Appl, Vol 143, No 1, pp 1–8
83. Szczerba Z, Lubosny Z, Zajczyk R, Siodelski A, Klucznik J, Malkowski R, Pochyluk R, Wrycza M (2001) Concept of wind turbines equipped with asynchronous generators cooperating with the power system (in Polish). Department of Electrical Power Engineering Report, Gdansk University of Technology, Gdansk
84. Szczesny R (1992) Computer simulation of power electronic systems (in Polish). Monograph, Gdansk University of Technology Press, Gdansk. Elektryka LXXII, No 488
85. Szczesny R (1999) Computer simulation of power electronic devices (in Polish). Monograph, Gdansk University of Technology Press, Polish Academy of Science, Gdansk
86. Tapia A, Tapia G, Ostolaza JX, Saenz JR, Criado R, Berasategui JL (2001) Reactive power control of a wind farm made up with doubly fed induction generators (I). In: Proc IEEE Porto Power Tech Conf, Sept. 10–13, Porto, Portugal, DRT1-054
87. Tapia A, Tapia G, Ostolaza JX, Saenz JR, Criado R, Berasategui JL (2001) Reactive power control of a wind farm made up with doubly fed induction generators (II). In: Proc IEEE Porto Power Tech Conf, Sept. 10–13, Porto, Portugal, DRT1-055
88. Technical description of the Optislip® feature in VESTAS wind turbines. (2000) February, Vetsas Wind Systems A/S, Ringkobing
89. Technical description NM 900/52, ©2000 NEG Micon A/S, Randers
90. Tunia H, Winiarski B (1987) Power electronic basics (in Polish). WNT, Warsaw
91. Venikov VA (1985) Electromechanical transient processes in electric power systems (in Russian). Izdanie 4. Moskwa: Wyzszaja Szkola
92. Wierzejski M, Bogalecka E (1995) Problems of doubly-fed induction machine control during the generation mode (in Polish). In: II Conference SENE'95, Lodz, Poland, Nov. 15–17, pp 636–657
93. Windtest Kaiser-Wilhelm-Koog GmbH (2001) Power quality measurement on the Vestas V80–2 MW. Final Report
94. Wiszniewski A (1990) Algorithms of digital measurement in power automatics (in Polish). WNT, Warsaw
95. Zajczyk R (1996) Generating node control during transient states. Monograph, Gdansk University of Technology Press, Gdansk. Elektryka LXXXI, No 542

# Index

Printed in the United States
212580BV00006B/1/A

9 783540 403401